W. Salomons U. Förstner (Eds.)

Chemistry and Biology of Solid Waste

Dredged Material and Mine Tailings

With 83 Figures

Springer-Verlag
Berlin Heidelberg New York
London Paris Tokyo

Dr. WIM SALOMONS
Delft Hydraulics Laboratory
Institute for Soil Fertility
P.O. Box 30003
NL-9750 RA Haren (Gr)
The Netherlands

Professor Dr. ULRICH FÖRSTNER
Arbeitsbereich Umweltschutztechnik
Technische Universität Hamburg-Harburg
Harburger Schloßstraße 20
Postfach 90 14 03
D-2100 Hamburg 90
Fed. Rep. of Germany

ISBN 3-540-18231-4 Springer-Verlag Berlin Heidelberg New York
ISBN 0-387-18231-4 Springer-Verlag New York Berlin Heidelberg

Library of Congress Cataloging-in-Publication Data. Chemistry and biology of solid waste ; dredged material and mine tailings. Includes index. 1. Mineral industries–Waste disposal–Environmental aspects. 2. Tailings (metallurgy)–Environmental aspects. 3. dredging spoil–Environmental aspects. 4. Geochemistry. 5. Ecology. I. Salomons, W. (Willem), 1945-. II. Förstner, Ulrich. TD899.M47C48 1988 628.4'4 88-2015

Printed in the United States of America.

Typesetting: Overseas Typographers, Inc., Makati, Philippines
2132/3130-543210

Preface

Mine tailings and dredged materials are solid wastes, which are produced at single sites in huge quantities. Costs are dramatically increasing both for the installation of new disposal sites and for the restoration of former deposits, which had been planned and operated in an inappropriate manner. The problems and questions these waste materials pose with regard to safe environmental deposition are similar: aquatic versus terrestrial disposal, revegetation, leaching of contaminants. Larger projects in the fields of both mine tailings reclamation and dredged material disposal are increasingly requiring a multidisciplinary team approach. Scientists with many different backgrounds – engineers, hydrologists, geologists analytical chemists, biologists, ecologists, environmental planners and economists – are searching for long-term solution, which involve minimum harm to nature, but can still be paid by society.

A typical project (and one of the motives for this book) was the planning of the "sludge island" off Rotterdam. After one decade of preparatory research, which was continuously interacted by public discussions, the Port of Rotterdam and the Netherlands Waterways Administration has now started to construct a containment for approximately 150 million m^3 of moderately to strongly polluted dredged sludge from the harbor area. The various "ecological" advantages will be discussed in several contributions to the present work. This solution also seems to be economically competitive with the old inland sites and even to the sea disposal of dredged materials (!). It is noteworthy, however, that the Municipality of Rotterdam has definitely stated that there is no intention of creating further large-scale sites after 2002. Measures have to be undertaken to improve the quality of the sediments, particularly by municipal and industrial dischargers in the Rhine River catchment area.

Mine tailings, compared to the more recent environmental implications with dredged materials, have been recognized as problematic materials for a long time. A report from 1868 of the River Pollution Commission in Britain described the situation in mid-Wales: "All these streams are turbid, whitened by the waste of the lead mines in their course; and flood waters bring down poisonous slime which, spreading over the adjoining flats, either befoul or destroy grass, and thus injure cattle and horses grazing on the dried herbage, or, by killing the plants whose roots have held the land together, render the shores more liable to abrasion and destruction on the next occasion of high water". Until now, problems arising from acid mine effluents are not satisfactorily solved in any part of the world. The major emphasis devoted to this aspect in the present book clearly demonstrates that the prediction and prevention of acid mine drainage are key elements of a strategy to control pollution from mining operations.

A major part of mineral reserves are in less-developed countries. A disproportionate fraction of resource development is expected to take place in the areas where environmental protection measures may be limited. Such experience will imply far-going demands from the host countries: (1) Reclamation should be carried out, as far as possible, during the life of the mine; (2) technology to ameliorate long-term effects should be as self-supporting as possible; (3) simple, reliable, low-energy techniques for minimizing deleterious effects of mining should be developed. The latter requirement can be summarized in the short expression "working with and not against nature". Initial efforts will be described in this book.

With 30 contributions, the two-volume book is mainly addressed to two major groups of potential users: to environmental chemists, biologists and geochemists working for mining companies, consultant agencies and universities, and to managers and planners in both industry and governmental agencies.

Chemistry and Biology of Dredged Materials and Mine Tailings, is introduced by review articles on three scientific disciplines, which seem to be particularly relevant for the present subject: solution/solid interactions of metal (A. Bourg), microbial processes (O.H. Tuovinen) and behavior of vegetation (W.H.O. Ernst). These reviews are followed by in-depth contributions on biological and chemical characteristics of the two types of solid wastes. They include case histories as well as laboratory (experimental) assessment of actual or potential environmental impact of both organic and inorganic priority pollutants.

Environmental Management of Dredged Materials and Mine Tailings will certainly fulfill expectations of a wide spectrum of practitioners, in that latest results are presented from management plans and decision-making processes in both fields. Examples are given from new mining operations in both developed and developing countries (I. Ritchie; R. Higgins), and the most advances approaches to dredged material handling by the Municipality of Rotterdam and the U.S. Government (H. Nijssen; C.R. Lee and R.K. Peddicord). With respect to the future development in these areas, the reader should particularly refer to the articles on "ecological engineering" (K. Kalin and R.O. van Everdingen), "biological engineering" (D.V. Ellis and L. Taylor), use of "integrated biological systems" (H. von Michaelis). Several other contributions describing methods of "geochemical engineering" emphasize the increasing efforts of using natural resources available at the disposal site for reducing negative environmental effects of all types of solid waste materials.

Last but not least we would like to express our gratitude to all contributors for their enthusiastic and cooperative response to this project. We are very thankful to Dr. Engel, Springer-Verlag, for his constant encouragement, and for the much appreciated assistance of the publisher in the preparation of these volumes.

W. Salomons
U. Förstner

Contents

Part I Chemical and Biological Principles

Part II Biological and Geochemical Assessment

Part I
Chemical and Biological Principles

Metals in Aquatic and Terrestrial Systems: Sorption, Speciation, and Mobilization

A.C.M. BOURG[1]

1 Introduction

Water is a major carrier of metal pollutants which can be transported from a disposal site both in dissolved and particulate form. The particles can be original components of the waste as well as newly formed material or existing solids contaminated during leachate flow through the receiving ecosystem. Water is also the medium in which the metal pollutants undergo biological and chemical reactions. Some of these reactions induce phase changes resulting in variations in the rate of metal transport. It is therefore important to understand which phenomena are responsible for retarding the metal fluxes and which are, in contrast, capable of increasing the metal mobility in order to select disposal sites which provide conditions to provoke minimum pollutant dispersion from the point source.

In this chapter a general overview of the various geochemical processes involved in the control of trace metals in aquatic and terrestrial systems is presented. Their relevance to the mobility of metal pollutants originating from mine tailings and dredged materials is discussed.

In spite of their different origin, mine tailings and dredged materials present strong similarities in behavior and environmental impact. The biogeochemistry of these waste materials is discussed in detail in the following chapters, but at this point one should emphasize that metal sulfide minerals are significant solid phases in both waste types and that pollutant remobilization will be strongly dependent on the redox reactions taking place after disposal (Salomons 1985). Under oxidizing conditions solid sulfides are dissolved and trace metals are no longer controlled by sulfide equilibria but rather by adsorption/desorption or, in limited cases, by other precipitation processes. The oxidation of sulfides generates acid and, if the buffer capacity of the neighboring, receiving environment is not sufficient, acidic conditions may develop which, for some metals such as Cd, may provoke an increasing mobility as a side effect.

The mobility of metal pollutants in the environment at large depends on a more or less elaborate network of interactions between transport processes and heterogeneous phenomena (Fig. 1). Adsorption and precipitation tend to delay the metal transfers. But the solubility, and thus the convective transport of metals, can be increased if complexation with inorganic or organic species takes place in the aqueous phase. A detailed understanding of the mobility of metal pollutants

[1]Water Resources Department, National Geological Survey, BRGM, BP 6009, F-45060 Orléans Cedex, France

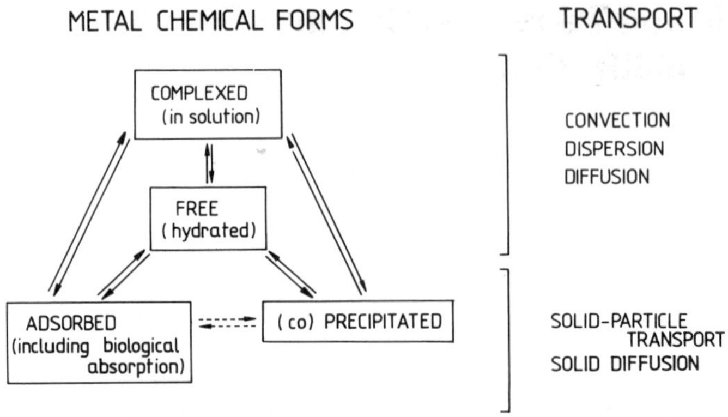

Fig. 1. Geochemical speciation and mobility of metals

therefore necessitates a thorough knowledge of sorption processes (adsorption and precipitation), of dissolved speciation, and of their kinetics in relation to fluid flow.

Since it is the extent of the heterogeneous reactions which controls the retention of elements in the various compartments of the hydrocycle, a significant difference between aquatic and terrestrial environments is the much larger (specific surface of solid)/(volume of liquid) ratio encountered in the latter (Table 1).

Table 1. Relative significance of the solid-solution interface (typical range of values)

System	Specific surface of solids (m^2g^{-1})	Concentration of solids ($g\ l^{-1}$)	Available solid surface (m^2l^{-1})
Aquatic (rivers, lakes, oceans)	10–20	0.00001–0.05	0.001–1
Terestrial (soil, subsoil)	0.1–1	2 000–10 000	1 000–2 000
Bottom sediments	5–20	1 000–5 000	20 000–25 000

2 Sorption by Solid Particulate Matter

The role of solid particles in regulating the composition of natural waters is now widely recognized (Whitfield and Turner 1979; Li 1981; Sigg et al. 1984). Natural solids can dissolve, increasing the dissolved load of the surrounding water phase. New particles can form because of super-saturation, removing or scavenging dissolved components from the aqueous phase. Dissolved elements can adsorb on newly formed or existing solid particles. "Particle surfaces, because of their ability to rival with solute complex formers in tying-up metal ions, are important scavengers and sinks for heavy metal ions in natural water systems" (Sigg et al. 1984). Moreover, if the sorption processes are, even only partially, reversible, natural solids will act as buffers of the dissolved trace metal content of aquatic and terrestrial ecosystems.

The uptake of trace metals by a variety of suspended aquatic particulates, bottom sediments, and soils has been extensively investigated. These studies, often consisting of the determination of sorption isotherms, are, however, usually of little conceptual value and thus they cannot be applied to the quantitative understanding of other systems or situations. It is well known, for example, that adsorption isotherms are strongly dependent on pH (Fig. 2a). The composition of the medium can also significantly affect the extent of adsorption. A tenfold increase in Ca concentration reduces by a factor of 3 the adsorption of Cd on a sandy loam (Fig. 2b). If linear or, better, curved adsorption isotherms have been used with success for modeling the mobility of trace metals in soil columns (Christensen 1985a), these calculations are not capable of accounting for the effects of variations in the major and minor element chemistry of the water phase.

A major difficulty in understanding adsorption by natural solids arises in the characterization of the surface properties of associations of aggregates of composite nature. Bulk composition should be modulated by specific surface contributions. For example, iron oxyhydroxides exert a much larger adsorptive action than could be predicted from their dry weight fraction (Fig. 3). The actual relative effect will be even larger than that reflected by Fig. 3b as these oxyhydroxides often occur as coatings.

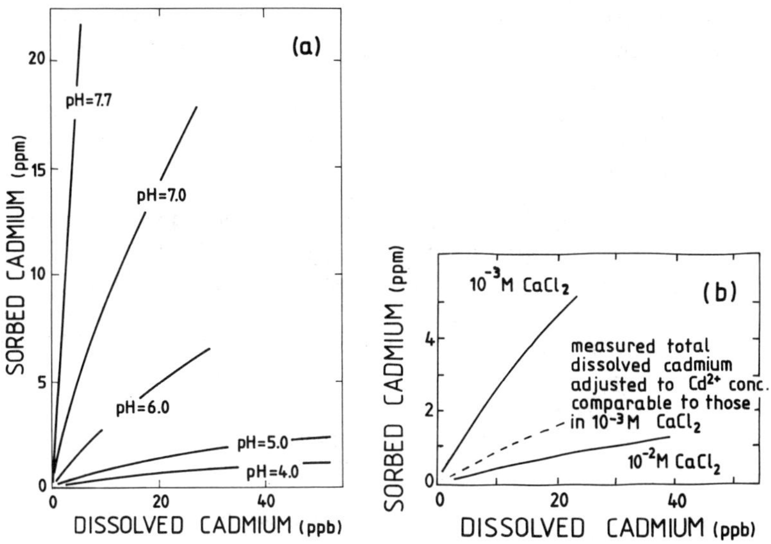

Fig. 2a,b. Sorption isotherms of Cd in the presence of a sandy loam: **a** in 10^{-3} M $CaCl_2$; **b** at pH 6.0 (After Christensen 1984a)

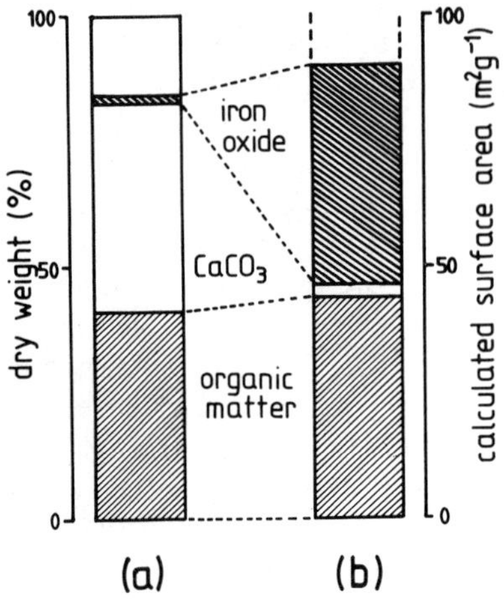

Fig. 3a,b. Composition of settling particles in Lake Zurich in terms of dry weight and calculated surface area. The surface areas were calculated from the specific surface area of amorphous iron hydroxide for the iron oxide and from the size and surface of typical algae for the organic (biological) material (After Sigg 1986).

3 Adsorption by Natural Solids

3.1 Adsorption Chemical Models

The various surface models which have been derived to describe the adsorption of metals on hydrous oxide surfaces can be broadly classified into two categories:

1. The chemical approach, in which dissolved species react chemically with specific surface entities (by ion exchange or by coordination reactions). Physical phenomena (e.g., electrostatic forces or steric considerations) are included only as correction factors.
2. The physical approach, in which the adsorption is interpreted in terms of a combination of electrostatic interactions between ions and the surface charge and ion-solvent interactions.

Both types of models can provide a good fit of experimental data [as shown, for example, by Vuceta (1978) for the adsorption of Cu(II) and Pb(II) on α-quartz in the absence and presence of carbonate, citrate, and EDTA]. The chemical approach is, however, preferred by many scientists because cations and anions seem to form inner-layer, surface complexes (Stumm et al. 1976, 1980). Moreover, the adsorption of metal complexes can be interpreted, similar to coordination chemistry, in terms of the formation of ternary (surface) complexes (Bourg 1983b; Bourg and Schindler 1978, 1988, Bourg et al. 1979).

The various chemical models differ mainly by:

1. The set of surface species and surface reactions [i.e., the fact that adsorption of heavy metals often occurs at pH values where hydrolysis also starts, made several researchers assume that hydrolysis was a prerequisite for adsorption, but it definitely need not be so (Stumm et al. 1976)];

2. The mathematical expression of the "mass action law" as a function of surface-site concentration; and
3. The formulation of the coulombic term.

Surface-adsorption reactions are different from reactions among simple solutes. They should actually be considered as reactions with polymers where the free-energy change must be subdivided into an intrinsic adsorption term (no effect of the neighboring site) and the electrostatic energy of interaction:

$$\Delta G_{adsorption} = \Delta G_{intrinsic} + \Delta G_{coulombic}. \tag{1}$$

Details of the theoretical background can be found elsewhere (Stumm et al. 1976; Stumm and Morgan 1981; Anderson and Rubin 1981). The final choice of surface species and reactions of a given model is always a compromise between simplicity and the necessity to fit experimental data (Morel et al. 1981). A simplified presentation of a commonly accepted chemical model, the surface-complex formation concept, is presented here. Only the basic fundamental aspects, necessary to the application of adsorption studies on simple, solid hydrous oxides to geochemical and environmental problems will be discussed, keeping in mind the waste management objective of the present volume.

3.2 The Surface Complexation Model

Hydrous oxides as well as all natural solids found in aquatic and terrestrial environments contain ionizable functional groups (-OH, -COOH, $-NH_2$, or even $-OPO_3H_2$, -SH). In the presence of water these surface groups are capable of accepting or giving out protons, depending on the pH of their surroundings. Surfaces, therefore, behave like amphoteres.

$$\overline{SOH} \rightleftharpoons \overline{SO}^- + H^+ \tag{2}$$
$$\overline{SOH} + H^+ \rightleftharpoons SOH_2^+, \tag{3}$$

where the superscript describes surface species. Surfaces are thus positively charged at low pH and negatively charged at high pH. The pH of zero point of charge, the pH for which the overall surface charge is equal to zero, is a convenient reference to predict the surface-charge dependence on pH (Parks 1967, Table 2).

The adsorption of metal cations (M^{Z+}) can be interpreted in terms of competition with protons for surface sites, according to the following equation (Hohl and Stumm 1976; Schindler et al. 1976):

$$(\overline{SOH})_n + M^{Z+} \rightleftharpoons (\overline{SO_n M^{(z-n)+}} + nH^+, \tag{4}$$

where, for steric reasons, n is equal to only 1 or 2 (Fig. 4). It has also been suggested by Davis and Leckie (1978) that Eq. (4) should be written as:

$$\overline{SOH} + M^{z+} \rightleftharpoons \overline{SO\ M(OH)_n{}^{(z-n)+}} + nH^+. \tag{5}$$

It is, however, usually not possible to differentiate between Eqs. (4) and (5) because they liberate the same number of protons per metal adsorbed and the concen-

Table 2. Zero point of charge[a]

Solid	pH_{ZPC}	Reference
SiO_2	2.0	Stumm and Morgan 1981
Montmorillonite	2.5	Parks 1967
MnO_2	2–4.5	Stumm and Morgan 1981
Kaolinite	5.5	Bourg 1983a
αFeOOH	7.0	Sigg and Stumm 1981
Calcite	8.1	Stumm and Morgan 1981
$Fe(OH)_3$ amorphous	8.5	Stumm and Morgan 1981
γ-Al_2O_3	8.8	Bourg 1983a
Sediment, Coloma Creek, California	4.3	Brown and Hem 1984
Suspended matter, Gironde Estuary, France	4.8	Bourg 1983a
Sediment, Garonne R., France	5.4	Bourg 1983a
Sediment, Meuse R., Belgium	7.3	Mouvet and Bourg 1983
Sediment, Rhone R., France	9.4	Bourg 1983a

[a]The values were obtained by different methods and are therefore not necessarily directly comparable.

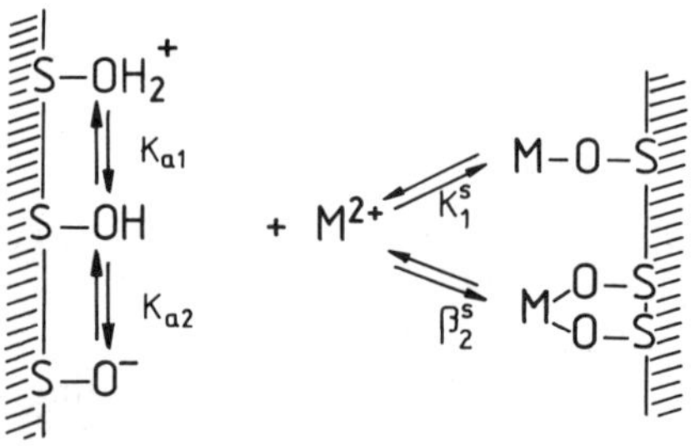

Fig. 4. The surface complexation model

tration of $\overline{(SOH)}_2$ is, in the absence of steric restrictions and for low surface coverage, equal to the concentration of single $\overline{SOH}$ surface groups.

The various surface reactions described by Eqs. (2) to (5) are similar to solution coordination reactions to the restriction that surface sites should be treated as polymers or associations of polymers since a single-surface adsorption site is not independent of other neighboring sites. The equilibrium constants (for surface acidity and surface complexation) should, as in Eq. (1), be split into two terms:

$$\beta_n^{surface} = \beta_{n(intrinsic)}^{surface} \cdot \exp\left(\frac{F\Psi\Delta z}{RT}\right), \quad (6)$$

where Ψ is the electrical potential at the plane of adsorption and Δz is the net change in charge number of the surface species due to the adsorption reaction. The value of Ψ cannot be measured directly. For acid-base surface reactions of simple hydrous oxides, the intrinsic constants can be evaluated by extrapolating the values of the microscopic constants measured for various surface charges to conditions of zero surface charge (Stumm et al. 1980). Overall constants are then derived from the pH dependence of the surface charge. For surface-complex formation reactions the coulombic term is often taken equal to unity (Hohl and Stumm 1976; Schindler et al. 1976) or assumed to be constant at a given ionic strength (and is thus included in the value of the intrinsic constants, which is therefore an apparent constant). A more rigorous treatment has been used for iron oxyhydroxide in sodium nitrate (Davis et al. 1978; Davis and Leckie 1978, 1980), for goethite in model seawater (Balistrieri and Murray 1981), and lake particles in model lake water (Baccini et al. 1982), but being cumbersome it can be avoided, especially for environmental management purposes.

Aquatic and terrestrial solids show the same adsorption pattern (Lion et al. 1982; Elliott 1983; Mouvet and Bourg 1983; Brown and Hem 1984; Kuo and McNeal 1984; Balistrieri and Murray 1984; Cavallaro and McBride 1984) as simple hydrous oxides. Typical examples of metal adsorption on various surfaces are presented in Fig. 5. The adsorption curves present a sharp increase within a restricted 1 to 2 unit pH range. This part of the curve, called the pH-adsorption edge by Benjamin and Leckie (1980), is characteristic of each metal, and also for a given metal, of adsorbent and, even if much less, of adsorbate concentrations (Benjamin and Leckie 1981; Leckie et al. 1980) and ionic strength (Bourg 1983a). According to the mass balance law of Lechatelier, a tenfold increase in surface-site concentration should shift the adsorption edge by 1 pH unit towards the acidic range (Fig. 6).

The consequences of the strong pH dependence on adsorption and of the influence of environmental parameters on the position of the adsorption edge are:

1. Observations such as "cadmium adsorbs onto iron oxide at pH 7" are meaningless if the adsorbate concentration and the chemical characteristics of the system considered are not given (Benjamin et al. 1982);
2. The pH variations (diurnal, seasonal, and spatial) typical of natural systems are thus potentially significant with respect to the adsorption (particle scavenging)/desorption (remobilization) of any trace metal; and

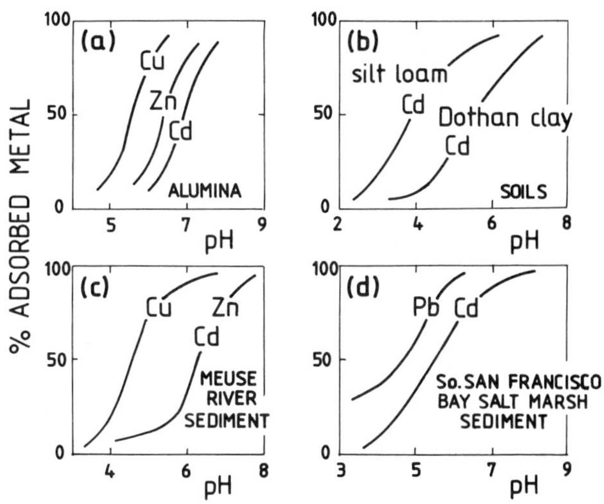

Fig. 5a-d. Adsorption of trace metals on various solids: **a** from Bourg (1983b); **b** from Elliott (1983); **c** from Mouvet and Bourg (1983); **d** from Lion et al. (1982)

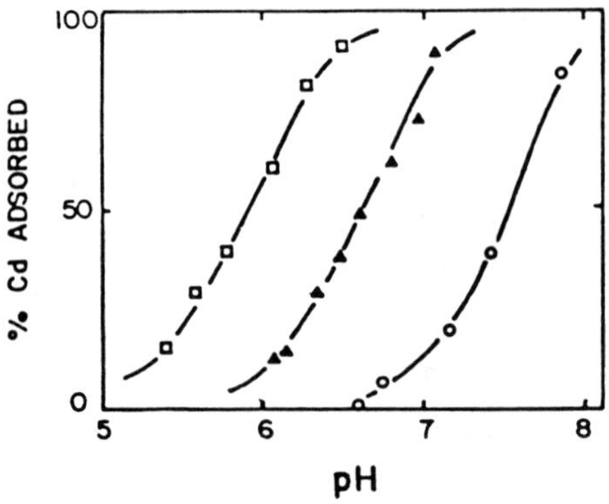

Fig. 6 Adsorption of cadmium on amorphous $Fe(OH)_3$ for various concentrations of solids (□: 1.3×10^{-2}; ▲: 1.3×10^{-3}; ○ 1.3×10^{-4} mol Fe dm^{-3}) (Benjamin and Leckie 1981)

3. All other parameters being unchanged, an increase in dissolved complexation (larger concentrations of the complexing agent or the same concentration of a stronger complex former) results in a shift toward higher pH of the adsorption edge.

3.3 Adsorption of Metal Complexes

It was assumed until recently that the chemical speciation of trace metals was the result of competition between adsorption and solution complexation. Increasing amounts of complexing agents could therefore remobilize adsorbed trace metals. There is now, however, substantial evidence that adsorption not only involves free (hydrated) ions, but also inorganic and organic complexed forms. For example, as seen in Fig. 7, the presence of equimolar concentrations of Cu and organic molecules does not necessarily prevent adsorption of the metal on amorphous silica. The cases of EDTA and oxalate correspond to the hypothesis mentioned above, of competition between adsorption and complexation. The EDTA is a

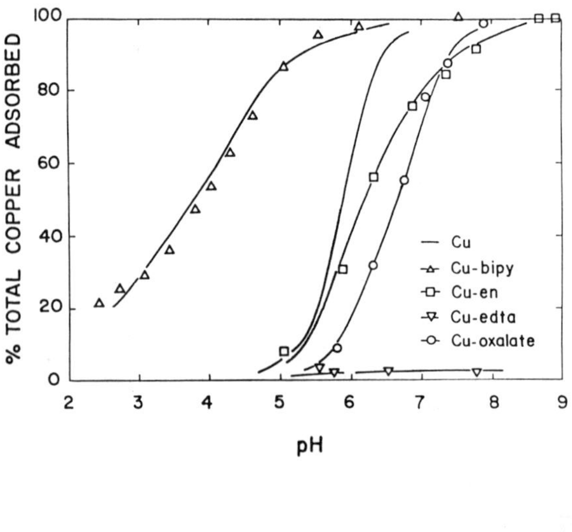

Fig. 7. Adsorption of Cu on amorphous silica in the presence of equimolar concentrations of complexing agent (Bourg and Schindler 1988)

strong enough chelate to prevent the uptake of Cu by silica in the whole pH range of Cu adsorption. Oxalate is a weaker chelate and at pH values greater than 6 it cannot compete successfully with the surface sites. The adsorption edge for the Cu-oxalate system in Fig. 7 corresponds to the uptake of "free" Cu. However, in the presence of 2,2'bipyridyl (bipy) and ethylenediamine (en) the behavior of Cu cannot be explained without the adsorption of metal complexes (Bourg et al. 1979; Bourg and Schindler 1978, 1988). The adsorption of the metal complexes is different from that of the free metal but it is still metallike (i.e., the quantity adsorbed increases with pH). Such an effect was also observed for inorganic complexes by Benjamin and Leckie (1982). The adsorption patterns of Cd on $Fe_2O_3.H_2O$(am), α-SiO_2, γ-Al_2O_3, and γ-FeOOH in the presence of Cl and SO_4 were consistent with a model in which chloro- and sulfato-cadmium complexes are assumed to adsorb, with the metal closer to the surface. The ternary surface complexes mentioned until now are of the form:

Surface-Metal-Ligand.

Complexed metals can also adsorb with the ligand closer to the surface:

Surface-Ligand-Metal.

The adsorption behavior of the metal complex is now completely different from that of the free metal. The quantity of metal adsorbed decreases with pH, the adsorption is ligandlike (Fig. 8). Similar adsorption patterns were observed for various organic and inorganic (S_2O_3) complexes of Cu, Zn, Cd, and Ag (Davis and Leckie 1978; Elliott and Huang 1979, 1980, 1981; Benjamin and Leckie 1982; Bowers 1982; Bourg 1983b; Rhoads and Huang 1983).

The speciation and thus the mobility of trace elements in natural systems is not an easy matter to predict even qualitatively. All significant parameters must be included in the evaluation. Of special importance is the relative contribution of the various reactions. The complexity of natural aquatic and terrestrial systems can easily be illustrated by the study of much simpler systems: one metal, one surface and one organic ligand (Fig. 9). In Fig. 9a as the total organic matter increases, the

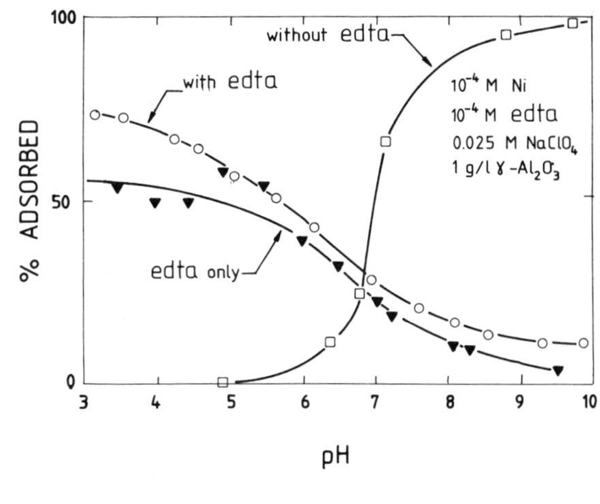

Fig. 8. A comparison of EDTA, Ni-EDTA, and free-Ni adsorption on γ-Al_2O_3 (Bowers 1982)

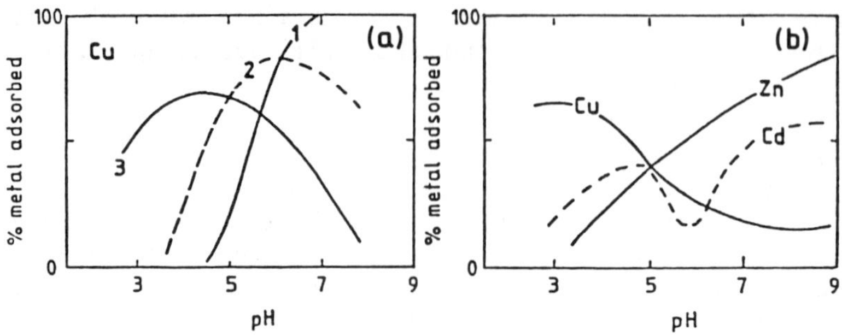

Fig. 9a,b. Adsorption of trace metals in ternary systems (metal, inorganic solid, organic matter). **a** Adsorption of Cu (10^{-6} M) on alumina (10 g l^{-1}). Organic carbon added: (1) none; (2) 9.4 mg l^{-1}; (3) 46.8 mg l^{-1} (After Davis 1984b); **b** adsorption of trace metals (5×10^{-6}M) on alumina (1 g l^{-1}) in the presence of fulvic acid (80 mg organics l^{-1}) (After Bourg and Schindler 1985)

adsorptive behavior of Cu changes from metallike to organic matterlike (Tipping et al. 1983; Davis 1984b). In Fig. 9b the behavior of Cu and Zn are ligand- and metallike, respectively, while that of Cd is intermediate (Bourg 1983a).

3.4 Adsorption Kinetics

Adsorption reactions are usually rapid with a possible slower second step of uptake (Bunzl et al. 1976; Benjamin and Leckie 1981; Bourg 1983b; Christensen 1984a; Aringhieri et al. 1985). Longer equilibration times, of up to more than 100 days, have been reported for marine sediments (Balistrieri and Murray 1983, 1984; Nyffeler et al. 1984; Santschi et al. 1984) and calcium carbonate (Davis 1984a). This time dependence can be interpreted either by evolution with time of the adsorbing particle or by adsorption-desorption equilibrium followed by a slow migration in the lattice structure.

3.5 Adsorption Reversibility

Desorption can be caused in two ways: (1) a variation in the chemistry of the water phase (lower pH, increasing content in dissolved complexing agents, or in adsorption competitors) and (2) for a constant water chemistry, a decrease in the dissolved fraction of the metal of interest. The former is, in fact, a combination of side reactions and disequilibrium imposed on the law of mass action in Eq. (4). The effect of changes in water chemistry can only be truly understood if the intrinsic desorption is characterized.

The reversibility of adsorption has been studied by several investigators. However, most of the results are not really usable because it is often overlooked that (1) desorption experiments must be performed with the same solid to solution ratio as the corresponding adsorption investigation since adsorption is a distribution between particulate (surface) phase and dissolved phase resulting from a mass-action law (or several mass-action laws if significant dissolved complexation occurs) and (2) chemical extraction by complexation and ion displacement may significantly modify the system, and the metal remobilization is the result of processes other than the previously mentioned surface/dissolved distribution.

Another experimental procedure consists in equilibrating the metal-containing solution with the solid at the appropriate pH for approximately 90% metal adsorption, as determined by prior experiments (determination of the adsorption edge) and in subsequently lowering the pH of the system to that appropriate for 20% adsorption. Separate experiments performed with Pb and Cd on sediments from the oxidized layer of mudflats in the South San Francisco Bay estuary indicated slow release of 85% of predicted desorbable Cd after 96 h, whereas less than 15% of predicted Pb was released after 11 days (Lion et al. 1982). Using the Tessier et al. (1979) sequential extraction procedure for an operationally defined distinction of associations of Pb and Cd with major solid phases, it was observed that a substantial fraction of total extractable Cd (50%) was associated with organic phases, while about 70% of the total extractable Pb was remobilized during the destruction of carbonate and Fe/Mn oxides. Moreover, the adsorption edge of Cd on extraction residuals shifted to a higher pH value (ca. 0.5 pH units) after the organic extraction step, while the edge of Pb shifted to a higher pH value (0.5 pH units) after the combined extraction of carbonate and Fe-Mn oxides. The kinetically poor reversibility of Pb adsorption can thus be attributed to incorporation into either or both of these solid components, most likely the carbonates. Additional problems may arise because of the changing particle characteristics with pH history. Significant changes in their aggregating characteristics may arise, thus affecting particle size, which in turn could significantly alter adsorption reversibility.

The approach of Mayer (1978) provides results which are directly applicable to the prediction of dissolved metal concentration in aquatic and terrestrial environments. The desorption experiments are carried out analogously to the adsorption experiments with the restriction that the added solute is initially free of metal. The solid and its adsorbed metal are the product of a previous adsorption experiment (after separation from the corresponding aqueous phase by centrifugation or filtration). Mayer (1978) observed full reversibility for an acid forest soil loaded with low Cd concentrations.

Using this same approach with Cd concentrations in the micromolar range on two Danish soils, Christensen (1984b) observed full adsorption reversibility for a loamy sand in 10^{-3} M $CaCl_2$ at pH 6.0, while under the same conditions, a partial reversibility of 1 μg Cd g^{-1} was exhibited in a sandy loam. This partial retention cannot be explained by any strong binding since no reversibility was observed in 10^{-2} M $CaCl_2$ at the same pH. Aging of the soil at 1°C, after exposure to Cd, for up to 67 weeks, did not produce any changes with respect to desorption of Cd at pH 6.0 in 10^{-1} M $CaCl_2$.

In similar experiments, Calvet (1981) observed a partial irreversibility, more or less marked depending on the soils studied. The strongest irreversibility seemed to be correlated to high carbonate-containing samples.

3.6 Adsorption Competition

Studies of single-metal retention on a given solid are not necessarily relevant to management objectives since competitive effects, for example for adsorption sites, may in some cases be significant. The presence of an initial concentration of 40 ppm Cu had no effect on the adsorption of 20 ppm Zn on an acid mineral soil (McBride and Blasiak 1979). On the other hand, the sorption of copper on various Californian soils was not significantly affected by zinc addition of up to 250 $\mu g\ l^{-1}$, though 15 $\mu g\ l^{-1}$ copper completely prevented the sorption of zinc in one low CEC soil (Kurdi and Doner 1983). The presence of competing metals increased the general availability of heavy metals, especially of Cd, in the presence of an orthic luvisol (Schmitt and Sticher 1986). Lead and copper were also, but to a lesser extent, affected by the presence of other metals. A strong decrease of Cd adsorption in the presence of competing metals was also observed by Christensen (1985b) on a sandy loam and a loamy sand.

4 Sorption by Carbonates

Trace metals can also be removed from the aqueous phase by reacting with solid carbonates, the phenomenon involved being incorporation in the solid lattice rather than purely surface adsorption. This is demonstrated by the stronger adsorption of Zn^{2+} on magnesian carbonates than on calcite because of the better spatial match with Mg than with Ca atoms in the $MgCO_3$ crystal lattice (Jurinak and Bauer 1956) and also by the long-term increase of Cd uptake by calcite, indicating diffusion of Cd in the $CaCO_3$ particles (Davis 1984a). Long-term equilibrations showed that equal quantities of coarse and fine $CaCO_3$ adsorbed Cd at greatly different rates (McBride 1980). For the fine $CaCO_3$ system, equilibrium was attained within several minutes, while in the case of the coarse calcite, the Cd dissolved concentrations kept decreasing even after the 10-day study.

Precipitation of $CdCO_3$ as a separate solid is unlikely to occur at levels of Cd encountered in most aquatic and terrestrial systems (McBride 1980; Bourg 1983b).

5 Dissolution/Precipitation

Precipitation/dissolution reactions are known to control the dissolved composition of many major and minor cations in aquatic systems resulting from geological weathering. Can this be also true of trace metals especially in mine tailing or dredged material disposal sites and in their vicinity?

The dissolution of a solid M_iL_j corresponds to the following equation:

$$M_iL_j(s) \overset{Ksp}{\rightleftharpoons} iM^{z+} + jL^{y-}, \tag{7}$$

where M and L are a metal ion of charge z+ and a ligand ion of charge y-, i and j are stoichiometric coefficients and Ksp is the solubility product.

The identification of a solid phase together with knowledge of its solubility and its rates of dissolution and precipitation would provide sufficient evidence of the potential control of trace-metal concentrations in aquatic and terrestrial aqueous systems. However, the mineralogical evidence for the existence of pure trace-metal precipitates is scarce.

Solid minerals of the common heavy metals have rarely been identified, most often because they were not looked for, which is usually justified when the equilibrium solubility is not overcome and, even so, such minerals are difficult to identify directly because of their small contribution to the total solid mass. A number of Pb phosphate minerals of the plumbo-gummite family have been identified (Norrish 1968). The basic copper minerals malachite $Cu_2(OH)_2CO_3$, brochaulite $Cu_4(OH)_6SO_4$, and paratacamite $Cu_4(OH)_6Cl_2$ occur widely in natural aquatic environments. They are found as corrosion products of copper and its alloys, in the oxidized portions of mineral deposits, as coatings on rocks in areas with minimal copper concentrations, and floating in the oceans (Jedwab 1979; Woods and Garrels 1986).

In the absence of positive identification of solid minerals of trace metals, the indirect method of equilibrium solubility calculation is used. The equilibrium dissolved speciation of the water studied is calculated (see Sect. 6) thus giving the free ionic activity of the species involved in the formation of the possibly existing minerals. Their saturation status is then calculated by means of the saturation index S.I.:

$$\text{S.I.} = \log \frac{\text{Ionic Activity Product}}{\text{Solubility Product}}. \tag{8}$$

The dissolved concentration of a metal is controlled by the precipitation/dissolution of a given mineral if the S.I. of this mineral is equal to zero. Positive S.I. values indicate supersaturation corresponding to either slow precipitation or the nonexistence of the mineral under the specified geochemical conditions. Negative S.I. values indicate undersaturation corresponding to slow dissolution, if the mineral of interest is present, or to another geochemical control.

This method is limited (Mattigod et al. 1981) by the following facts:

1. All possible minerals should be known and investigated.
2. Mixed minerals or, in other words, coprecipitation cannot be considered. Mixed solid formation is common (Table 3). For example, when minerals such as montmorillonite and Fe or Mn oxides precipitate from soil solutions, they are characterized by a wide range of isomorphous substitutions in which metal cations in their usual structures can be partly replaced by trace metal ions of the

Table 3. Coprecipitated trace metals in secondary soil minerals (After Sposito 1983)

Mineral	Coprecipitated trace metals
Fe oxides	Ni, Cu, Zn, Mo, Mn
Mn oxides	Fe, Co, Ni, Zn, Pb
Ca carbonates	Mn, Fe, Co, Cd
Clays	Fe, Mn, Ni, Co, Cr, Zn, Cu, Pb

same charge and of similar size (Sposito 1983). Ni, Co, Mn, and Cu can form minerals of widely ranging compositions with siderite, $FeCO_3$ (Thornber and Nickel 1976).

3. There is a lack of equilibrium solubility data (see point 1) of many minerals of trace metals. This problem can be circumvented to some degree by using empirical methods to estimate Ksp values as yet unavailable (see references in Mattigod et al. 1981).
4. The computation of ion activities can be inaccurate, especially for high ionic strength. Direct measurements are difficult. All complexes (especially organics) and ion pairs must be considered. Major and minor cations and anions must all be analyzed to permit these calculations.
5. It is an equilibrium method. For example, the composition of soil solutions is often controlled by fast-forming metastable solid phases. Consequently, all solids that are likely to occur, stable or metastable, must be considered.

6 Speciation

Chemical speciation can be defined as the distribution of a given element or compound into the various chemical forms which together make up the total concentration of that element or compound in a sample.

It is now well recognized that it is not the total amount of a metal pollutant which is most significant for the understanding of geochemical fluxes. One of the keys to the description of the mobility of a metal is its speciation. This is especially important for trace metals because they are largely influenced by the major ion chemistry of their aqueous environment:

$$iM^{z+} + jL^{y-} \overset{K_{M_iL_j}}{\rightleftharpoons} M_iL_j^{(iz-jy)+}, \quad (9)$$

where $K_{M_iL_j}$

is the formation constant of the complex. Let us note that one advantage of the surface complexation model described earlier is that surfaces can be considered as classical (polymer-type) ligands. The activity of any complexed species can be written as:

$$(M_iL_j) = K_{m_iL_j} (M^{z+})^i (L^{y-})^j. \quad (10)$$

Two types of effects can thus modify the speciation of M:

1. Direct effect: an increase in the concentration of L causes M to be complexed more.
2. Indirect effect: an increase in the concentration of a competitor of M for complexation causes a decrease in (L^{y-}) and thus M is less complexed.

In practice, the significant ligands for trace metals are OH^-, Cl^-, CO_3^{2-}, HCO_3^-, SO_4^{2-}, organic molecules and macromolecules, surface sites and, to a lesser extent, phosphorus and silicate forms and NO_3^-. The chemical forms of trace metals in soils and aquatic systems have been reviewed by Florence and Batley (1980), Sposito (1983), and Salomons and Förstner (1984).

Analytical in situ determination of the dissolved speciation can provide at best very limited information. A significant effort has been made in the last few years to determine the complexing capacity of natural waters (Kramer and Duinker 1984), but while everyone agrees that the dissolved organic carbon is a major complexing agent, there is no consensus yet as to the complexing intensity or the density of complexing sites.

The conditional constants for the complexation of trace metals with organic matter vary largely with the pH of the medium in which they were determined. In a first approximation, the simple linear relationship between the logarithmic values of the constant and the pH, observed by Buffle (1980), for Cu can be extended to other metals (Bourg and Vedy 1986).

The effect on the speciation of trace metals of some of the most significant types of parameters (pH for hydrolysis and carbonate complexes, chloride for inorganic complexation, NTA for organic complexation, and adsorbing surface for adsorption) is illustrated in Fig. 10. The consequences for the mobility of heavy metal will be discussed later.

Since the appearance of the first paper on chemical speciation of a natural water (Garrels and Thompson 1962), the use of computers to solve chemical equilibrium speciation calculations has expanded enormously (Nordstrom et al. 1979; Nordstrom and Ball 1984).

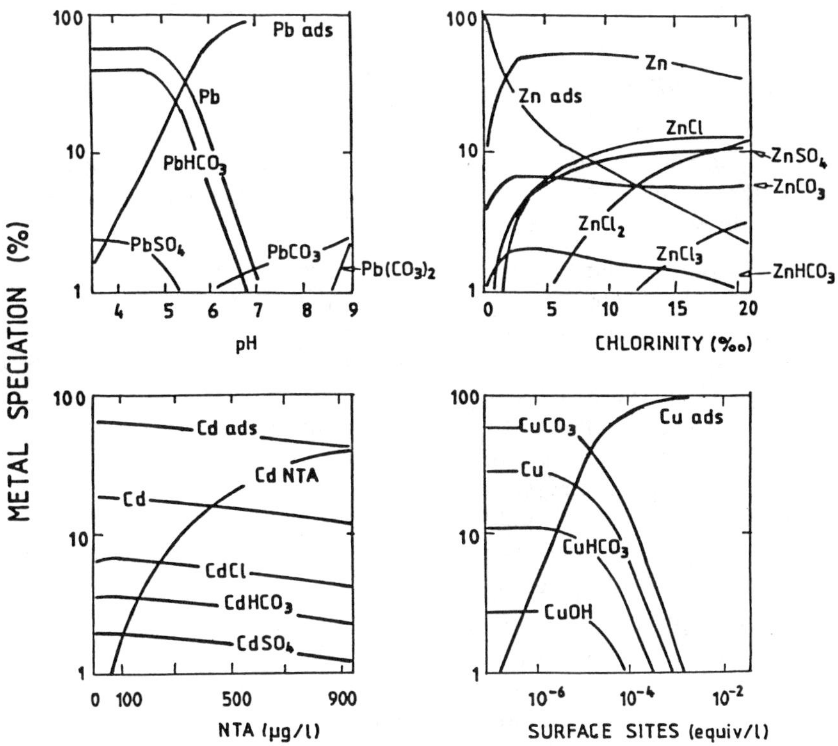

Fig. 10. Speciation of trace metals as a function of various parameters. (NTA diagram from Salomons et al. 1988, others from Bourg 1983b)

In the absence of a dynamic geochemical model, equilibrium calculations can provide useful qualitative information on potential fluxes between several compartments of a given ecosystem. For example, an equilibrium speciation computer code with the weakest reported binding constants between trace metals and dissolved organic matter was used to investigate expected forms of dissolved trace metals in the gravitational water of two acid-soil profiles (Bourg and Vedy 1986). In spite of the weak constants used, organic complexes were predicted to be significant species of trace metals in an acid brown earth and a humo-ferrugineous podzol. It was concluded that the fate of trace metals associated with the dissolved organics in these soils should follow the dynamics and evolution of their organic vectors. The acid brown soil A_1 horizon organic trap is effective, as shown, for example, by the strong accumulation of Pb in the organic surface layer of an acid brown soil from the Solling area, FRG (Heinrichs and Mayer 1980). On the other hand, even though they are known to retain organics, podzol B_h horizons do not seem to concentrate trace metals (Cheshire et al. 1981).

Other uses of equilibrium speciation calculations are numerous (Zirino and Yamamoto 1972; Sibley and Morgan 1975; Long and Angino 1977; Mantoura et al. 1978; Vuceta and Morgan 1978; Wilson 1978; Lindsay 1979; Dyrssen and Wedborg 1980; Jenne et al. 1980; Mattigod 1981; Turner et al. 1981; Chapman et al. 1983; Mouvet and Bourg 1983; Sposito 1983; Salomons 1984; Bourg 1984, 1988). Some of these will be described in the next section on control of dissolved trace metals.

7 Redox Conditions

The dispersion of metal pollutants from mine tailings and dredged materials is influenced by redox reactions at two stages during their release from the deposition area to the surrounding environment:

1. The disposal on land or in oxic aquatic environments of dredged material, the resuspension of contaminated sediments, the disposal of mining wastes, and the operation of mines all provoke drastic changes from reducing to oxidizing conditions. These changes are not due to the introduction in the global environment, but rather to the physical transfer of pollutants concentrated by given solids (mineral deposits, sediments) through natural processes and suddenly exposed to new chemical conditions.
2. Metal pollutants released from the deposition area because of the transition from anoxic to oxic conditions can later, in specific areas of the receiving surroundings, be exposed to further redox variations (daily, seasonal, or spatial) capable of retarding or even blocking their migration.

The weathering of mine tailings can be exemplified by the oxidation of pyrite, in which the side effect of oxidation is the release of acidity (Fig. 11). After the oxidizing reaction is initiated, the ferric ion is capable of taking over the role of oxygen to produce additional Fe(II) and acidity. The end product is soluble Fe^{2+} or solid $Fe(OH)_3$ depending on the pH and Eh buffering capacity of the mining

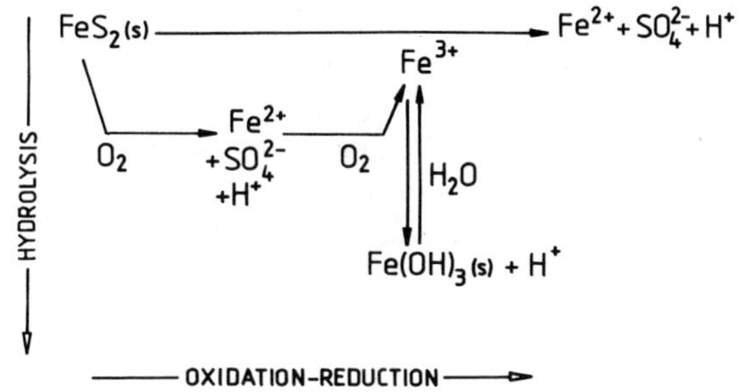

Fig. 11. Schematic description of the oxidation of FeS_2 (After Stumm and Morgan 1981)

waste and immediate surroundings as well as the dissolved organic content of the leachate.

For dredged material, organic matter is a significant component of the oxidation process (Fig. 12). The organic compounds formed by the microbial decay of organic matter can inhibit the oxidation of Fe(II). Moreover, the solubility of Fe(III) can be drastically enhanced by the formation of dissolved organic complexes.

A natural analog of the oxidation of dredged material is the fate of trace metals in anoxic areas of aquatic systems. In the estuary of the Scheldt River (Holland) anoxic conditions are encountered until the chlorinity reaches 5%. Upon entering the more oxic estuarine water, trace metals are released from the suspended matter (Salomons and Eysink 1981). A similar behavior was observed in an experiment simulating the dumping of dredged material in the North Sea. When anoxic sediments were brought in contact with oxic seawater in a flow-through system for 4 weeks, a large release of cadmium (50% of the total cadmium present in the original sediment) and smaller releases of other trace metals were observed (Salomons et al. 1982). A control experiment carried out in fresh water produced only a little release. An adsorption was even observed for some metals. The transition from anoxic to oxic conditions can be explained by the following consecutive reactions:

1. An oxidation of metal sulfides;
2. The formation under oxic conditions of new species, such as dissolved chloro-complexes in marine and estuarine environments or adsorbed compounds in fresh water.

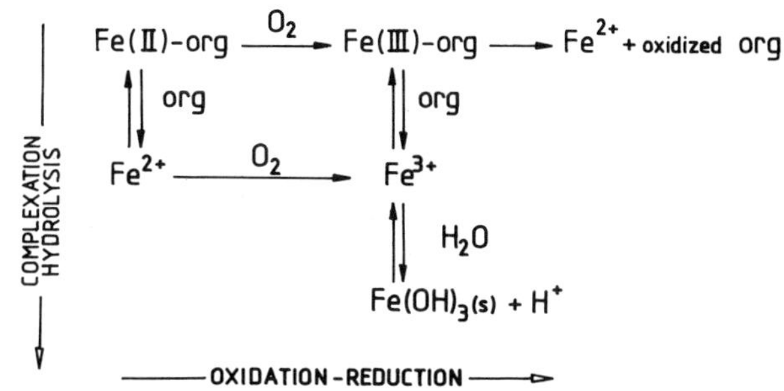

Fig. 12. Schematic description of the redox behavior of iron in the presence of organic matter (After Theis and Singer 1974)

The latter depends on the chemistry of the immediate environment of the metal remobilized during the former. As mentioned earlier, pH and complexing agents have a significant influence on the dissolved metal speciation (Fig. 10). Moreover, these two parameters as well as the availability of solid surfaces will affect the metal adsorption (Fig. 10).

The solubility and mobility of trace metals are influenced by the redox status of their environment, even though they are usually not directly involved in oxidation-reduction reactions. The oxidation of sulfide into sulfate will increase the solubility of heavy metals by lowering the ionic activity product of solid metal sulfides. The acidity produced during the reaction will also affect the solubility of trace metals (see Sect. 3). The redox chemistry of iron and manganese, ubiquitously present as coatings on natural solids (Jenne 1968), will also tremendously affect the mobility of trace metals (Fig. 13). Oxidizing conditions provoke the formation of coprecipitating and later adsorbing amorphous oxyhydroxides of Fe and Mn. Slightly reducing conditions, such as a decrease in soil O_2 levels resulting from temporary waterlogging or from high biological activity, induce the reduction of Mn^{4+} to Mn^{2+} and Fe^{3+} to Fe^{2+} thereby solubilizing their associated trace metals.

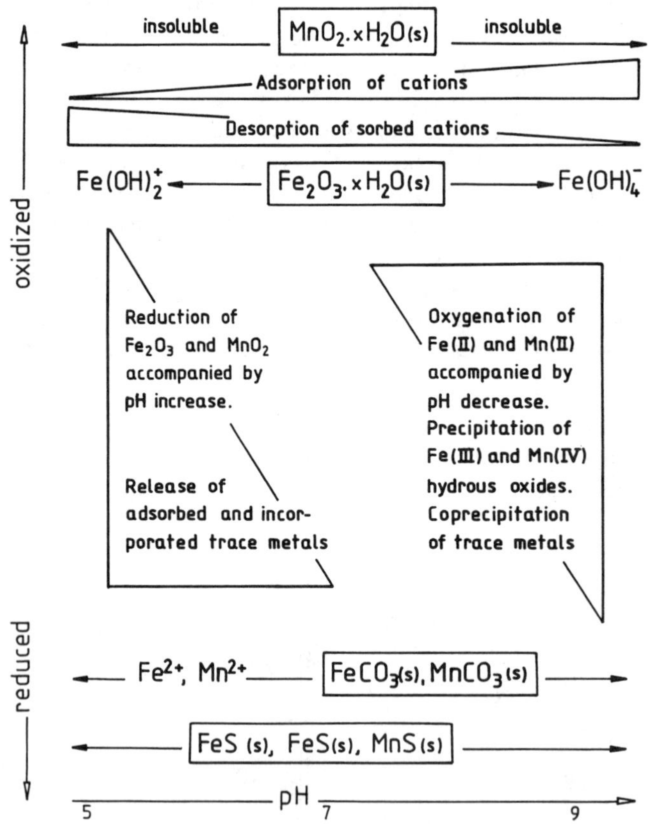

Fig. 13. Major features of the redox chemistry of Fe and Mn (After Stumm and Morgan 1970) in the absence of organic matter

The solubilization of iron and manganese in reducing environments is well documented (Lyons et al. 1979; Ellaway et al. 1980; Sholkovitz and Copland 1982; Smock and Kuenzler 1983). The water-soluble Fe and Mn of a waterlogged silt loam increases with decrease in pH and Eh (Fig. 14). Therefore, not only will the adsorbing capacity of the soil be lowered because of the dissolution of the strongly adsorbent Fe and Mn oxides, but Fe^{2+} and Mn^{2+} will also be competitors with trace metals for adsorption sites.

Khalid et al. (1975, 1979) investigated the uptake of Cd^{109} on a salt marsh sediment and of Hg on Mississippi River sediment (Fig. 14). In the cadmium study the low metal recovery at low Eh coincided with high concentrations of sulfides (>500 μg S g^{-1} solids) indicating cadmium sulfide precipitation. the low Cd recovery at high pH for the whole range of redox potentials is probably due to adsorption. In the mercury investigation the addition of 2.4 μg of Hg^{203} g^{-1} sediment resulted in recoveries always lower than 0.5%. The addition of 60 μg of Hg g^{-1}

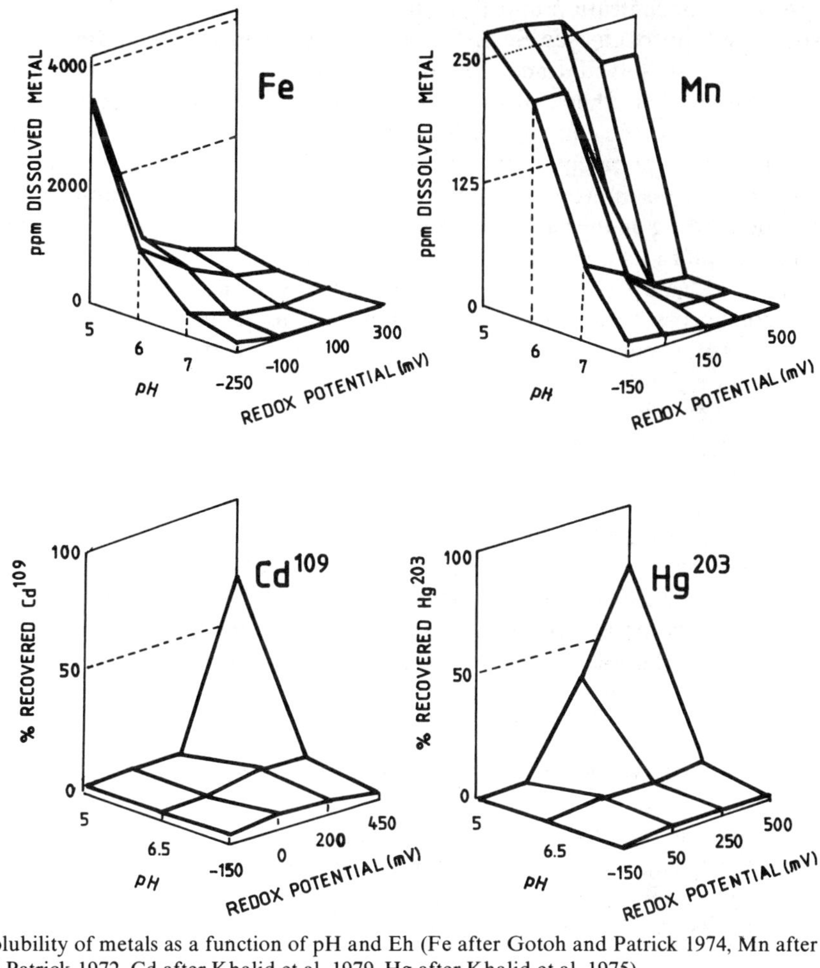

Fig. 14. Solubility of metals as a function of pH and Eh (Fe after Gotoh and Patrick 1974, Mn after Gotoh and Patrick 1972, Cd after Khalid et al. 1979, Hg after Khalid et al. 1975)

sediment produced a trend similar to that of cadmium. Under reduced and moderately oxidized conditions, essentially all the mercury added was sorbed by the sediment. A greater decomposition of organic matter under oxidized conditions may partially explain the lower uptake of Hg. A more realistic explanation for the increase in solubility with decrease in pH is the use of HCl to maintain pH at 6.5 and 5.0. Complexation with chloride could also explain the high Cd^{109} recovery for low pH and high Eh.

8 Control of Dissolved Trace Metals

The retention of trace metals in aquatic and terrestrial systems is due to heterogeneous geochemical processes and to physical deposition or filtering of natural solids. Which of the sorption processes, adsorption or precipitation, controls the dissolved concentration and thus the mobility of trace metals? The equilibrium solubility approach (see Sect. 5) has been used in many studies. Under some conditions, dissolved trace metals are regulated by precipitation phenomena (Table 4). Under others, adsorption is (Mouvet and Bourg 1983; Bourg 1984) or is suggested to be the control mechanism (Table 5).

Equilibrium calculations performed with available solubility products and adsorption constants on Meuse River sediments (Mouvet and Bourg 1983) suggest that depending on the chemical characteristics of the water phase, both adsorption and precipitation are regulating mechanisms of the dissolved heavy metal content of natural waters (Fig. 15). For pH values lower than 8, adsorption is likely to control the dissolved Cu of rivers with low Cu load. If the total Cu were to be higher, the adsorption curves of Fig. 15 would be shifted toward the top of the diagram. The malachite solubility would then be controlling the dissolved Cu at lower pH values.

Further evidence of the importance of adsorption as a control mechanism is provided (1) by the equilibrium model of Hunter (1983) for the scavenging of reactive metals from the deep oceans by suspended particles covered with an organic film and (2) by the simple black box adsorption model of Salomons (1983) for the regulation of dissolved Zn in the IJsselmeer, Holland.

Adsorption/desorption reactions are usually faster than many of the precipitation/dissolution reactions and they can thus well dominate during periods of high heavy metal input in any compartment of the hydrocycle.

The most important characteristics of suspended or deposited sediments, soils, and subsoils for the retention of metal pollutants are:

1. pH;
2. Content of hydrous oxides of Fe and Mn;
3. Organic matter content;
4. Oxidation/reduction potential;
5. Surface properties (cation-exchange capacity);
6. Nature and concentrations of constituents of the aqueous phase;
7. Quantity and flow rate of solution which moves through the aquatic or terrestrial system.

Table 4. Possible control of dissolved trace metals by precipitation/dissolution

Metal	Mineral	System	Reference
Pb	Pb jarosite, $PbFe_6(SO_4)_4(OH)_{12}$ (small amount found in sediments)	Acid mine drainage stream 1 (thick hydrous iron oxide crust on stream bed)	Chapman et al. 1983
Cu	$Cu_2(OH)_2CO_3$	Acid mine drainage stream 2	Chapman et al. 1983
Zn	$ZnSiO_3$	Acid mine drainage stream 2	Chapman et al. 1983
Pb	$PbFe_6(SO_4)_4(OH)_{12}$	Acid mine drainage stream 2	Chapman et al. 1983
Cu	$Cu_2(OH)_3CO_3$	Garonne River and Gironde Estuary (3 stations out of 6)	Bourg 1984
Zn	Zn_2SiO_4	U.S. rivers	Hem 1972
Cd, Zn	Cd, Zn carbonates	U.S. rivers (24 out of 80)	Hem 1972
Pb	$Pb(OH)_2$, $Pb_3(PO_4)_2$, $Pb_4O(PbO_4)_2$, $Pb_5(PO_4)_3OH$	Non calcareous soil (depending on pH)	Santillan-Medrano and Jurinak 1975
Pb	idem plus $PbCO_3$	Calcareous soil (depending on pH)	Santillan-Medrano and Jurinak 1975
Cd	$Cd_3(PO_4)_2$ and/or $CdCO_3$	Soil, high Cd concentration	Santillan-Medrano and Jurinak 1975
Pb	$Pb_5(PO_4)_3Cl$	Sewage treatment plant outfall	Bailey et al. 1975
Pb	$Pb_5(PO_4)_3$ Cl	Mining drainage	Jenne et al. 1980

Table 5. Possible control of dissolved trace metals by adsorption/desorption

Metal	System	Reference
Zn, Cd	U.S. rivers	Hem 1976
Cd	Soils	Santillan-Medrano and Jurinak 1975
Cd, Zn, (Cu)	Garonne River and Gironde Estuary	Bourg 1984
Zn, Cu	Acid mineral soil	McBride and Blasiak 1979
Zn	Soils	Brümmer et al. 1983
Cu, Zn, Cd	Sewage treatment plant outfall	Bailey et al. 1975
Cd	Hydrous ferric oxide (in phosphate free and phosphate enriched)	Kuo and McNeal 1984
Cu, Cd	Acid and calcareous soils	Cavallaro and McBride 1978
Cd, Cu, Ni, Pb, Zn	Meuse River	Bourg and Mouvet 1988

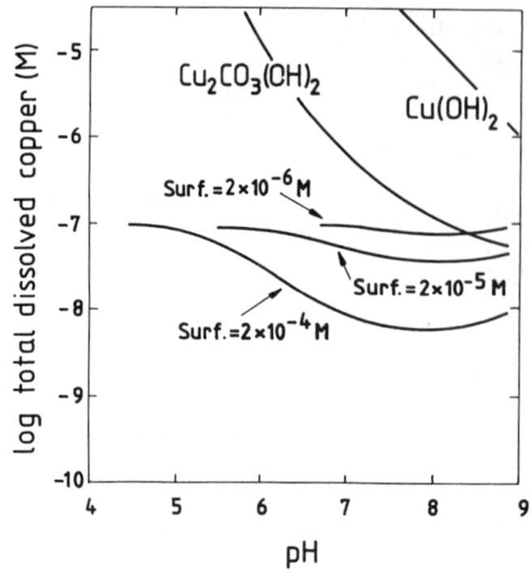

Fig. 15. Geochemical control of dissolved Cu in average river water (Bourg and Mouvet 1988)

Let us simply point out that the influence of clays on the total adsorbing strength of a natural solid is negligible compared to the contribution of organic matter and oxides of Fe and Mn (MacLaren et al. 1981).

9 Mobility of Trace Metals

A number of ions or compounds and processes can be involved in keeping heavy metals in solution or in remobilizing them from temporary retention substrates.

It is quite clear that complexation with chloride ions is a mobilizing factor.

An example of the migration of Cd, Cu, and Ni in the presence of Cl^- and ClO_4^- ions is provided by the soil-column experiments of Doner (1978). The perchlorate experiments were used as references since significant complexing of ClO_4^- with Cd, Cu, and Ni does not occur. Chloride complexation increases the metal mobility (Fig. 16). The effect of chloride on metal solubility followed the order of the log K values of the metal chloro-complexes:

$$Cd \gg Cu > Ni.$$

Mercury migrated through a sand column much faster in the presence of chloride (Behra 1986). Moreover, mercury retained by the sand column when introduced as the nitrate salt was completely remobilized when the supporting electrolyte of the percolating fluid was shifted to sodium chloride (Behra et al. 1988).

The presence of chloride also decreased the adsorption of Hg on sediments (Reimers and Krenkel 1974) and on silica (MacNaughton and James 1974), of Cd on soils (Christensen 1983b), on alumina (Bourg 1983b), and on montmorillonite (Garcia-Miragaya and Page 1976; Egozy 1976) and of Cd, Cu, and Zn on sediments (Bourg 1983b).

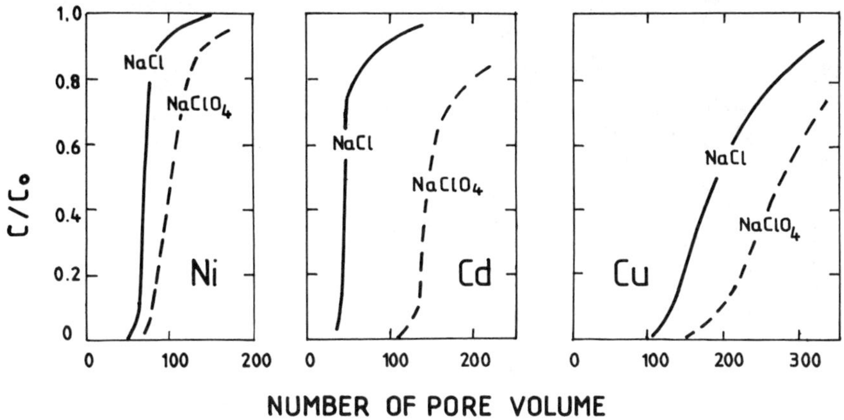

Fig. 16. Elution curves of Ni, Cd, and Cu as affected by Cl^{-1} and ClO_4^- ions (After Doner 1978)

The mobility of Cd, Ni, Hg, and Pb in the presence of a humic acid, NO_3^- or Cl^-, in an organic matter-free sandy soil was studied by Gomez and Juste (1982). A large fraction of the metal added as inorganic salts was removed by leaching during the first week of the experiment, while metal added as a humic compound was strongly immobilized with the exception of Ni.

Under acid leaching (pH 3) by dilute solution (0.025 M) of $AlCl_3$ and $FeCl_2$, the important soil properties describing the remobilization of metals in ten natural soils were the total metal originally present, the total Mn, and the free iron oxide content (Korte et al. 1975). These experimental conditions are close to that expected from acid-mining waste leachates. Even though they are not representative of natural conditions (soil pH values between 4.2 and 7.8) or of neutral or alkaline pH waste leachates, these results indicate the significance of Fe and Mn oxides in the retention of trace metals. As they are dissolved (between 6 and 68% of Mn originally present is dissolved during the leaching) Fe and Mn oxides (most likely coatings) liberate their associated trace metal load. However, only small fractions of metals like Cu and Pb are solubilized because either they were associated with the soil organic matter or, if not, after leaching, they can complex with solid organic matter or adsorb on other surfaces.

If inorganic complexing agents tend to remobilize heavy metals or maintain them in solution, the behavior of organic chelates is harder to ascertain. Strong synthetic organic chelators such as EDTA or NTA can reduce the extent of adsorption (Means et al. 1978; Bourg and Schindler 1979; Salomons and Van Pagee 1981; Cleveland and Rees 1981; Christensen 1984b). Natural organic matter is, however, subject to precipitation-coagulation phenomena and it is therefore susceptible to inhibit the mobility of some heavy metals such as Cu and Pb (Bourg and Schindler 1985).

Natural municipal solid waste leachates individually enriched with metals were percolated through soil columns under anaerobic saturated conditions at flow rates ranging from 1 to 15 ml h^{-1} (Alesii et al. 1980). In general, the slower the rate

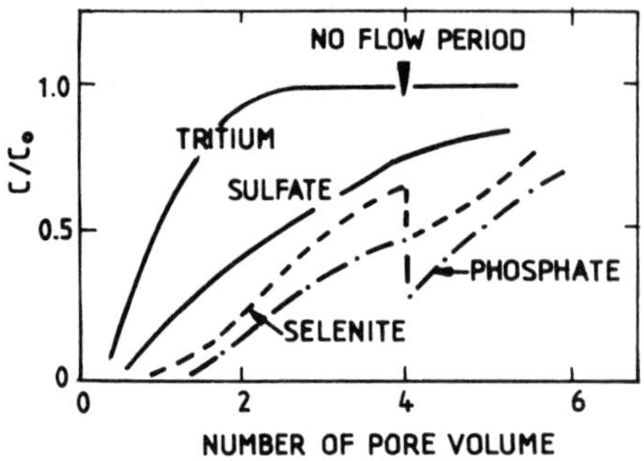

Fig. 17. Elution curves showing the effect of a 24-h no-flow period (After Murali and Aylmore 1980)

of flow, the greater the attenuation of the metal in solution percolating out of the column. The leachate velocity significantly influenced the retention of Cr(VI) and Fe(II), but was of little importance for Cd, Ni, and Zn.

This need to investate the inclusion of sorption dynamics for the adequate description of practical situations, particularly those involving intermittent flow is further supported by the experiments of Murali and Aylmore (1980). The breakthrough curves in Fig. 17 for solutions of 2.5 mM phosphate, 1.0 mM selenite, and 1.0 mM sulfate and tritium-labeled water percolated through a soil column at a flow rate of 0.3 pore volume per hour show an interesting pattern for a 24-h no-flow equilibrium period. The soil used was in a finely ground state and the depressions in outflow concentrations for phosphate and selenite following the no-flow period cannot be attributed to intra-aggregate diffusion. This is verified by the absence of a measurable depression for sulfate, despite the measurable adsorption of this ion, and for tritium.

10 Conclusions

This brief introductory chapter of a volume on the environmental impact and management of mine tailings and dredged materials was intended to present briefly the general fundamental aspects of the processes involved in the retention and mobility of trace metals in aquatic and terrestrial environments. As summarized in Fig. 1, the mobility of metals is governed by the competition between the reactions of sorption (adsorption and/or precipitation, including co-precipitation) and of solubilization (dissolved complexation). Moreover, there is some, even if limited, evidence that the participation of metals in these processes may be affected by the flow rates of the vectors of transport (water, solid particles).

Further theoretical work, in tight connection with field investigations, should be conducted on the characterization of adsorption by natural solids, of complexation with natural organic matter as well as on the kinetics of the various processes mentioned and especially for the redox reactions involving the solubilization and precipitation of Fe and Mn.

References

Alesii BA, Fuller WH, Boyle MV (1980) Effect of leachate flow rate on metal migration through soil. J Environ Qual 9:119–126

Anderson MA, Rubin AJ (eds) (1981) Adsorption of Inorganics at Solid-Liquid Interfaces. Ann Arbor Sci Publ, Ann Arbor: 357 pp

Aringhieri R, Carrai P, Petruzzelli G (1985) Kinetics of Cu^{2+} and Cd^{2+} adsorption by an italian soil. Soil Sci 139:197–204

Baccini P, Grieder E, Stierli R, Goldberg S (1982) The influence of natural organic matter on the adsorption properties of mineral particles in lake water. Schweiz Z Hydrol 44:99–116

Bailey SM, Helz GR, Harris RL (1975) Investigation of the transport of metals and orthophosphate away from a sewage treatment plant outfall. Environ Lett 10:159–169

Balistrieri LS, Murray JW (1981) The surface chemistry of goethite (αFeOOH) in major ion seawater. Am J Sci 281:788–806

Balistrieri LS, Murray JW (1983) Apparent equilibrium constants for metal-solid interactions in the marine environment. Geochim Cosmochim Acta 47:1091–1098

Balistrieri LS, Murray JW (1984) Marine scavenging: trace metal adsorption by interfacial sediment from MANOP site H[1]. Geochim Cosmochim Acta 48:921–929

Behra P (1986) Migration or retention of mercury (II) salts when percolating through a porous medium constituted of a natural quartz sand. Proc 2nd Internat Conf Environmental Contamination, Sept 1986, Amsterdam. pp 318–320

Behra P, Bourg ACM, Sardin M (1988) Influence of the chemical composition of the liquid phase upon the migration of mercury through sand. In preparation

Benjamin MM, Leckie JO (1980) Adsorption of metals at oxide interfaces: effects of the concentrations of the adsorbate and competing metals. In: Baker RA (ed) Contaminants and Sediments, Ann Arbor Sci, Ann Arbor: pp 305–322

Benjamin MM, Leckie JO (1981) Multiple site adsorption of Cd, Cu, Zn and Pb on amorphous iron oxyhydroxide. J Colloid Interface Sci 79:209–211

Benjamin MM, Leckie JO (1981) Effects of complexation by Cl, SO_4, S_2O_3 on adsorption behavior of Cd on oxide surfaces. Environ Sci Technol 16:162–170

Benjamin MM, Hayes MM, Leckie JO (1982) Removal of toxic trace metals from power-generation waste streams by adsorption and co-precipitation. J Water Pollut Control Fed 54:1472–1481

Bourg ACM (1983a) Role of fresh water/sea water mixing on trace metals adsorption phenomena. In: Wong CS, Burton JD, Boyle E, Bruland K, Goldberg ED (eds) Trace Metals in Sea Water, Plenum, New York, pp 195–208

Bourg ACM (1983b) Modélisation du Comportement des Métaux Traces à l'Interface Solide-Liquide dans les Systèmes Aquatiques. Document du B.R.G.M. n°62, Editions of the Bureau de Recherches Géologiques et Minières, Orléans, France, 171 pp

Bourg ACM (1984) Geochemical control of dissolved cadmium, copper and zinc in the low salinity zone of the Gironde Estuary, France. Proc Internat Symp Interactions between Sediments and Water, Aug 1984, Geneva, pp 249–252

Bourg ACM (1988) Physico-chemical speciation of trace elements in oxygenated estuarine waters. In: Trace Elements in Natural Waters. Internat Union Pure and Applied Chem. In press

Bourg ACM, Mouvet C (1988) Adsorption as a control mechanism of dissolved trace metals in the Meuse River. In preparation

Bourg ACM, Schindler PW (1978) Ternary surface complexes. (1) Complex formation in the system silica-Cu(II)-ethylenediamine. Chimia 32:166–168

Bourg ACM, Schindler PW (1979) Effect of EDTA on the adsorption of Cu(II) at amorphous silica. Inorg Nucl Chem Lett 15:225–229

Bourg ACM, Schindler PW (1985) Control of trace metals in natural aquatic systems by the adsorptive properties of organic matter. Proc Internat Conf Heavy Metals in the Environment, Sept 1985, Athens, Vol 1, pp 97–99

Bourg ACM, Schindler PW (1988) Adsorption of Cu(II) complexes on amorphous silica: ternary surface complexes. In preparation

Bourg ACM, Vedy JC (1986) Expected speciation of trace metals in gravitational water of acid soil profiles. Geoderma. 38:279–292

Bourg ACM, Joss S, Schindler PW (1979) Ternary surface complexes. (2) Complex formation in the system Silica - Cu(II) - 2,2'bipyridyl. Chimia 33:19–21

Bowers AR (1982) Adsorption characteristics of various heavy metals at the oxide-solution interface: effect of complex formation. PhD Thesis, Univ of Delaware, Newark

Brown DW, Hem JD (1984) Development of a model to predict the adsorption of lead from solution on a natural streambed sediment. US Geological Survey Water-Supply Paper 2187, 35pp

Brümmer G, Tiller KG, Herms U, Clayton PM (1983) Adsorption-desorption and/or precipitation-dissolution processes of zinc in soils. Geoderma 31:337–354

Buffle J (1980) A critical comparison of studies of complex formation between copper (II) and fulvic substances of natural waters. Anal Chim Acta 118:29–44

Bunzl K, Schmidt W, Sansoni B (1976) Kinetics of ion exchange in soil organic matter IV. Adsorption and desorption of Pb^{2+}, Cu^{2+}, Cd^{2+}, Zn^{2+} and Ca^{2+} by peat. J Soil Sci 27:32–41

Calvet R (1981) Etude du comportement physico-chimique de métaux lourds et des pesticides et de leur transport dans les sols. Report Institut Nat Rech Agronomique, Paris, grant 77.53 Ministère Environnement 104 pp

Cavallaro N, McBride MB (1978) Copper and cadmium adsorption by acid and calcareous soils. Soil Sci Soc Am J 42:550–555

Cavallaro N, McBride MB (1984) Zinc and copper sorption and fixation by an acid soil clay: effect of selective dissolutions. Soil Sci Soc Am J 48:1050–1054

Chapman BM, Jones DR, Jung RF (1983) Processes controlling metal ion attenuation in acid mine drainage streams. Geochim Cosmochim Acta 47:1957–1973

Cheshire ML, Berrow ML, Goodman BA, Mundie CM (1981) The nature and origin of the organic matter-metal complexes of soil. In: Migration Organo-Minérales dans les Sols Tempérés (Colloques Internationaux du CNRS n° 303) pp 241–246

Christensen TH (1984a) Cadmium soil sorption at low concentrations: I. Effect of time, cadmium load, pH and calcium. Water, Air and Soil Pollut 21:105–114

Christensen TH (1984b) Cadmium soil sorption at low concentrations: II. Reversibility, effect of changes in solute composition and effect of soil aging. Water, Air and Soil Pollut 21:115–125

Christensen TH (1985a) Cadmium soil sorption at low concentrations: III. Prediction and observation of mobility. Water, Air and Soil Pollut 26:255–264

Christensen TH (1985b) Heavy metal competition for soil sorption sites at low concentrations. Proc Internat Conf Heavy Metals in the Environment, Sept 1985, Athens, Vol 2, pp 394–397

Cleveland JM, Rees TF (1981) Characterization of plutonium in Maxey Flats radioactive trench leachates. Science 212:1506–1510

Davis JA (1984a) personal communication

Davis JA (1984b) Complexation of trace metals by adsorbed natural organic matter. Geochim Cosmochim Acta 48:679–691

Davis JA, Leckie JO (1978) Surface ionization and complexation at the oxide-water interface: (2) Surface properties of amorphous iron oxyhydroxide and adsorption of metal ions. J Colloid Interface Sci 67:90–107

Davis JA, Leckie JO (1980) Surface ionization and complexation at the oxide-water interface: (3) Adsorption of anions. J Colloid Interface Sci 74:32–43

Davis JA, James RO, Leckie JO (1978) Surface ionization and complexation at the oxide-water interface: (1) Computation of electrical double layer properties in simple electrolytes. J Colloid Interface Sci 63:480–499

Doner HE (1978) Chloride as a factor in mobilities of Ni(II), Cu(II) and Cd(II) in soil. Soil Sci Soc Am J 42:882–885

Dyrssen D, Wedborg M (1980) Major and minor elements, chemical speciation in estuarine waters. In: Olausson E, Cato I (eds) Chemistry and Biogeochemistry of Estuaries, Wiley, New York, pp 71–119

Egozy Y (1980) Adsorption of cadmium and cobalt on montmorillonite as a function of solution composition. Clays Clay Miner 28:311–318

Ellaway M, Beckett R, Hart B (1980) Behaviour of iron and manganese in the Yarra Estuary, Aust J Mar Freshwater Res 31:597–609

Elliott HA (1983) Adsorption behavior of cadmium in response to soil surface charge. Soil Sci 136:317–321

Elliott HA, Huang CP (1979) The adsorption characteristics of Cu(II) in the presence of chelating agents. J Colloid Interface Sci 70:29–45

Elliott HA, Huang CP (1980) Adsorption of some copper (II)-amino acid complexes at the solid-solution interface: Effects of ligand and surface hydrophobicity. Environ Sci Technol 14:87–93

Elliott HA, Huang CP (1981) Adsorption characteristics of some Cu(II) complexes on aluminosilicates. Water Res 15:849–855

Florence TM, Batley GE (1980) Chemical speciation in natural waters. Crit Rev Anal Chem 9:219–296

Garcia-Miragaya J, Page AL (1976) Influence of ionic strength and inorganic complex formation on the sorption of trace amounts of Cd by montmorillonite. Soil Sci Soc Am J 40:658–663

Garrels RM, Thompson ME (1962) A chemical model for seawater at 25°C and one atmosphere total pressure. Am J Sci 260:57–66

Gomez A, Juste C (1981) Mobilité du cadmium, du mercure, du nickel et du plomb associés à des acides humiques ou à des anions minéraux dans des colonnes de sable. In: Migrations Organo-Minérales dans les Sols Tempérés (Colloques Internationaux du CNRS n° 303) pp 291–296

Gotoh S, Patrick Jr WH (1972) Transformation of manganese in a waterlogged soil as affected by redox potential and pH. Soil Sci Soc Am Proc 36:738–742

Gotoh S, Patrick Jr WH (1974) Transformation of iron in a waterlogged soil as influenced by redox potential and pH. Soil Sci Soc Am Proc 38:66–71

Heinrichs H, Mayer R (1980) The role of forest vegetation in the biogeochemical cycles of heavy metals. J Environ Qual 9:111–118

Hem JD (1972) Chemistry and occurrence of cadmium and zinc in surface water and groundwater. Water Resour Res 8:661–679

Hohl H, Stumm W (1976) Interaction of Pb^{2+} with hydrous γAl_2O_3. J Colloid Interface Sci 55:281–288

Hunter KA (1983) The adsorptive properties of sinking particles in the deep ocean. Deep Sea Res 30:669–675

Jedwab J (1979) Copper, zinc and lead minerals suspended in ocean waters. Geochim Cosmochim Acta 43:101–110

Jenne EA (1968) Controls on Mn, Fe, Co, Ni, Cu and Zn concentrations in soils and water: The significant role of hydrous Mn and Fe oxides. Am Chem Soc Symp Ser 93:337–387

Jenne EA, Ball JW, Matter Burchard J, Vivit DV, Barks JH (1980) Geochemical modeling apparent solubility controls on Ba, Zn, Cd, Pb and F in waters of the Missouri Tri-State mining area. Proc 14th Conf Trace Substances in Environmental Health, June 2–5, 1980, Columbia, Missouri, pp 353–361

Jurinak JJ, Bauer N (1956) Thermodynamics of zinc adsorption on calcite, dolomite and magnesite type minerals. Soil Sci Soc Am Proc 20:466–471

Khalid RA, Gambrell RP, Patrick Jr WH (1975) Sorption and release of mercury by Mississippi River sediment as affected by pH and redox potential. In: Biological implications of metals in the environment, ERDA series 42. Available from Hanford Life Sciences Symp, Richland, Washington; CONF-750929

Khalid RA, Gambrell RP, Patrick Jr WH (1979) Chemical mobilization of cadmium in an estuarine sediment as affected by pH and redox potential. Proc Internat Conf Heavy Metals in the Environment, Sept 1979, London: 320–324

Korte NE, Skopp J, Niebla EE, Fuller WH (1975) A baseline study on trace metal elution from diverse soil types. Water, Air, Soil Pollut 5:149–156

Kramer CJM, Duinker JC (eds) (1984) Complexation of trace metals in natural waters. Martinus Nijhoff/Dr W Junk Publ, The Hague, 448 pp

Kuo S, McNeal BL (1984) Effects of pH and phosphate on cadmium sorption by a hydrous ferric oxide. Soil Sci Soc Am J 48:1040–1044

Kurdi F, Doner HE (1983) Zinc and copper sorption and interaction in soils. Soil Sci Soc Am J 47:873–876

Leckie JO, Benjamin MM, Hayes K, Kaufman G, Altman S (1980) Adsorption/coprecipitation of trace elements from water with iron oxyhydroxide. Final Report, EPRI RP-910, Electric Power Res Inst, Palo Alto, Calif

Li YH (1981) Ultimate removal mechanisms of elements from the ocean. Geochim Cosmochim Acta 45:1659–1664

Lindsay WL (1979) Chemical equilibria in soils. Wiley, New York, 449 pp

Lion LW, Altman RS, Leckie JO (1982) Trace-metal adsorption characteristics of estuarine particulate matter: evaluation of contributions of Fe-Mn oxide and organic surface coatings. Environ Sci Technol 16:660–666

Long DT, Angino EE (1977) Chemical speciation of Cd, Cu, Pb and Zn in mixed freshwater, seawater and brine solutions. Geochim Cosmochim Acta 41:1183–1191

Lyons WB, Gaudette HE, Armstrong PB (1979) Evidence for organically associated iron in nearshore pore fluids. Nature 282:202–203

MacLaren RG, Swift RS, Williams JG (1981) The adsorption of copper by soil materials at low equilibrium solution concentrations. J Soil Sci 32:247–256

MacNauthton MG, James RO (1974) Adsorption of aqueous mercury (II) complexes at the oxide/water interface. J Colloid Interface Sci 47:431–440

Mantoura RFC, Dickson A, Riley JP (1978) The complexation of metals with humic materials in natural waters. Estuar Coastal Mar Sci 6:387–408

Mattigod SV (1981) Speciation of heavy metals in soils. Proc. Internat Conf Heavy Metals in the Environment, Sept 1981, Amsterdam, pp 721–727

Mattigod SV, Sposito G, Page AL (1981) Factors affecting the solubilities of trace metals in soils. In: Dowdy RH, Ryan JA, Volk VV, Baker DE (eds) Chemistry in the soil environment, Am Soc Agronomy and Soil Sci Soc Am, Madison, pp 203–221

Mayer R (1978) Z Pflanzenernährung Bodenk 141:11–20, cited in Christensen (1984b)

McBride MB (1980) Chemisorption of Cd^{2+} on calcite surfaces. Soil Sci Soc Am J 44:26–28

McBride MB, Blasiak JJ (1979) Zinc and copper solubility as a function of pH in an acid soil. Soil Sci Soc Am J 43:866–870

Means JL, Grerar DA, Duguid JO (1978) Migration of radioactive wastes: radionuclide mobilization by complexing agents. Science 200:1477–1481

Morel FMM, Westall JC, Yeasted JG (1981) Adsorption models: a mathematical analysis in the framework of general equilibrium calculations. In: Anderson MA, Rubin AJ (eds) Adsorption of inorganics at solid-liquid interfaces. Ann Arbor Sci Publ, Ann Arbor, pp 263–294

Mouvet C, Bourg ACM (1983) Speciation (including adsorbed species) of copper, lead, nickel and zinc in the Meuse River. Observed results compared to values calculated with a chemical equilibrium computer program. Water Res 17:641–649

Murali V, Aylmore LAG (1980) No-flow equilibration and adsorption dynamics during ionic transport in soils. Nature 283:467–469

Nordstrom DK, Ball JW (1984) Chemical models, computer programs and metal complexation in natural waters. In: Kramer CJM, Duinker JC (eds) Complexation of trace metals in natural waters, M Nijhoff, Dr W Junk Publ, The Hague pp 149–164

Nordstrom DK, and eighteen co-authors (1979) A comparison of computerized chemical models for equilibrium calculations in aqueous systems. In: Jenne EA (ed) Chemical modelling in aqueous systems. Am Chem Soc Symp Ser 93:857–892

Norrish K (1968) Some phosphate minerals of soils. Trans 9th Int Congr Soil Sci Amer, Elsevier, New York, 2:713–723

Nyffeler UP, Li YH, Santschi PM (1984) A kinetic approach to describe trace-element distribution between particles and solution in natural aquatic systems. Geochim Cosmochim Acta 48:1513–1522

Parks GA (1967) Aqueous surface chemistry of oxides and complex oxide minerals; isoelectric point and zero point of charge. In: Equilibrium concepts in natural water systems, Adv in Chemistry Series n° 67 ACS, Wash, DC, pp 121–159

Reimers RS, Krenkel PA (1974) Kinetics of mercury adsorption and desorption in sediments. J Wat Poll Control Fed 46:352–365

Rhoads EA, Huang CP (1983) The adsorption of zinc (II) onto synthetic aluminosilicates. Report, Grant CEE 8104728, US National Science Foundation

Salomons W (1983) Trace metal cycling in a polluted lake (IJsselmeer, the Netherlands). Delft Hydraulics Laboratory Report S 357/EEC contract no 199-7-1 ENV N. 50 pp

Salomons W (1984) Contaminants in sediment: Out of sight out of mind? Proc Internat Conf Environmental Contamination, July 1984, London, pp 766–774

Salomons W (1985) Dredged material and mine tailings: Similarities in behavior and impact. Proc Internat Conf Heavy Metals in the Environment, Sept 1985, Athens, Vol 2, p 23

Salomons W, Eysink W (1981) Pathways of mud and particulate trace metals from rivers to the Southern North Sea. In: Nio SD, Schuettenhelm RTE Weering TCE van (eds) Holocene Marine Sedimentation in the North Sea Basin. Spec Publ Int Assoc Sedimentol 5:429–450

Salomons W, Foerstner U (1984) Metals in the hydrocycle, Springer, Berlin, 349 pp

Salomons W, Van Pagee JA (1981) Prediction of NTA levels in river systems and their effects on metal concentration. Int Conf. Heavy metals in the environment, Amsterdam: 694–697

Salomons W, van Driel W, Kerdijk H, Boxma R (1982) Help! Holland is plated by the Rhine (environmental problems associated with contaminated sediments). Effects of waste disposal on groundwater. Proc Exeter Symp IHAS 139:255–269

Salomons W, de Rooij NM, Kerdijk H, Bril J (1988) Sediments as a source for contaminants. In press

Santillan-Medrano J, Jurinak JJ (1975) The chemistry of lead and cadmium in soil: Solid phase formation. Soil Sci Soc Amer Proc 39:851–856

Santschi PH, Nyffeler UP, Li YH, O'Hara P (1984) Radionuclide cycling in natural waters: Relevance of sorption kinetics. Proc 3rd Internat Symp Interactions between sediments and water, Aug 1984, Geneva: 18–27

Schindler PW, Fürst B, Dick R, Wolf PU (1976) Ligand properties of surface silanol groups: Surface complex formation with Fe^{3+}, Cu^{2+}, Cd^{2+} and Pb^{2+}. J Colloid Interface Sci 55:469–475

Schmitt HV, Sticher H (1986) Long-term trend analysis of heavy metal content and translocation in soils. Geoderma, 38:145–208

Sholkovitz ER, Copland D (1982) The chemistry of suspended matter in Esthwaite Water, a biologically productive lake with seasonally anoxic hypolimnion. Geochim Cosmochim Acta 46:393–410

Sibley TH, Morgan JJ (1975) Equilibrium speciation of trace metals in freshwater: seawater mixtures. Proc Internat Conf. Heavy metals in the environment, Oct 1975, Toronto, Vol 1:323–338

Sigg L (1986) Surface chemical aspects of the distribution and fate of metal ions in lakes. Lecture given on Jan 24, 1986, Workshop on Aquatic Surface Chemistry, EAWAG, Ermatingen, Switzerland, (abstract) pp 67–70

Sigg L, Stumm W (1981) The interaction of anions and weak acids with the hydrous goethite (α-FeOOH) surface. Colloids and Surfaces 2:101–117

Sigg L, Stumm W, Zinder B (1984) Chemical processes at the particle-water interface; implications concerning the form of occurrence of solute and adsorbed species. In: Kramer CJM, Duinker JC (eds) Complexation of trace metals in natural waters, M Nijhoff/Dr. W Junk Publ, The Hague: 251–266

Smock LA, Kuenzler E (1983) Seasonal changes in the forms and species of iron and manganese in a seasonally inundated floodplain swamp. Water Res 17:1287–1294

Sposito G (1983) The chemical forms of trace metals in soils. In: Thornton I (ed) Applied environmental geochemistry, Academic Press, London: 123–170

Stumm W, Morgan JJ (1970) Aquatic chemistry, 1st ed, Wiley, New York, 583 pp

Stumm W, Morgan JJ (1981) Aquatic chemistry, 2nd ed, Wiley, New York, 780 pp

Stumm W, Hohl H, Dalang F (1976) Interaction of metal ions with hydrous oxide surfaces. Croat Chem Acta 48:491–504

Stumm W, Kummert R, Sigg L (1980) A ligand exchange model for the adsorption of inorganic and organic ligands at hydrous oxide interfaces. Croat Chem Acta 53:291–312

Tessier A, Campbell PGC, Bisson M (1979) Sequential extraction procedure for the speciation of particulate trace metals. Anal Chem 51:844–851

Theis TL, Singer PC (1974) Complexation of iron (II) by organic matter and its effect on iron (II) oxygenation. Environ Sci Technol 8:569–572

Thornber MR, Nickel EH (1976) Supergene alteration of sulfides. II. The composition of associated carbonates. Chem Geol 17:45–72

Tipping E, Griffith JR, Hilton J (1983) The effect of adsorbed humic substances on the uptake of copper (II) by goethite, Croat Chem Acta 56:613–621

Turner DR, Whitfield M, Dickson AG (1981) The equilibrium speciation of dissolved components in freshwater and seawater at 25°C and 1 atm pressure. Geochim Cosmochim Acta 45:885–881

Vuceta J (1978) Adsorption of Pb(II) and Cu(II) on quartz from aqueous solutions: influence of pH, ionic strength and complexing ligands. PhD Thesis, California Institute of Technology, Pasadena, California

Vuceta J, Morgan JJ (1978) Chemical modeling of trace metals in fresh waters: role of complexation and adsorption. Environ Sci Technol 12:1302–1309

Whitfield M, Turner DR (1979) Water-rock partition coefficients and the composition of seawater and river water. Nature 278:132–137

Wilson DE (1978) An equilibrium model describing the influence of humic materials on the speciation of Cu^{2+}, Zn^{2+} and Mn^{2+} in fresh waters. Limnol Oceanog 23:499–507

Woods TL, Garrels RM (1986) Use of oxidized copper minerals as environmental indicators. Applied Geochemistry 1:181–187

Zirino A, Yamamoto S (1972) A pH-dependent model for the chemical speciation of copper, zinc, cadmium and lead in seawater. Limnol Oceanog 17:661–671

Microbiological Oxidations of Minerals in Mine Tailings

B. C. KELLEY[1] and O. H. TUOVINEN[2]

1 Introduction

Industrially significant metal sulfides are relatively stable in their natural crystalline form. As a result of mining activities, these minerals are exposed and interact with water, oxygen, carbon dioxide, and soluble chemical species, factors which collectively enhance the mineral dissolution. The very nature of mining practices such as milling and grinding, which are designed to maximize metal recovery, also ensure maximum surface area exposure of mine tailings and waste materials to subsequent oxidative processes.

An important prerequisite prior to oxidative dissolution is the chemical dissociation of the mineral sulfide (Sato 1960). The rate of metal dissolution is proportional to the solubility product of the respective metal sulfide (Torma and Bosecker 1982), although the concentrations of holes in the sulfide valence band, electron extraction by Fe^{3+}, and the role of protons (Tributsch and Bennett 1981a,b) are also important considerations.

Following mineral sulfide dissociation, the chemical oxidation of the reduced valence state ionic species is greatly enhanced by the catalytic activity of the ubiquitous mixed communities of microorganisms associated with sulfide ore wastes and tailings. A major contributor to the microbial consortia is the chemolithoautotrophic bacterium *Thiobacillus ferrooxidans*. For a more comprehensive listing of the sulfur- and iron-oxidizing bacteria, the reader is directed to the excellent reviews by Ralph (1979) and Norris and Kelly (1982). Under field conditions, the degradation of these mineral materials is a dynamic process involving a succession of microbial populations (Ralph 1979; Lundgren and Malouf 1983) which develop according to the prevailing environmental conditions. These, in turn, are controlled by a complex array of physicochemical conditions which are generally site-specific.

The oxidative dissolution of a sulfide mineral is commonly associated with an increasing acidification of the surrounding medium. This alone provides a heavy selection pressure on the development of the microbial succession since a range of pH conditions are encountered. Variations in pH also affect the development and viability of microbial populations through the availability of electron donors such as ferrous iron, the oxidation of which is sensitive to pH.

[1]Biotechnology Australia Pty. Ltd., P.O. Box 20, Roseville, NSW 2069, Australia

[2]Department of Microbiology, The Ohio State University, 484 West 12th Avenue, Columbus, OH 43210, USA

In a field environment, however, mineral sulfide degradation is affected by a far more complex interaction of physicochemical parameters. These have been previously discussed (Ralph 1979, 1985; Lundgren and Malouf 1983) and include the availability of oxygen, carbon dioxide, water, and suitable nutrients at the reaction site. Climatic conditions also play an important role, particularly temperature and rainfall. The mineralogical composition of the tailings or waste material influences both mineral sulfide dissociation and bacterial establishment. Highly siliceous or carbonaceous gangue associations consume acid, thereby displacing conditions beyond the pH range suitable to many leaching organisms. The reduction in proton levels can also reduce the rate of chemical dissociation of many metal sulfides thus reducing their availability as bacterial substrates (Tributsch and Bennett 1981a,b). Gangue minerals may also contain metal species such as arsenic, silver, and molybdenum, which on being rendered soluble, may be toxic to the chemoautotrophic bacteria (Tuovinen et al. 1971; Lundgren and Silver 1980). Other aspects of gangue mineralogy and its associated effects have been discussed by Lundgren and Malouf (1983) and Ralph (1985). The type and relative abundance of sulfide minerals will vary according to the nature of the mineral deposit and the efficiency of the primary metal extraction process. Since the various mineral sulfides differ in their free energies of formation, their abilities to act as substrates for bacterial oxidative processes vary quite considerably (CL Brierley 1978).

It is apparent that tailings dams and waste deposits represent extremely complex and variable environments. All mineral deposits differ in composition and extensive mineralogical and grade variations can exist within any one deposit. In mining practice, attempts are often made to blend feed ores in an effort to maintain uniform grades. Despite this, variations in the nature of the tailings occur between different sections of the same dam as a result of changes in the grade of the run-of-mine feed, fluctuations in metallurgical extraction efficiency, and physical characteristics of the dam including water availability and depth. The residence time of the deposited material must also be considered. Such complexity does not permit uniform modeling of sulfide mineral dissolution and bacterial oxidation of mine tailings. Nevertheless, significant progress has been made in our understanding of the fundamental chemical and biological processes associated with these materials. This must contribute to our attempts to industrially exploit the sizeable metal values contained within these low-grade deposits and to minimize the possibilities of pollution.

2 Oxidation of Sulfide Minerals and Fe(II)

2.1 The Oxidation of Sulfide Minerals

The dissolution of a large number of metal sulfides has been shown to be biologically assisted. Very few natural metal sulfides have been studied as purified mineral fractions and, although the stoichiometry of bacterial oxidation can be predicted, incomplete oxidations, side reactions, by-products, and impurities complicate the understanding of the microbiology of mineral leaching. For discussions on the leaching of various metal sulfides, together with appraisals of industrial-scale bacterial leaching methods, the reader is directed to several reviews (Tuovinen and Kelly 1974; Torma 1977; CL Brierley 1978; Kelly et al. 1979; Manchee 1979; Ehrlich 1981; Lundgren and Malouf 1983; Karavaiko 1985; Ralph 1985).

Iron, both in its divalent and trivalent state, plays a central role in sulfide mineral degradation. Attempts to demonstrate direct bacterial oxidation of these sulfides must necessarily exclude iron. This is extremely difficult to accomplish as most naturally occurring mineral sulfides are associated with various iron minerals. Our understanding of the detailed mechanisms of bacterial attack remains obscure. Ralph (1979, 1985) has considered the mechanisms of their oxidative process in some detail. Electrochemical processes may be important in providing substrates for bacterial metabolism. The cathodic reaction, the reduction of oxygen on the mineral sulfide surface, is the rate-limiting step owing to the formation of a passive sulfur layer over the surface of the mineral. The presence of this layer could be oxidized to sulfuric acid by bacterial action. In an elegant series of experiments, Vanselow (1976) was further able to demonstrate that certain strains of *Thiobacilli* could oxidize the mineral while physically separated from it. He concluded that while physical attachment of the bacterium to the mineral sulfide surface is not an essential component of the oxidative mechanism, it would be of importance in reducing diffusion distances at low substrate concentrations. These findings were supported by Torma and Bosecker (1982), who suggested that substrate dissociation from the mineral surface rather than bacterial attachment is the prerequisite for bacterial oxidation of the insoluble metal sulfides. In an iron-free environment, the rate of metal dissolution is proportional to the solubility product of the metal sulfide. Thus, metal sulfide dissociation at the mineral surface provides the sulfide moiety as an oxidizable substrate for the bacterial metabolic system (Torma and Bosecker 1982):

$$MS \rightleftarrows M^{2+} + S^{2-}. \quad (1)$$

(M = bivalent metal)

The bacteria remain proximal to the mineral sulfide surface and the resulting bacterial oxidation shifts the dissociation equilibrium in favor of further mineral sulfide dissolution.

Tributsch and Bennett (1981a,b) also demonstrated that metal sulfides with high solubility products are good substrates for bacterial leaching. However, this is not the only condition for bacterial leaching. The amenability of minerals to bacterial leaching is also critically dependent on the composition, structure, and impurities of metal sulfides (Tributsch and Bennett 1981a,b). The presence of lattice imperfections and impurities has been shown to effect the susceptibility to bacterial degradation of a range of zinc sulfide minerals (Khalid and Ralph 1977).

Other conditions also apply, such as the presence of a high concentration of holes in the valence band of the semiconducting sulfide (p-type conduction), or when holes are generated through electron extraction from the sulfide surface by oxidants such as Fe^{3+} (Tributsch and Bennett 1981b). When ferric ion is present, the relative position between the energy levels of Fe^{3+} and the sulfide valence band determines the rate of hole injection. This will be discussed in more detail later in this section.

In the absence of iron, protons (H^+) play a central catalytic role in the mechanism of bacterial oxidation of metal sulfides (Tributsch and Bennett 1981a,b). This is based on the observation that electronically unfavorable systems (e.g., ZnS) with large energy gaps leach faster than more electronically favorable sulfides (Tributsch and Bennett 1981a; Vaughan 1984). There appears to be no direct correlation with the energy difference between the position of the electron transport system of the bacterium and the highest filled energy band of the sulfides. Tributsch and Bennett (1981a,b) have, therefore, postulated that protons are important in chemically breaking surface bonds, including shifting electron levels of sulfides with large energy gaps into the energy range of the forbidden energy gap, where they can participate in electron transfer reactions. The proton catalyzed reaction produces weakly bonded $-SH\Delta^-$ groups which are in turn removed from the sulfide surface through complex formation with an unidentified molecular carrier used by the bacterium (Tributsch and Bennett 1981a,b). This proposed mechanism is not applicable to sulfides in which the valence band of the semiconductor is derived from metal orbitals instead of from sulfur orbitals. MoS_2 and WS_2 are given as examples (Tributsch and Bennett 1981a).

T. ferrooxidans has been shown to oxidize the sulfide moiety of a range of nonferrous metal sulfides such as ZnS, CuS, NiS, and CoS:

$$\begin{aligned} ZnS + 2.5O_2 + 2H^+ &\rightarrow Zn^{2+} + SO_4^{2-} + H_2O; \\ CuS + 2.5O_2 + 2H^+ &\rightarrow Cu^{2+} + SO_4^{2-} + H_2O; \\ NiS + 2.5O_2 + 2H^+ &\rightarrow Ni^{2+} + SO_4^{2-} + H_2O; \\ CoS + 2.5O_2 + 2H^+ &\rightarrow Co^{2+} + SO_4^{2-} + H_2O. \end{aligned} \tag{2}$$

Copper selenide has also been reported to be oxidized by *T. ferrooxidans* (Torma and Habashi 1972; Torma 1977) according to the following reaction:

$$CuSe + 0.5O_2 + 2H^+ \rightarrow Cu^{2+} + Se^0 + H_2O. \tag{3}$$

Synthetic gallium sulfide is reputed to be directly oxidized by *T. ferrooxidans* (Torma 1978):

$$Ga_2S_3 + 6O_2 \rightarrow Ga_2(SO_4)_3. \tag{4}$$

While the biological oxidation of Fe^{2+} is extremely well documented and will be discussed later, the fate of other reduced metals such as Cu(I), Sn(II), and U(IV) has received far less attention. Demonstrations of the coupling of these substrate oxidations with energy conservation in *T. ferrooxidans* have in part been complicated by the presence of iron, particularly in the natural leaching systems. DiSpirito and Tuovinen (1984) have reviewed the current state of knowledge with respect to oxidation of these metals. The coupling of Cu(I) and Sn(II) oxidation to oxygen uptake and carbon dioxide fixation awaits confirmation. The oxidation of uranous compounds in the absence of iron has been linked to both oxygen uptake and carbon dioxide fixation (DiSpirito and Tuovinen 1981, 1982a,b), although growth dependency on U(IV) compounds as sole sources of energy has yet to be demonstrated. The kinetics of uranous ion oxidation differed markedly to those for

Fe^{2+} and uranous ion was found to be a competitive inhibitor of iron oxidation (DiSpirito and Tuovinen 1982b). The addition of ferrous iron at concentrations of up to 0.5 μM failed to stimulate uranous ion oxidation, suggesting that under these conditions, trace levels of iron do not directly affect the oxidation of U^{4+} (DiSpirito and Tuovinen 1982b). These results have been confirmed and extended under conditions designed to further minimize the presence of Fe^{2+} (DiSpirito and Tuovinen 1984).

As already indicated, in natural environments iron is nearly always associated with the leaching system. The kinetics of mixed substrates (e.g., FeS_2, $CuFeS_2$, Cu_2S) are not well understood. The (direct) biological oxidation of chalcopyrite is represented as follows:

$$2CuFeS_2 + 8.5O_2 + 2H^+ \rightarrow 2Cu^{2+} + 2Fe^{3+} + 4SO_4^{2-} + H_2O. \quad (5)$$

However, chalcopyrite displays both chemical and microbiological recalcitrance and is typically oxidized at a slow rate in comparison to secondary copper sulfides such as chalcocite and covellite. Its incomplete oxidation in leaching systems is a common problem and produces secondary copper sulfides and elemental sulfur.

One of the most important naturally occurring growth substrates for *T. ferrooxidans* is the iron sulfide mineral, pyrite (FeS_2), which is often associated with other, more valuable minerals. The direct biological oxidation of pyrite would require metal sulfide dissociation as a prerequisite as already discussed. The bacteria probably oxidize pyrite by the following reaction:

$$2FeS_2 + 7.5O_2 + H_2O \rightarrow 2Fe^{3+} + 4SO_4^{2-} + 2H^+. \quad (6)$$

Other iron sulfide minerals differ in their degree of recalcitrance (CL Brierley 1978). Pyrrhotite ($Fe_{1-x}S$) leaches more readily than pyrite (Ahonen et al. 1986), while marcasite (FeS_2) is slightly more recalcitrant than pyrite.

Iron can be brought into solution by the aqueous oxidation of pyrite by molecular oxygen. In an excellent review, Lowson (1982) identified three reaction paths for pyrite oxidation, namely bacterial, chemical, and electrochemical, restricting discussion to the latter two. The chemical oxidation path is a three-step process:

1. The oxidation of pyrite by molecular oxygen;

$$2FeS_2 + 7O_2 + 2H_2O \rightarrow 2Fe^{2+} + 4H^+ + 4SO_4^{2-}. \quad (7)$$

2. The oxidation of ferrous iron to ferric iron by molecular oxygen, which is thought to be the rate-limiting step;
3. The oxidation of pyrite by ferric iron. Thermodynamic data predict sulfate formation (Dutrizac and MacDonald 1974; Lowson 1982) as shown in the following reaction:

$$FeS_2 + 14Fe^{3+} + 8H_2O \rightarrow 15Fe^{2+} + 16H^+ + 2SO_4^{2-}. \quad (8)$$

The electrochemical pathway (Lowson 1982) is the summation of two half-cell reactions as follows:

1. The anode:

$$FeS_2 + 8H_2O \rightarrow Fe^{3+} + 2SO_4^{2-} + 16H^+ + 15e^-. \tag{9}$$

2. The cathode:

$$O_2 + 4H^+ + 4e^- \rightarrow 2H_2O. \tag{10}$$

Both the chemical and electrochemical reactions make important contributions to localized iron environments in nature (Lowson 1982). In a natural environment, the large number of physicochemical parameters involved would greatly influence the kinetics and direction of pyrite dissolution.

The diversity of chemical and biological reactions, coupled with the frequent presence of iron sulfides, ensures that the chemoautotrophic bacteria are exposed to reduced iron and sulfur compounds as energy sources for growth:

$$\begin{aligned} &4Fe^{2+} + O_2 + 4H^+ \rightarrow 4Fe^{3+} + 2H_2O; \\ &2S° + 3O_2 + 2H_2O \rightarrow 4H^+ + 2SO_4^{2-}. \end{aligned} \tag{11}$$

The products of these reactions, ferric sulfate and sulfuric acid, provide an excellent means of dissolving a variety of mineral sulfides by what is often termed the "indirect" mechanisms of bacterial attack:

$$MS + Fe_2(SO_4)_3 \rightarrow MSO_4 + 2FeSO_4 + S°. \tag{12}$$

(M = bivalent metal)

The ferrous sulfate and elemental sulfur thus produced are further oxidized in mirobiologically active systems. Elemental sulfur has a passivation effect due to the formation of a diffusion barrier between the mineral surface and bulk solution, a problem particularly associated with oxidative dissolution of chalcopyrite. The role of bacteria in the formation and oxidation of the sulfur layer surrounding the mineral surface is poorly understood and not well documented. As a phenomenon, sulfur accumulation indicates the predominance of ferric ion-mediated leaching. Essentially no information is available on soluble sulfur intermediates (sulfo-oxyanions) in mineral bioleaching systems. Compounds such as thiosulfate and polythionates are extremely metastable in acid, metal-containing solutions. Trithionate has been detected as an intermediate in bacterial pyrite oxidation (Basaran and Tuovinen 1987), but its pathway is not presently known.

Unlike the direct mechanism of bacterial attack, when ferric iron is present, there is less need to break chemical bonds in the mineral sulfide surface by proton interaction (Tributsch and Bennett 1981a,b), although the proton concentration is critical to maintain ferric iron solubility. Ferric iron may directly break chemical bonds by extracting electrons from the valence band of the sulfide provided that energetic constraints are met. This provides an efficient means of sulfide oxidation.

In the natural leaching system, it is likely that all of these parameters including the solid state, chemical and electrochemical properties, and electronic structures of the mineral sulfide, the presence of iron, the proton concentration, and the physicochemical environment are all important in the relationship between *T. ferrooxidans* and the mineral sulfide.

The physical characteristics of the tailings dam or waste material are of prime importance in the microbial oxidation of sulfide minerals. These include factors such as galvanic effects associated with mixtures of sulfide minerals (Mehta and Murr 1982, 1983; Natarajan and Iwasaki 1983) and the particle size distribution. In most tailings deposits, the gangue associated with the mineral sulfide represents the bulk of the material present and therefore largely contributes to the environment in which the sulfide minerals are degraded. The gangue may consist of silicates, clays, and limestone-type materials which are highly acid-consuming. Acid removal can bring about the precipitation of ferric iron which in turn reduces the diffusion of water, nutrients, and dissolved gases to sites of microbial activity. The physical nature of the gangue may also bring about similar problems. The adsorption characteristics of the gangue must also be considered. The large buffering capacity of the gangue minerals will affect factors such as heat dissipation. Thus, the thermal conductivity and heat capacity of these materials is of importance. As pointed out by Ralph (1979), the composition of the water associated with tailings deposits will also influence the microbial oxidation of sulfide minerals. Factors such as nutrient status, pH, redox status, heavy metal ion concentrations, iron concentration, and the presence of potentially toxic materials will interact and affect sulfide mineral oxidation. These in turn are also influenced by prevailing climatic conditions including rainfall, the length of dry periods, and ambient temperatures. Changes in mining practices are obviously also of importance.

Acid mine drainage has prompted the development of some models to describe the problem particularly in coal strip mine areas (Cathles and Apps 1975; Jaynes et al. 1984a,b). Such models incorporate several interactive components including aspects of bacterial dynamics. However, the sheer complexity of these environments dynamically affects the qualitative and quantitative distribution of their microbial populations. Such complex interactions have thus far defied attempts to accurately and comprehensively model the naturally occurring processes associated with the microbial oxidations in mine tailings.

2.2 The Oxidation of Fe(II)

The oxidation of Fe^{2+} by *T. ferrooxidans* has been reviewed by Ingledew (1982) and Cox and Brand (1984) and will be briefly summarized here. During aerobic growth on ferrous iron at pH 2.0, *T. ferrooxidans* couples ferrous iron oxidation and the reduction of oxygen by an electron transport chain. The half-reactions which operate are as follows:

$$2Fe^{2+} \rightarrow 2Fe^{3+} + 2e^-$$

$$2e^- + 0.5O_2 + 2H^+ \rightarrow H_2O$$

$$\text{Sum } 2Fe^{2+} + 0.5O_2 + 2H^+ \rightarrow 2Fe^{3+} + H_2O. \quad (13)$$

The steady state E_h of the Fe^{3+}/Fe^{2+} couple during growth is about +770 mV, while that of the $0.5O_2/H_2O$ couple at the higher cytosolic pH (pH 5.5–6.0), the site of oxygen reduction, is +890 mV (Ingledew et al. 1977; Ingledew 1982). There is therefore little energy available to the cell from this oxidative reaction. Chemolithotrophic bacteria that use these types of high redox potential substrates which are energetically unfavorable, typically rely on a high throughput of electrons along the electron transport chain in order to satisfy their energy requirements for growth. To achieve these high electron throughputs, high oxidation rates, up to ten times those reported for beef heart mitochondria (Ingledew 1982), are achieved as the cells contain elevated concentrations of some of the electron transport components. As discussed by Ingledew (1982), the copper protein rusticyanin, which is the proposed primary electron acceptor of the respiratory chain, accounts for up to 5% of the total cell protein, while for cytochrome *c*, the figure more closely approximates 10%. The components of the electron transport chain have received considerable study (Cobley and Haddock 1975; Ingledew and Cobley 1980; Ingledew 1982).

Since the Fe^{3+}/Fe^{2+} couple is so electropositive (+770 mV), ferrous ion does not directly reduce NADP ($NADP^+/NADPH$, E_0 = -320 mV). Reduced pyridine nucleotides are essential as a source of reducing power for cellular processes such as carbon dioxide fixation. Therefore, in *T. ferrooxidans*, the mechanism of NADPH production is assumed to be via an energy-dependent reversal of electron flow. Such a process has been generally documented in chemolithotrophs (Ingledew 1982), but has been observed infrequently in *T. ferrooxidans* (Aleem et al. 1963). Ingledew (1982) provides an excellent discussion on the thermodynamics and possible mechanisms of reversed electron transport in this organism.

The nature of the electron transport chain operating during iron oxidation can now be considered. Most of the electron flux occurs between Fe^{2+} and O_2, while less than 10% of the electrons flow "uphill" from Fe^{2+} to $NADP^+$ via the reversed electron flow (Ingledew 1982). The electron transport chain has been modeled (Ingledew et al 1977; Ingledew 1982) and can be summarized as follows. Fe^{2+} is oxidized outside of the cell membrane under physiologically favorable pH conditions (pH 2.0). Cytosolic oxidation of Fe^{2+} would not be expected to occur since the cytosolic pH (5.5–6.0) would favor rapid autooxidation of the Fe^{2+} substrate and chemical precipitation of the Fe^{3+} product. Both rusticyanin, the probable primary electron acceptor, and cytochrome *c* are considered as components of the periplasmic space (Ingledew et al. 1977). Rusticyanin is an acid-stable protein and it would not be surprising to find unusual acid stability also for the first cytochrome *c* type redox carrier in view of the organization of these components in the periplasmic space. Ingledew et al. (1977) consider that one of the multiple forms of cytochrome *c* present may be bound to the cell membrane as is also the terminal oxidase, cytochrome *a*, where O_2 reduction occurs at the inner or cytosolic surface of the cell membrane.

The relationship between electron flow and ATP generation in *T. ferrooxidans* has been discussed in terms of the chemiosmotic theory (Ingledew et al. 1977; Ingledew 1982; Cox and Brand 1984). The acidity of the external medium coupled with the neutrality of the cytosol provides a large and favorable chemical potential of H^+ into the cell. The organism thus uses the acidic environment to its advantage, generating a transmembrane pH gradient by internal proton consumption (reduction of oxygen at the cytosol-cell membrane interface) and using this gradient for the synthesis of ATP via the reversible ATPase system. This, therefore, represents a simple chemiosmotic model that does not postulate redox-coupled proton pumps, and is consistent with the environment in which the organism lives.

Leptospirillum ferrooxidans is also an acidophilic iron-oxidizing bacterium, but is incapable of mineral sulfide oxidation in the absence of iron. The cell shape varies from curved rods to spirals and is distinctly different from the straight rod shape of *T. ferrooxidans*. The bacterium has been isolated from exposed ore deposits and acid mine waters which are typical environments also for *T. ferrooxidans*. Some differences in growth rates, acid tolerance, and metal resistance patterns between *L. ferrooxidans* and *T. ferrooxidans* have been reported (Eccleston et al. 1985; Hutchins et al. 1986) which may be of some use in the selective enrichment and isolation of *L. ferrooxidans*. In general, however, the isolation of *L. ferrooxidans* is difficult because similar mineral salts media are used for both types of bacteria and because *T. ferrooxidans* tends to be abundant in natural samples of *L. ferrooxidans*. Nucleic acid and serological studies and protein profiles suggest a high degree of unrelatedness between these two bacteria (Harrison 1984, 1986a; Huber et al. 1986; Jerez et al. 1986). Both *L. ferrooxidans* and *T. ferrooxidans* constitute heterogeneous groups within which there is considerable diversity (Harrison and Norris 1985; Harrison 1986a,b) that may warrant further speciation in the future. The presence of a rusticyanin-type redox protein has not been determined in *L. ferrooxidans* and its metabolism is essentially unexplored at the present time.

3 Temperature as an Environmental Determinant

Measurements of ambient temperature fluctuations provide little insight into the temperature profile operating within the tailings deposit. Indeed, seasonal variations in temperature might be expected to affect only the surface layers of the tailings, depending on whether the material is covered by water, and if so, to what depth. Harries and Ritchie (1981) showed that surface temperature changes affected temperature distributions in a waste rock dump down to a depth of approximately 6 m. Fluctuations in ambient temperature are more likely to affect airflow through waste materials thus influencing the oxygen level.

There have been few attempts to measure heat profiles in mine tailings. Temperature distributions have been measured in waste rock dumps to evaluate heat source distributors as indicators of mineral sulfide oxidation sites (Harries and Ritchie 1981, 1983). In this comprehensive study, oxidation sites were primarily located at the surface of the dump, although some sites were found down to depths of 15 m. Temperatures were generally in the range of 32° to 37°C. However, temperatures as high as 56°C were also detected (Harries and Ritchie 1981). The authors surmised that oxygen diffusion from the surface of the dump through the pore space to deep-seated oxidation sites would be the possible mechanism for O_2 transport.

The oxidation of pyrite, a common component of the mineralogical assemblages associated with mine tailings and waste materials, is an exothermic reaction as discussed by Harries and Ritchie (1981).

$$\begin{aligned}
&FeS_2 + 3.50_2 + H_2O \rightarrow Fe^{2+} + 2SO_4^{2-} + 2H^+ \\
&\qquad H = 1{,}440 \text{ kJ mol}^{-1} \\
&Fe^{2+} + 0.250_2 + H^+ \rightarrow Fe^{3+} + 0.5H_2O \qquad (14) \\
&\qquad H = 102 \text{ kJ mol}^{-1} \\
&FeS_2 + 14Fe^{3+} + 8H_2O \rightarrow 15Fe^{2+} + 2SO_4^{2-} + 16H^+ \\
&\qquad H = 11 \text{ kJ mol}^{-1}.
\end{aligned}$$

In field situations, the development of the microbial succession (Ralph 1979) involves the chemical and biological oxidation of sulfides and ferrous iron. Within the physical confines of tailings dams and waste deposits, there is little dissipation of the heat evolved during these oxidations due to the high buffering capacity of the low conductivity mining materials (Ralph 1979; Lundgren and Malouf 1983). Indeed, temperatures can increase to levels where purely chemical reactions involving sulfides are thought to dominate. Moderate temperatures of 56°C in the waste rock dump from a uranium operation in northern Australia (Harries and Ritchie 1981) and 59°C in copper-bearing waste (Murr and Berry 1979) have been reported. Similar temperatures have been recorded in low-grade copper ore which was tested at a large-scale facility in New Mexico (Murr and Brierley 1978). Even higher temperatures in excess of 80°C have been observed in low-grade copper dumps in Bulgaria (Groudev et al. 1978) and in the United States (Beck 1967).

High temperatures greater than 40°C would certainly limit the distribution of the mesophilic *Thiobacilli.* The optimum temperature for ferrous iron oxidation varies from 28° to 35°C and for metal sulfide leaching, values of 35°C have been reported in the literature. In evaluating microbial populations in waste rock dumps in northern Australia (Goodman et al. 1981), acidophilic iron oxidizers were found to be the most abundant organism associated with oxidation zones at temperatures of 45°C or higher.

It is now known that groups of specialized thermophilic chemolithotrophic bacteria may contribute to the biological oxidation of tailings and wastes at significantly higher temperatures than those favored by the mesophilic *Thiobacilli.* Mineral sulfide-oxidizing acidophilic organisms with an optimum temperature of approximately 50°C have now been isolated from a variety of environments (Brierley et al. 1980). These include copper leach dumps (JA Brierley 1978), a large-scale experimental copper waste leaching facility in New Mexico (Murr and Brierley 1978), coal spoil heaps (Marsh and Norris 1983a,b), and hot springs (Le Roux et al. 1977). Of the moderately thermoacidophilic iron-oxidizing organisms that have been extensively studied, some appear to require a source of reduced sulfur during growth on iron, e.g., tetrathionate, thiosulfate, or elemental sulfur (Norris and Barr 1985), although the mechanism of reduced sulfur incorporation is unknown. All of the isolates are capable of chemolithoheterotrophic growth on iron when supplied with yeast extract. In addition, most strains are capable of growing autotrophically (Norris and Barr 1985). Thus, the environments found at microsites within many tailings and waste deposits may be suitable for the establishment and growth of these types of organisms. Mixotrophic growth of some iron-oxidizing thermophiles has recently been described (Wood and Kelly 1984, 1985). The identification, distribution, and role of moderately thermophilic organisms in natural and industrial leaching environments have yet to be evaluated.

However, the isolation of these bacteria from such sites, coupled with their ability to degrade mineral sulfides in laboratory studies, suggests that they may make important contributions to the oxidation reactions in mine tailings at elevated temperatures.

Sulfobacillus thermosulfidooxidans, capable of growing with Fe(II) and sulfide minerals at temperatures up to 60 °C, was isolated from elevated temperature zones of a copper leaching dump (Golovacheva and Karavaiko 1978). It is also possible that extremely thermophilic archaebacteria of the genus *Sulfolobus* may be associated with elevated temperature zones up to 85 °C. *Sulfolobus acidocaldarius* was isolated and described by Brock et al. (1972). A similar organism was previously isolated from an acid-thermal region of Yellowstone National Park by Brierley and Brierley (1973). The organism is mixotrophic and uses inorganic substrates (e.g., Fe^{2+} and reduced sulfur compounds) and simple organic compounds (e.g., glucose and yeast extract) as sources of energy for growth. It is of interest that the organism prefers highly acidic environments (pH 1.5–2.5) and will grow at temperatures between 55° to 85 °C. These types of environments have been found in mineral sulfide leaching operations. In applying the organism to mineral sulfide substrates, it was found to be quite effective in the biological degradation of molybdenite (Brierley and Murr 1973; Brierley 1974) and copper wastes and ores (Brierley and Brierley 1978).

The extreme thermophiles of the genus *Sulfolobus* have largely been isolated from hot springs and thermal acid soils to date. The first isolation of *Sulfolobus* from a coal heap in the United Kingdom (Marsh and Norris 1983a) indicates the presence of these types of organisms in mining environments. Future studies may more clearly elucidate their presence and role in the microbial oxidation successions associated with mine tailings and waste materials.

The metabolic diversity of sulfur-dependent archaebacteria has been further demonstrated in two recent independent studies. Depending on the oxygen supply, *S. brierleyi* (=*Acidianus brierleyi*) could either grow anaerobically by reducing molecular sulfur, or aerobically by oxidizing the same element (Segerer et al. 1985, 1986). *S. brierleyi* can also anaerobically oxidize sulfur with molybdate as an electron acceptor. In a separate study, Zillig et al. (1985), working on a *Sulfolobus*-like isolate, also demonstrated the ability of this organism to grow anaerobically by reducing sulfur with hydrogen. The versatility of these organisms warrants further investigation. Studies on their ecology may also provide an insight into their ability to transform inorganic materials in natural systems at highly elevated temperatures.

Mine tailings have zones where the heat evolution is extremely slow with no noticeable increase in the ambient temperature. In certain geographic regions, a low temperature season constitutes a major phase in the annual cycle. Truly psychrophilic strains of *T. ferrooxidans* have not been reported in the literature. However, psychrotrophic isolates of *T. ferrooxidans* were characterized by Ferroni et al. (1986) from a uranium mine in the Elliot Lake district, Ontario. An Arrhenius plot of their growth rate constants displayed linearity in the range of 25° to 12 °C. The data also indicated that the minimal temperature for growth was close to 2 °C. An Arrhenius plot for a mesophilic *T. ferrooxidans* reference culture displayed considerable deviation from this profile and the minimal temperature for growth was around 6°C (Ferroni et al. 1986). Low temperature isolates from water samples of a copper mine in a comparable climatic region (Finland) were able to grow with ferrous ion at 4°C with a generation time between 56 and 72 h depending on the initial incubation temperature of isolation (P Hietanen, L Ahonen, OH Tuovinen, unpublished results). A 28°C isolate from the same mine had a generation time of 7 h at 28°C but it increased to 102 h when the culture was grown at 4°C. These and other temperature-switch experiments suggest that natural *T. ferrooxidans*-type populations are subject to temperature selection. In a mixed substrate system such as mine tailings, complex and substrate-specific temperature effects may exist which influence the thermodynamics and kinetics of microbiological oxidations as well as population dynamics in a manner yet to be explored.

4 Redox Potential and pH

Infiltration water from rain and snow promotes the formation of oxidized surface zones in tailings, typically high in soluble and insoluble oxidation products (Dubrovsky et al. 1985). This oxidized zone tends to expand as the oxidation proceeds. Ferric ion produced in these zones is a chemical oxidant in reactions which do not directly involve dissolved molecular oxygen.

Iron plays a central catalytic role in the chemical and biological oxidation of mineral sulfide mine tailings and its oxidation state is the main factor that determines the prevailing redox potential. The biogeochemistry of iron has been reviewed by Lundgren and Dean (1979) and Nealson (1983). The chemical stability of iron and its compounds is very sensitive to conditions of pH and Eh and will to a large degree determine the types of biological populations that develop. Iron oxidation is usually rapid and is sensitive to both pH and oxygen concentration, particularly at pH values above 3.5 (Ackman and Kleinmann 1984). After iron has been oxidized to the ferric oxidation state, it tends to hydrolyze in solution according to the following reactions:

$$
\begin{array}{l}
2Fe^{3+} + 2H_2O \rightarrow 2FeOH^{2+} + 2H^+ \\
2FeOH^{2+} + 2H_2O \rightarrow 2Fe(OH)_2^+ + 2H^+ \\
2Fe(OH)_2^+ + 2H_2O \rightarrow 2Fe(OH)_3 + 2H^+ \\
\hline
2Fe_2(SO_4)_3 + 6H_2O \rightarrow 2Fe(OH)_3 + 3H_2SO_4.
\end{array}
\qquad (15)
$$

This reaction leads to a net increase in acid formation in the environment. The ferric hydroxide can further interact with various sulfates to form Fe(III) hydroxy-sulfate and oxyhydroxide complexes (Nordstrom 1982; Ross et al. 1982; Brady et al. 1986). The buffering capacity of these compounds also influences the pH of the environment (Lundgren and Dean 1979). The precipitation of these materials may affect localized oxygen concentrations by creating diffusion barriers to various gases and solutions. Channeling effects can also occur, thus enhancing the complexity of field situations. The availability of iron may in addition be affected by chelation with a large number of organic molecules as has been discussed by Nealson (1983).

The chemical and biological dissolution of mineral sulfides associated with tailings and waste materials is generally characterized by an increase in acidity (H^+). It should be recognized that the oxidative dissolution of pyrite is an acid-yielding reaction, whereas pyrrhotite oxidation is initially an acid-consuming reaction. Protons are also consumed in the bacterial oxidation of ferrous ion and monosulfides (e.g., ZnS, Cu_2S, NiS). The oxidation of pyrite, often the most abundant iron sulfide, and the hydrolysis of ferric ion are the two major acid-generating systems. However, the net acidity and iron solubility are in turn affected by the physical and chemical characteristics of the gangue minerals which may consume acid and adsorb or precipitate components of the leaching system. The importance of the establishment of appropriate pH regimes can be seen from a study of the growth requirements of the iron-oxidizing bacterium *T. ferrooxidans*, the major biological component of many leaching systems. The optimum pH for growth of this organism on a range of mineral sulfides generally lies between pH 2–2.5 and the lower pH limit still permissive for growth probably is in the pH range 1.2–1.3.

Such a level of acidity minimizes the chemical oxidation of ferrous ion, thus ensuring a ready supply of divalent iron as growth substrate. Indeed, the biological oxidation of Fe^{2+} is a proton-consuming reaction. The consumption of protons occurs in the reduction of oxygen at the cytosol-cell membrane interface and is a requirement for the chemiosmotic generation of ATP (Ingledew et al. 1977; Ingledew 1982; Cox and Brand 1984). The internal cellular consumption of protons thus facilitates the maintenance of the ATPase-mediated passage of protons into the cell and ensures that the cytosolic pH is maintained. Thus, the growth of *T. ferrooxidans* is uniquely adapted to and dependent on highly acidic environments. Protons also play an important role in the chemical dissociation of mineral sulfides which is a prerequisite step prior to bacterial oxidation. The protons disrupt chemical bonds in the sulfide valence band shifting electronic states into positions more closely related to the bacterial metabolic system (Tributsch and Bennett 1981a,b). Protons influence the solubility of ferric ion and, practically speaking, ferric iron has an extremely low solubility in sulfate-containing environments at pH>2.5 and would not effectively participate as an oxidizing agent in the chemical or "indirect dissolution of mineral sulfides. The solubility of Fe^{3+} can be expressed as its activity in relation to the pH, numerically as $\log a_{Fe^{3+}} = 2.4-3\,pH$ (Miller 1980).

Nealson (1983) considers that the growth of autotrophic iron bacteria is largely dependent on the establishment of appropriately acidic environments. These may be best generated through the sulfur-oxidizing cocapacity of the iron-oxidizing bacteria, or by associated miroorganisms. Thus, in the microbial oxidation of mine tailings, the sulfur metabolism of the *Thiobacilli* must be considered as more significant than iron oxidation, since the acid produced is required to stabilize the ferrous ion substrate and permit proton-dependent growth of *T. ferrooxidans*. This would have particular significance in the initial colonization of mine tailings and in the initiation and development of the microbial succession. *T. thiooxidans* is native with *T. ferrooxidans* in acid mineral environments and can thrive at the pH range 0.5–1 which is prohibitive to *T. ferrooxidans*.

The redox properties of the tailings are largely determined by the qualitative and quantitative distribution of the mineral sulfides and their associated gangue minerals, pH, and oxygen content. Particularly pH and Eh are extremely important parameters affecting the establishment, growth, and metabolism of chemoautotrophic bacteria. The diversity of metabolic functions of these organisms further enhances their contribution to the range of reactions occurring in mine tailings and wastes. Considering the complexity of these environments as influenced by a variety of parameters, including Eh and pH, we are still far from a comprehensive understanding of the overall processes involved.

5 Gases

There is little published data available on the gas composition of natural waste dumps and tailings. Oxygen is essential for the oxidation of sulfides to sulfates. Indeed, Harries and Ritchie (1985) consider that the transport of oxygen to the oxidation sites may be the rate-limiting process in dumps and tailings deposits. In a 2-year study of the atmospheric and temperature conditions within a reclaimed coal strip mine, Jaynes et al. (1983) have now provided evidence that gas transport in waste dumps from the surface to oxidation sites at depth occurs by diffusion, convection, and advection. The relative contribution of these processes to overall gas transport is dependent on a variety of parameters relating to the position within the dump, the component materials and minerals, and the way in which the dump has been constructed (Harries and Ritchie, 1985). Oxygen transport and pore gas composition have also been evaluated in coal refuse and spoils (Erickson et al. 1985).

Pore gas composition studies have also provided relevant data relating to microbiological requirements. Many sites characterized by low oxygen concentrations were strongly correlated with high carbon dioxide levels (Jaynes et al. 1983; Harries and Ritchie 1983) and would readily satisfy the autotrophic requirements of the *Thiobacilli* and other carbon dioxide-fixing organisms (Harries and Ritchie 1985). The enrichment of carbon dioxide in these environments is not well understood, although it could in part relate to carbonate gangue neutralization of acid produced from pyrite oxidation (Jaynes et al. 1983). The results of Harries and Ritchie (1985) clearly provide evidence of how waste dump design may influence localized environmental conditions and oxidative reactions within the material, and therefore the development of microbial populations.

Both pH and Eh are critical parameters in the establishment of any one group of organisms. The metabolic activity of these populations in turn changes the conditions of pH and Eh, thus contributing to the observed succession of microbial communities associated with the oxidation of mine tailings until some type of steady state is achieved. The iron- and sulfur-oxidizing *Thiobacilli* are generally regarded as highly aerobic organisms and a reduction in oxygen concentration affects the rate of substrate oxidation. Some strains of *T. ferrooxidans* are also known to fix atmospheric dinitrogen under microaerophilic conditions (Mackintosh 1978; Stevens and Tuovinen 1986; Stevens et al. 1986). Low levels of oxygen have been found in the pore spaces associated with a reclaimed coal strip mine (Jaynes et al. 1983) and waste rock dumps (Harries and Ritchie 1985). Under the acidic conditions associated with mineral sulfide oxidation, the solubility of dissolved gases such as oxygen is also lowered, and therefore the rate of mass transfer of the gas to its reaction site is highly important. The mechanisms of gas transport have been discussed by Harries and Ritchie (1985). The distribution and activity of the chemoautotrophic bacteria under the microaerophilic or anaerobic conditions associated with these types of environments has received little attention. There are only scattered data available on the distribution and numbers of bacteria in tailings and dumps. The activity and number of bacteria are primarily determined by physical factors such as moisture content, aeration, and availability of electron donors. It is apparent that the microbiologically active zone is site-specific. Bhappu et al. (1969) reported a decline in *T. ferrooxidans* numbers of four orders of magnitude upon an increasing depth from the very surface layer to below 2.4 m in an active leach dump. However, in active leach mines, the bacterial distribution has been primarily determined in leach liquor samples (Tuovinen et al. 1981; DiSpirito et al. 1982) and little information is available on the bacterial colonization at various depths and zones of exposed ore materials or tailings.

The effect of elevated carbon dioxide concentrations on the leaching of mineral sulfides has been discussed by Goodman et al. (1983). *T. ferrooxidans* has been shown to oxidize a zinc-iron sulfide mineral ore under anaerobic conditions in the presence of CO_2 (Goodman et al. 1983). Leaching rates increased at elevated carbon dioxide concentrations. Such conditions have been found to exist in field situations (Jaynes et al. 1983; Harries and Ritchie 1983, 1985) and apparently can support significant populations of *T. ferroxidans* (Goodman et al. 1983). Other studies have also shown that this organism catalyzed inorganic transformations under microaerophilic or anaerobic conditions (Pugh and Umbreit 1966). Under anaerobic conditions, *T. ferrooxidans* couples the oxidation of inorganic sulfur to the reduction of Fe^{3+} to Fe^{2+} (Brock and Gustafson 1976; Sugio et al. 1985), but this is not coupled with growth. Both *Sulfolobus brierleyi* and *S. acidocaldarius* can couple the oxidation of sulfur to the reduction of hexavalent molybdenum (Brierley and Brierley 1982). The hydrogen-dependent anaerobic reduction of sulfur by sulfur-dependent archaebacteria has also recently been reported (Segerer et al. 1985; Zillig et al. 1985).

6 Concluding Remarks

The formation of sulfate and ferric ion are good indicators of bacterial activity on sulfide minerals, but it is presently recognized that mine tailings support complex microbial consortia. Although the bulk substrate is practically inorganic, heterotrophic bacteria, yeasts, and filamentous fungi can sometimes be readily isolated from samples of mine waters or tailings. it is generally assumed that there is an interplay between chemoautotrophic *Thiobacilli* and heterotrophic satellites, based on the excretion of organic metabolites by *Thiobacilli* and their subsequent utilization by the heterotrophs. However, the organic carbon and nutrient dynamics are not well understood in these habitats. It should be noted that tailing waters also contain a low chemical residual from the mill and organic compounds from the source of process water and as a result of leaching from the adjacent surface soil.

Several acidophilic, heterotrophic bacteria that live in coculture with *T. ferrooxidans* and in associated environments have been isolated and characterized. *T. acidophilus* (Guay and Silver 1975) and *T. organoparus* (Markosyan 1973) are sulfur oxidizers, but they do not oxidize Fe(II) compounds. Both organisms can utilize a large number of organic electron donors such as sugars and organic acids. Several species of *Acidiphilium* have also been discovered (Harrison 1981; Lobos et al. 1986; Wichlacz et al. 1986) which appear to be typical in acid environments associated with the mining industry and they are obligately dependent on organic compounds for growth.

Acid drainage and mobilization of metals from tailings due to microbiological activity are well-recognized problems and cause major undesirable chemical and biological changes in soil and water. Environmental problems associated with uranium tailings are particularly aggrevated due to radionuclides. Recent lysimeter and column studies at CANMET have started to unravel the complex environmental problem of uranium tailings by also addressing the significant microbiological component and methods of inactivation (Ritcey and Silver 1982; Silver 1985; Silver and Ritcey 1985; Silver et al. 1985). The target nuclide is ^{226}Ra which is a decay product from ^{238}U. The solubility of ^{226}Ra is related to the chemical composition of the water, which in turn is greatly influenced by microbiological oxidations in tailings. Bacterial oxidation of pyrite enhances the dissolution of uranium and associated radionuclides from tailings. ^{226}Ra forms insoluble complexes with jarosites and it can also be effectively precipitated as a sulfate complex with barium chloride treatment. The microbiological production of sulfate promotes ^{226}Ra precipitation, but subsequent dissolution of sulfate complexes may occur upon continued application of water due to a sulfate elution effect.

Acid neutralization with lime is commonly practiced either at the source or downstream to control acid drainage emanating from sulfide-containing tailings. On-site liming does not completely cease microbiological production of acid due to the microenvironments and void spaces inaccessible by liming. Therefore, repeated liming is required and the development of acidic hot spots is not an uncommon phenomenon. Such problems complicate further efforts of reclamation of tailings areas. Liming of the effluent, on the other hand, requires continuous treatment because it does not arrest the acid production at the source. Downstream effluent treatment is also possible with aeration basins and contacting devices which utilize attached iron-oxidizing bacteria for the microbiological conversion of Fe^{2+} to Fe^{3+} (Ishikawa et al. 1983), followed by neutralization and iron precipitation. Again, these treatments require continuous operation because the source is not treated. Besides sulfuric acid and ferric iron production, bacterial oxidations in tailings may render extremely toxic elements soluble which pose a serious health hazard in surface and groundwater. For example, the leaching of arsenic from tailings is such a potential health problem that some mining companies may not undertake processing of arsenic minerals because low-cost technology has not been developed to produce arsenic-free tailings or to prevent the solubilization of arsenic from waste dumps into groundwater supplies.

The microbiological problems are fundamentally similar in tailings produced in the mining of sulfide minerals and in refuse and cob piles associated with coal mining. *Thiobacilli,* as any microorganism, can be inactivated by both chemical and physical treatments. Several chemicals have proved effective in laboratory studies of inhibition of bacterial iron oxidation and acid production. With a view to curtailing *Thiobacilli* and their activities at the source, laboratory studies have demonstrated anionic surfactants and organic acids to be effective treatments (Kleinmann et al. 1981; Erickson et al. 1982; Dugan and Apel 1983; Onysko et al. 1984; Sand 1985; Shellborn and Rastogi 1984; Dugan 1987). Their use in the treatment of tailings by spray application is unlikely to provide a long-lasting effect because both surfactants and organic acids are readily degraded by heterotrophic microorganisms at the site. The introduction of controlled slow-release technology (Fox and Rastogi 1983) shows promise as a long-term treatment for the abatement of microbiological activity at the source of acid production, but requires further field trials to evaluate efficacy and site specificity.

Acknowledgment. Special thanks are due to Laurie A. Haldeman for the typing of the manuscript.

References

Ackman TE, Kleinmann RLP (1984) In-line aeration and treatment of acid mine drainage. US Bur Mines Rep Invest 8868. US Dep Interior

Ahonen L, Hiltunen P, Tuovinen OH (1986) The role of pyrrhotite and pyrite in the bacterial leaching of chalcopyrite ores. In: Lawrence RW, Branion RMR, Ebner HG (eds) Fundamental and applied biohydrometallurgy. Elsevier, Amsterdam, pp 13–22

Aleem MIH, Lees H, Nicholas DJD (1963) Adenosine triphosphate-dependent reduction of nicotinamide adenine dinucleotide by ferrocytochrome *c* in chemoautotrophic bacteria. Nature (London) 200:759–761

Basaran AH, Tuovinen OH (1987) Iron pyrite oxidation by *Thiobacillus ferrooxidans*: sulfur intermediates, soluble end products, and changes in biomass. Coal Preparat 3 (in press)

Beck JV (1967) The role of bacteria in copper mining operations. Biotechnol Bioeng 9:487–497

Bhappu RB, Johnson PH, Brierley JA, Reynolds DH (1969) Theoretical and practical studies on dump leaching. Trans Soc Min Eng AIME 244:307–320

Brady KS, Bigham JM, Jaynes WF, Logan TJ (1986) Influence of sulfate on Fe-oxide formation: comparisons with a stream receiving acid mine drainage. Clays Clay Mineral 34:266–274

Brierley CL (1974) Molybdenite-leaching: use of a high-temperature microbe. J Less Common Met 36:237–247

Brierley CL (1978) Bacterial leaching. Crit Rev Microbiol 6:207–262

Brierley CL, Brierley JA (1973) A chemoautotrophic and thermophilic microorganism isolated from an acid hot spring. Can J Microbiol 19:183–188

Brierley CL, Brierley JA (1982) Anaerobic reduction of molybdenum by *Sulfolobus* species. Zentralbl Bakt Hyg I Abt Orig C3:289–294

Brierley CL, Murr LE (1973) Leaching: use of a thermophilic and chemoautotrophic microbe. Science 179:488–490

Brierley CL, Brierley JA, Norris PR, Kelly DP (1980) Metal-tolerant micro-organisms of hot, acid environments. In: Gould GW, Corry JEL (eds) Microbial growth and survival in extremes of environment. Academic Press, London New York, pp 39–51

Brierley JA (1978) Thermophilic iron-oxidizing bacteria found in copper leaching dumps. Appl Environ Microbiol 36:523–525

Brierley JA, Brierley CL (1978) Microbial leaching of copper at ambient and elevated temperatures. In: Murr LE, Torma, AE, Brierley JA (eds) Metallurgical applications of bacterial leaching and related microbiological phenomena. Academic Press, London New York, pp 477–490

Brock TD, Gustafson J (1976) Ferric iron reduction by sulfur- and iron-oxidizing bacteria. Appl Environ Microbiol 32:567–571

Brock TD, Brock KM, Belly RT, Weiss RL (1972) *Sulfolobus*: a new genus of sulfur-oxidizing bacteria living at low pH and high temperature. Arch Mikrobiol 84:54–68

Cathles LM, Apps JA (1975) A model of the dump leaching process that incorporates oxygen balance, heat balance, and air convection. Metall Trans 6B:617–624

Cobley JG, Cox JC (1983) Energy conservation in Acidophilic Bacteria, Microbiol Rev 47:579–595

Cobley JG, Haddock BA (1975) The respiratory chain of *Thiobacillus ferrooxidans:* the reduction of cytochromes by Fe^{2+} and the preliminary characterization of rusticyanin, a novel "blue" copper protein. FEBS Lett 60:29–33

Cox JC, Brand MD (1984) Iron oxidation and energy conservation in the chemoautotroph *Thiobacillus ferrooxidans*. In: Strohl WR, Tuovinen OH (eds) Microbial chemoautotrophy. Ohio State Univ Press, Columbus, pp 31–46

DiSpirito AA, Tuovinen OH (1981) Oxygen uptake coupled with uranous sulfate oxidation by *Thiobacillus ferrooxidans* and *T. acidophilus.* Geomicrobiol J 2:275–291

DiSpirito AA, Tuovinen OH (1982a) Uranous ion oxidation and carbon dioxide fixation by *Thiobacillus ferrooxidans.* Arch Microbiol 133:28–32

DiSpirito AA, Tuovinen OH (1982b) Kinetics of uranous ion and ferrous iron oxidation by *Thiobacillus ferrooxidans.* Arch Microbiol 122:33–37

DiSpirito AA, Tuovinen OH (1984) Oxidations of non-ferrous metals by *Thiobacilli.* In: Strohl WR, Tuovinen OH (eds) Microbial chemoautotrophy. Ohio State Univ Press, Columbus, pp 11–29

DiSpirito AA, Silver M, Voss L, Tuovinen OH (1982) Flagella and pili of iron-oxidizing thiobacilli isolated from a uranium mine in northern Ontario, Canada. Appl Environ Microbiol 43:1096–1200.

Dubrovsky NM, Cherry JA, Reardon EJ, Vivyurka AJ (1985) Geochemical evolution of inactive pyritic tailings in the Elliot Lake uranium district. Can Geotech J 22:110–128

Dugan PR (1987) Prevention of formation of acid mine drainage from high-sulfur coal refuse by inhibition of iron- and sulfur-oxidizing microorganisms. II. Inhibition of "run of mine" refuse under simulated field conditions. Biotechnol Bioeng 29:49–58

Dugan PR, Apel WA (1983) Bacteria and acidic drainage from coal refuse: inhibition by sodium lauryl sulfate and sodium benzoate. Appl Environ Microbiol 46:279–282

Dutrizac JE, MacDonald RJC (1974) Ferric ion as a leaching medium. Mineral Sci Eng 6:59–100

Eccleston M, Kelly DP, Wood AP (1985) Autotrophic growth and iron oxidation and inhibition kinetics of *Leptospirillum ferrooxidans.* In: Caldwell DE, Brierley JA, Brierley CL (eds) Planetary ecology. Van Nostrand Reinhold, New York, pp 263–272

Ehrlich HL (1981) Geomicrobiology. Dekker, New York

Erickson PM, Kleinmann RLP, Campion PSA (1982) Reducing oxidation of pyrite through selective reclamation practices. In: Symp Surface mining, hydrology, sedimentology and reclamation. Univ Kentucky, Lexington, pp 97–102

Erickson PM, Ladwig KJ, Kleinmann RLP (1985) Acid mine drainage from inactive eastern coal operations. Environ Geochem Health 7:16–25

Ferroni GD, Leduc LG, Todd M (1986) Isolation and temperature characterization of psychrotrophic strains of *Thiobacillus ferrooxidans* from the environment of a uranium mine. J Gen Appl Microbiol 32:169–175

Fox LA, Rastogi V (1983) Developments in controlled release technology and its application in acid mine drainage. In: Symp Surface mining, hydrology, sedimentology and reclamation. Univ Kentucky, Lexington, pp 447–455

Golovacheva RS, Karavaiko GI (1978) A new genus of thermophilic spore-forming bacteria, *Sulfobacillus.* Mikrobiologiya 47:815–822

Goodman AE, Khalid AM, Ralph BJ (1981) Microbial ecology of Rum Jungle 1: Environmental study of sulphidic overburden dumps, experimental heap leach sites and tailings dam area. Australian atomic energy commission, AAEC/E531

Goodman AE, Babij T, Ritchie AIM (1983) Leaching of a sulphide ore by *Thiobacillus ferrooxidans* under anaerobic conditions. In: Rossi G, Torma AE (eds) Recent progress in biohydrometallurgy. Assoc Min Sarda, Iglesias, It, pp 361–376

Groudev SN, Genchev FN, Gaidarjiev SS (1978) Observations on the microflora in an industrial copper dump leaching operation. In: Murr LE, Torma AE, Brierley JA (eds) Metallurgical applications of bacterial leaching and related microbiological phenomena. Academic Press, London New York, pp 253–274

Guay R, Silver M (1975) *Thiobacillus acidophilus* sp. nov.; isolation and some physiological characteristics. Can J Microbiol 21:281–288

Harries JR, Ritchie AIM (1981) The use of temperature profiles to estimate the pyritic oxidation rate in a waste rock dump from an open cut mine. Water Air Soil Pollut 15:405–423

Harries JR, Ritchie AIM (1983) The microenvironment within waste rock dumps undergoing pyritic oxidation. In: Rossi G, Torma AE (eds) Recent progress in biohydrometallurgy. Assoc Min Sarda, Iglesias, It, pp 377–392

Harries JR, Ritchie AIM (1985) Pore gas composition in waste rock dumps undergoing pyritic oxidation. Soil Sci 140:143–152

Harrison AP, Jr (1981) *Acidiphilium cryptum* gen. nov., sp. nov.: heterotrophic bacteria from acidic mineral environments. Int J Syst Bacteriol 31:327–332

Harrison AP, Jr (1984) The acidophilic thiobacilli and other acidophilic bacteria that share their habitat. Annu Rev Microbiol 38:265–292

Harrison AP, Jr (1986a) Characteristics of *Thiobacillus ferrooxidans* and other iron-oxidizing bacteria, with emphasis on nucleic acid analyses. Biotechnol Appl Biochem 8:249–257

Harrison AP, Jr (1986b) The phylogeny of iron-oxidizing bacteria. Biotechnol Bioeng Symp 16:311–317

Harrison AP, Jr, Norris PR (1985) *Leptospirillum ferrooxidans* and similar bacteria: some characteristics and genomic diversity. FEMS Microbiol Lett 30:99–102

Huber G, Huber H, Stetter KO (1986) Isolation and characterization of new metal-mobilizing bacteria. Biotechnol Bioeng Symp 16:239–251

Hutchins SR, Davidson MS, Brierley JA, Brierley CL (1986) Microorganisms in reclamation of metals. Annu Rev Microbiol 40:311–336

Ingledew JW (1982) *Thiobacillus ferrooxidans*: the bioenergetics of an acidophilic chemolithotroph. Biochim Biophys Acta 683:89–117

Ingledew JW, Cobley JG (1980) A potentiometric and kinetic study on the respiratory chain of ferrous iron grown *Thiobacillus ferrooxidans*. Biochim Biophys Acta 590:141–158

Ingledew JW, Cox JC, Halling PJ (1977) A proposed mechanism for energy conservation during Fe^{2+} oxidation by *Thiobacillus ferrooxidans*: chemiosomotic coupling to net H^+ influx. FEMS Microbiol Lett 2:193–197

Ishikawa T, Murayama T, Kawahara I, Imaizumi T (1983) A treatment of acid mine drainage utilizing bacterial oxidation. In: Rossi G, Torma AE (eds) Recent progress in biohydrometallurgy. Assoc Min Sarda, Iglesias, It, pp 393–407

Jaynes DB, Rogowski AS, Pionke HB, Jacoby EK (1983) Atmosphere and temperature changes within a reclaimed coal strip mine. Soil Sci 136:164–177

Jaynes DB, Rogowski AS, Pionke HB (1984a) Acid mine drainage from reclaimed coal strip mines. 1. Model description. Water Resourc Res 20:233–242

Jaynes DB, Pionke HB, Rogowski AS (1984b) Acid mine drainage from reclaimed coal strip mines. 2. Simulation results of model. Water Resourc Res 20:243–250

Jerez CA, Peirano I, Chamorro D, Campos G (1986) Immunological and electrophoretic differentiation of *Thiobacillus ferrooxidans* strains. In: Lawrence RW, Branion RMR, Ebner HG (eds) Fundamental and applied biohydrometallurgy. Elsevier, Amsterdam, pp 443–456

Karavaiko GI (1985) Microbial processes for the leaching of metals from ores. Centre of International Projects (GKNT), Moscow, USSR

Kelly DP, Norris PR, Brierley CL (1979) Microbiological methods for the extraction and recovery of metals. Symp Soc Gen Microbiol 29:263–308

Khalid AM, Ralph BJ (1977) The leaching behaviour of various zinc sulphide minerals with three *Thiobacillus* species. In: Schwartz W (ed) Conference bacterial leaching. Chemie, Weinheim, pp 165–173

Kleinmann RLP, Crerar DA, Pacelli RP (1981) Biogeochemistry of acid mine drainage and a method to control acid formation. Min Eng 33:300–305

Le Roux NW, Wakerley DS, Hunt SD (1977) Thermophilic thiobacillus-type bacteria from Icelandic thermal areas. J Gen Microbiol 100:197–201

Lobos JH, Chisolm TE, Bopp LH, Holmes DS (1986) *Acidiphilium organovorum* sp. nov., an acidophilic heterotroph isolated from a *Thiobacillus ferrooxidans* culture. Int J Syst Bacteriol 36:139–144

Lowson RT (1982) Aqueous oxidation of pyrite by molecualr oxygen. Chem Rev 82:461–497

Lundgren DG, Dean W (1979) Biogeochemistry of iron. In: Trudinger PA, Swaine DJ (eds) Biogeochemical cycling of mineral-forming elements. Elsevier, Amsterdam, pp 211–251
Lundgren DG, Malouf EE (1983) Microbial extraction and concentration of metals. Adv Biotechnol Proc 1:223–249
Lundgren DG, Silver M (1980) Ore leaching by bacteria. Annu Rev Microbiol 34:263–283
Mackintosh ME (1978) Nitrogen fixation by *Thiobacillus ferrooxidans*. J Gen Microbiol 105:215–218
Manchee RJ (1979) Microbial mining. Trends Biochem Sci 4:77–80
Markosyan GE (1973) A new mixotrophic sulfur bacterium developing in acid media, *Thiobacillus organoparus* sp. nov. Dokl Akad Nauk SSSR 211:1205–1208
Marsh RM, Norris PR (1983a) The isolation of some thermophilic, autotrophic iron- and sulphur-oxidizing bacteria. FEMS Microbiol Lett 17:311–315
Marsh RM, Norris PR (1983b) Mineral sulphide oxidation by moderately thermophilic acidophilic bacteria. Biotechnol Lett 5:585–590
Mehta AP, Murr LE (1982) Kinetic study of sulfide leaching by galvanic interaction between chalcopyrite, pyrite, and sphalerite in the presence of *T. ferrooxidans* (30°C) and a thermophilic microorganism (55°C). Biotechnol Bioeng 24:919–940
Mehta AP, Murr LE (1983) Fundamental studies of the contribution of galvanic interaction to acid-bacterial leaching of mixed metal sulfides. Hydrometallurgy 9:235–256
Miller SD (1980) Sulfur and hydrogen ion buffering in pyritic strip-mine soil. In: Trudinger PA, Walter MR, Ralph BJ (eds) Biogeochemistry of ancient and modern environments. Springer, Berlin Heidelberg New York, pp 537–543
Murr LE, Berry VK (1979) Observations of a natural thermophilic microorganism in the leaching of a large, experimental, copper-bearing waste body. Metall Trans B 10:523–531
Murr LE, Brierley JA (1978) The use of large scale test facilities in studies of the role of microorganisms in commercial leaching operations. In: Murr LE, Torma AE, Brierley JA (eds) Metallurgical applications of bacterial leaching and related microbiological phenomena. Academic Press, London New York, pp 491–520
Natarajan KA, Iwasaki I (1983) Role of galvanic interactions in the bioleaching of Duluth copper-nickel sulfides. Separ Sci Technol 18:1095–1111
Nealson KH (1983) The microbial iron cycle. In: Krumbein WE (ed) Microbial geochemistry. Blackwell, Oxford, pp 159–190
Nordstrom DK (1982) Aqueous pyrite oxidation and the consequent formation of secondary iron minerals. In: Kittrick JA, Fanning DS, Hossner LR (eds) Acid sulfate weathering. Soil Sci Soc Am, Madison, Wisc, pp 37–56
Norris PR, Barr DW (1985) Growth and iron oxidation by acidophilic moderate thermophiles. FEMS Microbiol Lett 28:221–224
Norris PR, Kelly DP (1982) The use of mixed microbial cultures in metal recovery. In: Bull AT, Slater JH (eds) Microbial interactions and communities. Academic Press, London New York, pp 443–474
Onysko SJ, Kleinmann RLP, Erickson PM (1984) Ferrous iron oxidation by *Thiobacillus ferrooxidans*: inhibition with benzoic acid, sorbic acid, and sodium lauryl sulfate. Appl Environ Microbiol 48:229–231
Pugh LH, Umbreit WW (1966) Anaerobic CO_2 fixation by autotrophic bacteria. *Hydrogenomonas* and *Ferrobacillus*. Arch Biochem Biophys 115:122–128
Ralph BJ (1979) Oxidative reactions in the sulfur cycle. In: Trudinger P, Swaine DJ (eds) Biogeochemical cycling of mineral-forming elements. Elsevier, Amsterdam, pp 369–400
Ralph BJ (1985) Biotechnology applied to raw minerals processing. In: Robinson CW, Howell JA (eds) Comprehensive biotechnology, vol 4. Pergamon, Oxford, pp 201–234
Ritcey GM, Silver M (1982) Lysimeter investigations on uranium tailings at CANMET. CIM Bull 75 (10):134–143
Ross GJ, Ivarson KC, Miles NM (1982) Microbial formation of basic ferric sulfates in laboratory systems and in soils. In: Kittrick JA, Fanning DS, Hossner LR (eds) Acid sulfate weathering. Soil Sci Soc Am, Madison, Wisc, pp 77–94
Sand W (1985) The influence of four detergents on the substrate oxidation by *Thiobacillus ferrooxidans*. Environ Technol Lett 6:439–444
Sato M (1960) Oxidation of sulfide ore bodies II. Oxidation mechanisms of sulfide minerals at 25°C. Econ Geol 55:1202–1231

Segerer A, Stetter KO, Klink F (1985) Two contrary modes of chemolithotrophy in the same archaebacterium. Nature (London) 313:787–789

Segerer A, Neuner A, Kristjansson JK, Setter KO (1986) *Acidianus infernus* gen. nov., sp. nov., and *Acidianus brierleyi* comb. nov.: facultatively aerobic, extremely acidophilic sulfur-metabolizing archaebacteria. Int J Syst Bacteriol 36:559–564

Shellborn MA, Rastogi V (1985) Laboratory methods for determining the effects of bactericides on acid mine drainage. In: Symp Surface mining hydrology, sedimentology, and reclamation. Univ Kentucky, Lexington, pp. 77–81

Silver M (1985) Water leaching characteristics of uranium tailings from Ontario and northern Saskatchewan. Hydrometallurgy 14:189–217

Silver M, Ritcey GM (1985) Effects of iron-oxidizing bacteria and vegetation on acid generation in laboratory lysimeter tests on pyrite-containing uranium tailings. Hydrometallurgy 15:255–264

Silver M, Ritcey GM, Cauley MP (1985) A lysimeter comparison on the effects of uranium tailings deposition methods on the release of environmental contaminants. Hydrometallurgy 15:159–172

Stevens CJ, Tuovinen OH (1986) Ferrous ion oxidation, nitrogen fixation (acetylene reduction), and nitrate reductase activity by *Thiobacillus ferrooxidans*. In: McCready RGL (ed) Proc 2nd Annu Meet BIOMINET, CANMET, Ottawa, pp 37–45

Stevens CJ, Dugan PR, Tuovinen OH (1986) Acetylene reduction (nitrogen fixation) by *Thiobacillus ferrooxidans*. Biotechnol Appl Biochem 8:351–359

Sugio T, Domatsu C, Munakata O, Tano T, Imai K (1985) Role of ferric ion-reducing system in sulfur oxidation of *Thiobacillus ferrooxidans*. Appl Environ Microbiol 49:1401–1406

Torma AE (1977) The role of *Thiobacillus ferrooxidans* in hydrometallurgical processes. Adv Biochem Eng 6:1–37

Torma AE (1978) Oxidation of gallium sulfides by *Thiobacillus ferrooxidans*. Can J Microbiol 24:888–891

Torma AE (1983) Biotechnology applied to mining of metals. Biotech Adv 1:73–80

Torma AE, Bosecker K (1982) Bacterial leaching. Progr Ind Microbiol 16:77–118

Torma AE, Habashi F (1972) Oxidation of copper (II) selenide by *Thiobacillus ferrooxidans*. Can J Microbiol 18:1780–1781

Tributsch H, Bennett JC (1981a) Semiconductor-electrochemical aspects of bacterial leaching. 1. Oxidation of metal sulfides with large energy gaps. J Chem Technol Biotechnol 31:565–577

Tributsch H, Bennett JC (1981b) Semiconductor-electrochemical aspects of bacterial leaching. 2. Survey of rate-controlling sulfide properties. J Chem Technol Biotechnol 31:627–635

Tuovinen OH, Kelly DP (1974) Use of microorganisms for the recovery of metals. Int Metall Rev 19:21–31

Tuovinen OH, Niemelä SI, Gyllenberg HG (1971) Tolerance of *Thiobacillus ferrooxidans* to some metals. Antonie van Leeuwenhoek 37:489–496

Tuovinen OH, Silver M, Martin PAW, Dugan PR (1981) The Agnew Lake uranium mine leach liquors: chemical examinations, bacterial enumeration and composition of plasmid DNA of iron-oxidizing thiobacilli. In: Proc Int Conf Use of microorganisms in hydrometallurgy. Hung Acad Sci, Pécs, pp 59–69

Vanselow DG (1976) Mechanisms of bacterial oxidation of the copper sulfide mineral covellite. PhD thesis, Univ New South Wales, Sydney, 198 pp

Vaughan DJ (1984) Electronic structures of sulfides and leaching behavior. In: Bautista RG (ed) Hydrometallurgical process fundamentals. Plenum, New York, pp 23–40

Wichlacz PL, Unz RF, Langworthy TA (1986) *Acidiphilium angustum* sp. nov., *Acidiphilium facilis* sp. nov., and *Acidiphilium rubrum* sp. nov.: acidophilic heterotrophic bacteria isolated from acidic coal mine drainage. Int J Syst Bacteriol 36:197–201

Wood AP, Kelly DP (1984) Growth and sugar metabolism of a thermophilic iron-oxidizing mixotrophic bacterium. J Gen Microbiol 130:1337–1349

Wood AP, Kelly DP (1985) Autotrophic and mixotrophic growth and metabolism of some moderately thermoacidophilic iron-oxidizing bacteria. In: Caldwell DE, Brierley JA, Brierley CL (eds) Planetary ecology. Van Nostrand Reinhold, New York, pp 251–262

Zillig W, Yeats S, Holz I, Bock A, Groff F, Retterberger M, Lutz S (1985) Plasmid-related anaerobic autotrophy of the novel archaebacterium *Sulfolobus ambivalens*. Nature (London) 313:789–791

Response of Plants and Vegetation to Mine Tailings and Dredged Materials

W.H.O. ERNST[1]

1 Introduction

Mine tailings and disposal sites of dredged materials are new environments which have been superimposed on existing ones; at the same time open mining casts and the dredging sites themselves are disturbing existing ecosystems. Tailings from mineral operations and dredged materials can be stored as upland or underwater disposals including the backfilling of mined-out areas (Andrews 1975). The bare surfaces of above ground disposals are open to erosion by water and wind without a protective vegetation. Therefore, the final aim of managing these sites is primarily to diminish the environmental impact on the surrounding landscape by stabilizing these dumps by vegetation, because technical solutions have proved to be insufficient (Greszta 1973; Barner 1978). The possibilities for revegetation of surface and underwater disposal sites depend on the ability of the colonizing plant species to respond adequately with their roots and shoots to the physical structure and the chemical composition of these materials. With regard to the high diversity of structures and compounds the statement is appropriate: "No combination of environmental factors in mine tailings and dredged materials seems to be impossible" (Ernst and Joosse-van Damme 1983). The biological answer to combinations of extreme environmental factors is more restricted: "Not all is possible for plant growth". This contribution will present some general principals which have to operate in plant populations and vegetation if revegetation is to be a successful enterprise either on upland or underwater sites. No special attention will be given to the impact of disposal sites on the surroundings, because pollution by wind and water erosion causes similar stress on the affected ecosystems as on the vegetation of the disposal site itself.

2 General Aspects: Environmental Stress and the Selection of Convenient Stress Tolerators

It is convenient to define the presence of a high or very low concentration, amplitude, or frequency of an environmental factor as stress if it affects plant metabolism and plant performance (Levitt 1980). Stress, however, has to be judged

[1]Department of Ecology and Ecotoxicology, Biological Laboratories, The Free University, Postbus 7161, 1007 MC Amsterdam, The Netherlands

by the response of the organism to this external factor, and not by the observer of the organism. A plant which is growing and reproducing in a permanent flooded environment will be injured by a longer drought period, reducing dramatically the availability of water; it will be stressed by drought. In contrast, a plant from a sand dune will survive certain drought periods (Ashenden et al. 1975), but it will die under flooding.

The colonization of a new environment as mine tailings or dredged materials can start with all seeds or vegetative propagules which have entered the site either from the surrounding vegetation or after long-distance transport by wind or water. The success of these colonizers, however, will depend on their ability to cope with the stress factor(s) and on their capacity to achieve all necessary resources for growth and reproduction in adequate amounts. After successful vegetative or generative reproduction these stress-tolerant individuals will build up a population, i.e. a group of interbreeding individuals of the same plant species that coexist together in space and time (Hedrick 1984).

Due to the presence of heavy metals in all of these wastes, I will take these chemical elements to explain the basic biological principles of the evolution and selection of stress-tolerating plant species and plant populations. Old mine tailings in Europe and new ones in other continents are mostly located in the vicinity of the ore excavation. As far as these ore veins are cropping out at or near the soil surface, a vegetation with metal-tolerant plants is already present due to the selection processes over thousands of years. The offspring of these metal-tolerant plants has the highest potential for a successful natural revegetation of these mine tailings (Ernst 1974, 1982; Bradshaw 1975). Nevertheless, there is an enormous amount of bare tailings in old mining districts (Ernst 1974, 1981). These unvegetated areas indicate either that the colonizing species are not sufficiently tolerant to the high levels of heavy metals in the bare tailing sites or that they have to cope with additional environmental factors, such as low soil fertility and a low water-holding capacity. In addition, the angular shape of the tailings determines its physical stability and the danger of burying the colonizing species by wind and water erosion. The start of the natural revegetation process at the bottom of the tailings indicates that the adaptation of the metal-tolerant plants seems to be perhaps less inferior to the metalliferous material than their adaptation to drought, burial and nutrient deficiency (cf. Fig. 2).

In order to identify well-adapted genotypes of a plant species to certain environmental constraints, especially heavy metals in wastes and disposal material, Wilkins (1957; for a reappraisal of the technique: Wilkins 1978) has developed a technique to test individual plants for their resistance to heavy metals. This so-called rooting technique compares the response of root growth of the tiller of one individual in a solution with the metal under consideration to that of another tiller of the same individual in a solution without the metal. It is a very inexpensive, handsome, and therefore, often used technique of quantifying resistance to heavy metals and to other edaphic factors in grasses (Table 1), but it is less suitable for herbs due to their rooting behaviour (Ernst 1974, 1982). For herbs Repp (1963) has developed another technique, the so-called comparative protoplasmatology. In this case small pieces of leaf or shoot tissue are incubated in series with increasing metal concentrations; after 1 or 2 days the vitality of the tissue is analyzed (Gries

Table 1. Quantification of resistance to soil chemical factors in grasses by the rooting technique

Soil chemical factor	Grass species	Reference
Cu, Zn	*Agrostis canina*	Ernst et al. 1975
Cu, Zn, Pb, Ni	*Agrostis capillaris*	Gregory and Bradshaw 1965
Cu, Zn, Pb, Ni, Cd	*Agrostis capillaris*	Symeonidis et al. 1985
Cu, Zn, Ni	*Agrostis gigantea*	Hogan and Rauser 1979
Zu, Zn	*Agrostis stolonifera*	Wu and Antonovics 1975
Zn	*Anthoxanthum odoratum*	Gartside and McNeilly 1974
Cu, Zn, Pb, Ni	*Anthoxanthum odoratum*	Karataglis 1978
Zn	*Deschampsia cespitosa*	Godbold et al. 1983
Cu, Ni, Al	*Deschampsia cespitosa*	Cox and Hutchingson 1980
Pb	*Festuca ovina*	Urquhart 1971
Cu, Zn, Pb	*Festuca rubra*	Wong 1982
Fe, Mn	*Festuca rubra*	Dueck et al. 1984
Cd, Al	*Holcus lanatus*	McGrawth et al. 1980
Zn	*Molinia caerulea*	Dueck et al. 1984

1966; Rüther 1967; Ernst 1974, 1982). By applying both techniques to the appropriate plant groups it is possible to detect and select genotypes also for commercial breeding (Smith and Bradshaw 1979), thus creating sufficient plant material for specific revegetation demands (Fig. 1).

An analysis of the metal tolerance of plant species in a gradient from the bottom to the upper slopes of old mine tailings allowed the detection that individuals at the slope were more tolerant to heavy metals than those at the bottom, although the total amount of heavy metals did not significantly differ between sites. The chemical form of the water-soluble heavy metals, however, was quite different (Fig. 2). This change in metal speciation (cf. Bourg, this Vol.) is due to biotransformation of heavy metals by the metabolism of metal-adapted plants (Ernst 1975; Mathys 1977; Lolkema et al. 1984) and due to the complexation of heavy metals during decomposition and rearrangement of dead plant material in the process of soil formation to humic and fulvic acids (van der Werff 1981). During the above mentioned natural revegetation process, natural selection has favoured highly metal-tolerant genotypes on the sparsely vegetated slopes with predominant ionic metals, whereas at the highly vegetated bottom of the tailings the decreased toxicity of organo-metal complexes less tolerant genotypes survive.

When selection can operate so efficiently at such a small distance, it is obvious that the various metal-tolerant genotypes may differ in fitness; this means that there is a price to pay for tolerance. More than 30 years ago, it had already been indicated that zinc-tolerant plants of *Silene cucubalus* have a lower growth and biomass production than zinc-sensitive plants (Baumeister 1954). In the meantime this tendency has been shown for many plant populations with tolerance mechanisms against environmental hazards (Ernst 1983). The reduction of biomass production by tolerant plants compared to their sensitive counterparts varies between 20 and 40% and can even surmount this percentage (Table 2), although this phenomenon may have exceptions. As a consequence, plants with tolerances against many heavy metals, the so-called multiple metal-tolerant genotypes, have a lower biomass production than genotypes which are tolerant against only one heavy metal (Ernst

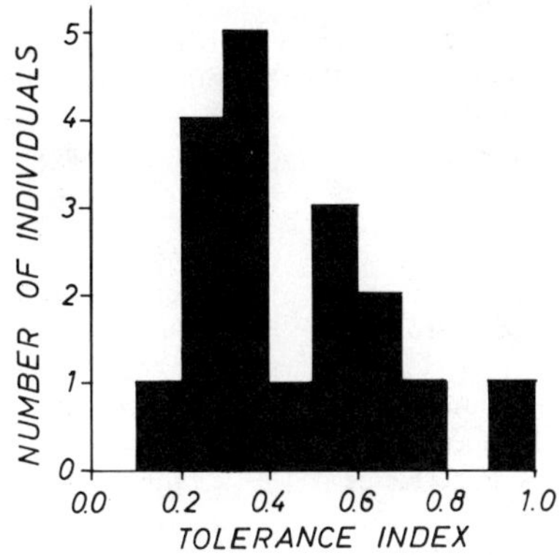

Fig. 1. Zinc tolerance in a population of the grass *Molinia caerulea* from uncontaminated birch wood. The tolerance index is given as the ratio of root elongation of tillers in a solution with 200 μM zinc compared to a treatment without zinc. This "normal" population already contains a high variability of zinc tolerance. Data are based on Dueck et al. (1984)

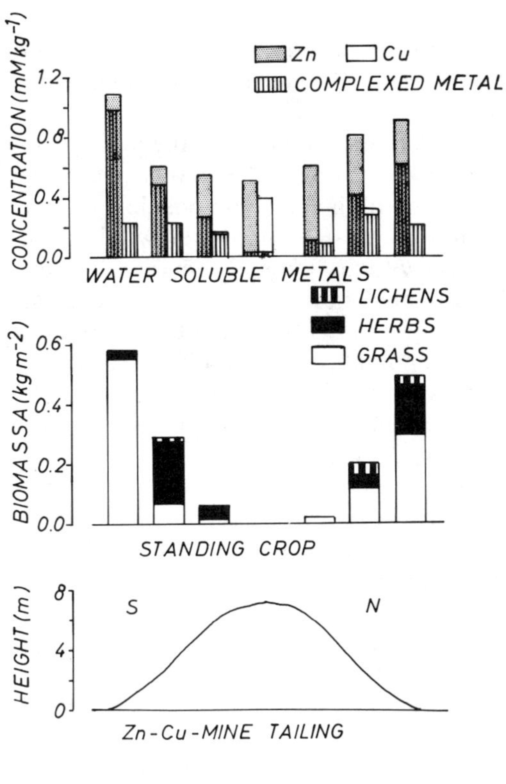

Fig. 2. Standing crop and water-soluble metals in a more than 100-year-old tailing in the vicinity of a copper and zinc mine in the Harz mountains. The analysis has followed a north-south gradient of the tailing. Productivity of the vegetation is hampered by the ionic status of the water-soluble heavy metals. Data are based on Ernst (1974, and unpublished results)

1974; Verkleij and Bast-Cramer 1985). It is hoped that ongoing experiments may give a firm foundation for a realistic calculation of the price which has to be paid for the evolution of tolerance mechanisms. A thorough knowledge of these energetic consequences will determine the possibilities of breeding for multiple metal resistance. Energy which is invested in the maintenance of tolerance levels is not available for growth and reproduction; therefore, hypertolerance will not only reduce the fitness of an individual in a population, but it will also diminish the

Table 2. Percentage decrease of biomass production in tolerant populations of various plant species in comparison to their sensitive equivalents (all references in Ernst 1983, except Lolkema et al. 1984)

Tolerance against	Plant species	Decrease in biomass production	Reference
Zinc	*Silene cucubalus*	-20%	Baumeister 1954
Nickel	*Festuca cinerea*	-25%	Sasse 1976
Azin herbicide	*Senecio vulgaris*	-25%	Conard and Radosevich 1979
Sodium chloride	*Festuca rubra*	-40%	Rozema 1978
Copper	*Silene cucubalus*	-40%	Lolkema et al. 1984
Azin herbicide	*Amaranthus retroflexus*	-40%	Conard and Radosevich 1979
Magnesium	*Agrostis stolonifera*	-45%-75%	Marrs and Proctor 1976

contribution of these plants to the production of organic material available for the soil formation. Colonization of disposal sites by hypertolerant plants will retard the amelioration of the tailings.

If the mine tailings and the disposal sites of dredged materials are erected in surroundings where the plant species have not experienced such a physical structure and chemical composition, it cannot be expected that all plant species and all individuals of a population will have the appropriate physiological answer to these new environmental demands. In such a situation only a few genotypes can be favoured by selection, or in certain situations a mutation may be the only answer to the new ecological demand. Both processes will rely on the genetic variability of plant populations. At this point, there are two very decisive questions, when tailings and disposal sites should be revegetated in every stochastically chosen site. Firstly, do all plant species have the same ability for the evolution of metal-tolerant genotypes? Secondly, are adaptations to more than one changed environmental factor compatible with other metabolic processes in a plant? The lack of many plant species in heavy metal-rich sites indicates that not all species have the same genetic potential to withstand this extraordinary ecological situation. Selection experiments with commercial grassland species have confirmed this observation on heavy metal sites (Gartside and McNeilly 1974; Lolkema 1985). There is a further important restriction. It is very obvious that the most productive grasses and herbs are not capable of evolving metal resistance. The consequence of this result is that revegetation is only supported by slowly growing species, i.e. it is a long-term process.

The answer to the second question on the physiological compatibility of the adaptation to several environmental stress factors is more difficult due to the paucity of information. As already mentioned, some plant species can simultaneously cope with many high concentrations of heavy metals, i.e. for *Silene cucubalus* on the mine tailings of the Harz mountains with zinc, copper, lead, cadmium and arsenic (Ernst 1974). The physiological tolerance mechanism against each of these heavy metals is very specific. Cellular detoxification takes place either by cellular compartmentation in cell walls as in the case of lead or in the vacuoles as in the case of zinc (Ernst and Weinert 1972; Ernst 1975) or by sequestration in organic compounds in the cytoplasm as in the case of copper (Lolkema et al. 1984). Obviously, these various tolerance mechanisms provide possibilities to breed plants for all combinations of metals and their concentration in mine tailings and disposal sites of dredged material. But remember, the more tolerance, however, desired, the less the biomass will be in these multiple-tolerant species.

In contrast to this obvious compatibility of several heavy metal tolerances in one plant, a combination of these tolerances with other abiotic factors may be incompatible. As already mentioned, grasses from very nutrient-rich soils, i.e. perennial ryegrass (*Lolium perenne*) and cocksfoot (*Dactylis glomerata*) have an essentially lower metal tolerance than grasses from areas poor in nitrogen and phosphorus, i.e. bentgrass (*Agrostis capillaris*) and sheep fescue (*Festuca ovina*). It will be premature to give a physiological explanation for these incompatibilities, but it may have a strong impact on the possibility of fertilizer application in these disposal sites.

In conclusion, the physical and chemical homogeneity or heterogeneity of tailings and dredged material will strongly influence the necessary amount of genotypes in a colonizing population, if a complete vegetation cover should be achieved. If a disposal site is very homogeneous, the propagation of a few genotypes of vegetatively reproducing, stress-tolerating plant species such as creeping bentgrass (*Agrostis stolonifera*) in the temperate regions (Wu et al. 1975) and of couch grass (*Cynodon dactylon*) in the subtropical regions (Hill 1978) will produce an acceptable plant cover on these substrates. If the mine tailings or disposal sites are very heterogeneous, a lot of genetic variation of the colonizing species must be present (Verkleij 1980), if the various niches in a mosaic environment should be filled up by plants.

3 Upland Disposal and its Problems to (Re)vegetation

Upland disposal of mine wastes and dredged materials are generally characterized by some properties which are hostile, at least in their combination to plant growth:

1. High levels of various heavy metals and other elements in toxic concentrations;
2. Low amounts of major plant nutrients;
3. Poor physical structures;
4. A variable access of oxygen to the upper and lower layers of the dumps.

For good vegetation development some way must be found to overcome or at least modify the hostility of some of these environmental factors to higher plants.

3.1 Heavy Metals and Other Elements in Toxic Concentrations

There is no doubt that the above mentioned evolution of metal tolerance by certain plant species is the basis for reclamation of metalliferous wastes. It has been demonstrated that indeed these tolerant plants offer a very promising way of overcoming certain degrees of metal toxicity in the revegetation process

Table 3. The effect of vesicular-arbuscular mycorrhizal (VAM) fungi on the growth and zinc concentration in three clones of *Calamagrostis epigeijos* with zinc exposure (After Visser, unpublished MSc Thesis). Results with different indices are significant at $P < 0.05$

Clone	Zinc tolerance index (according to Wilkins)	Biomass (g dry wt. per plant)				Zinc concentration of the shoot (μmol g^{-1} dry wt.)			
		-VAM -Zn	-VAM +Zn	+VAM -Zn	+VAM +Zn	-VAM -Zn	-VAM +Zn	+VAM -Zn	+VAM +Zn
I	0.38±0.14	2.62^{bc}	0.62^{f}	3.11^{a}	2.33^{c}	0.72^{A}	16.30^{D}	0.75^{A}	14.02^{D}
II	0.27±0.16	1.91^{cd}	0.88^{f}	2.49^{c}	1.50^{de}	0.60^{A}	25.68^{E}	0.68^{A}	22.22^{E}
III	0.20±0.04	1.65^{cde}	1.18^{ef}	2.91^{ab}	2.23^{cd}	0.91^{B}	16.74^{D}	0.90^{B}	11.00^{D}

(Lohmeyer 1964; Gadgil 1969; Gemmell 1977; Johnson et al. 1977; Smith and Bradshaw 1979). In addition to the metal tolerance of grasses and herbs, a further biological way of diminishing metal toxicity has been opened by the discovery of metal-tolerant mycorrhizal fungi (Gildon and Tinker 1981). Generally, most of the grasses and herbs in uncontaminated environments are infected by vesicular-arbuscular mycorrhizal (VAM) fungi. This symbiosis has a function in the mineral nutrition and water supply of the higher plants (Gianinazzi-Pearson and Gianinazzi 1983). The most important aspect of these VAM fungi seems to be their involvement in the biotransformation of heavy metals. Dueck and co-workers (1986) have demonstrated that in the presence of VAM fungi the toxicity of zinc for the grasses *Festuca rubra* and *Calamagrostis epigeios* has been diminished without affecting the zinc concentration of the grasses themselves (Table 3). These results give new perspectives for the accelerated establishment of vegetation on metalliferous wastes.

In the past it has been attempted to reduce the toxicity of heavy metals in mine tailings by the application of technical complexing agents such as ethylenediamine tetraacetic acid (EDTA, Ernst 1981) and of humic materials such as sewage sludges (Street and Goodman 1967). The organic wastes, however, decompose in a few years and demand, even as EDTA, an intermittent application, i.e. an expensive management. In contrast, the decomposition of organic materials from metal-tolerant vegetation is a slow process despite a certain metal tolerance of the microbial organisms (Williams et al. 1977), giving rise to an accumulation of organic matter and a metal detoxification during soil formation (Ernst 1966, 1974). At this point it should be mentioned that the uptake of complexed or otherwise organically bound heavy metals by higher plants is lower than that of ionic metals, regardless of whether the complexes are produced synthetically (Ernst 1968) or by natural processes such as humic or fulvic acids (Marquenie-van der Werff and Out 1981). If the application of organic material is to be a successful mitigatory step in the management of disposal sites, a thorough mixture of applied humic material with the metalliferous wastes will be necessary, if the organic material should complex the metals in the surface layer of the disposal site. It may be expected that this mixing process will hamper the decomposition of the organic material by its indigenous microbial populations (McCarvil and Macham 1985), but it will promote the development of an increased tolerance to heavy metals by microbial populations in the long term (Jordan and Lechevalier 1975; Nordgren et al. 1985; Kelley and Tuovinen, this Vol.). This retarded decomposition may be very valuable for the recolonization of metal-tolerant plants.

Another aspect of tailings is the release of acids due to the oxidation of pyrites (Pulford et al. 1983; Davé et al. 1985). As long as a neutralization of acids by carbonates can occur, plant roots will not be affected. If the neutralization capacity has been lost, a change of soil pH will increase the availability of heavy metals and that of aluminium to higher plants. In such a situation aluminium tolerance has to be present in colonizing plant populations. Up to now, only a few plant species are known to evolve such an aluminium tolerance in addition to metal tolerance. The most pronounced examples are *Deschampsia cespitosa* from the mining district of Sudbury (Cox and Hutchinson 1980) and *Vaccinium vitis-idaea* from a highly industrialized area in Upper Silesia (Czuchajowska and Straczek 1981). In some other grasses, i.e. *Agrostis tenuis* and *Festuca ovina* tolerances to heavy metals or aluminium have been described, however, not in combination. Despite the compatibility of both heavy metal and aluminium tolerances, the impact on the energy budget seems to be very high, if the slow revegetation process in the Sudbury area is considered.

3.2 Imbalances of Major Nutrients

Mine tailings and disposal sites of dredged materials have often a high imbalance of the major and minor nutrients (Ernst 1974, 1982; Smilde and van Driel 1977). In addition to the high amounts of those heavy metals which are minor nutrients, but which occur in concentrations typical for major nutrients, there is an acute shortage of major nutrients, particularly nitrogen and phosphorus. Nutrient shortage can be overcome, because metal-tolerant populations can grow under low nutrient conditions (Jowett 1959), but biomass production is very low. Can this problem be solved by natural processes during revegetation?

Nitrogen is the chemical element which has the highest concentration of soil-derived elements in living plants. Despite an enormous variability of the nitrogen demand of plant species and populations, nitrogen strongly influences the biomass production. This major nutrient can principally be enriched in waste materials by a natural process, the nitrogen fixation by free-living and symbiotic microorganisms, assuming that they can tolerate other abiotic hazards and achieve the necessary carbohydrates for the expensive process of nitrogen fixation (Stewart 1975). Already in less extreme environments the contribution of free-living, nitrogen-fixing bacteria to the nitrogen budget of ecosystems is very small. It may be expected that their activity will be insignificant under the conditions of tailings and disposal sites, even if they can evolve metal tolerances (Duxbury and Bicknell 1983; Kelley and Tuovinen, this Vol.). Another input of nitrogen may come from symbiotic, nitrogen-fixing bacteria. In the case of the *Rhizobium*-legume symbiosis it is known that a few legume species can tolerate moderate concentrations of heavy metals, i.e. *Lotus corniculatus, Anthyllis vulneraria* (Ernst 1974, 1976) and *Trifolium repens* (Rother et al. 1982) in European heavy metal soils, *Tephrosia* and *Indigofera* species in Africa (Ernst 1972). All the metal-resistant populations of these species have a certain amount of active root nodules, but the contribution of these legumes to the biomass and thus to the nitrogen budget of metal-contaminated areas is very low.

Disposal sites with small amounts of heavy metals offer more possibilities for symbiotic nitrogen fixation, as demonstrated for the China clay wastes in Cornwall by the group of Bradshaw (Bradshaw et al. 1975; Dancer et al. 1977; Palaniappan et al. 1979). In tailings with toxic metal concentrations the only realistic way to improve the nitrogen balance will be the application of fertilizer, as long as the chemical form of the nitrogen and its release from the fertilizer satisfy the demand of the metal-tolerant plant species.

In contrast to nitrogen, phosphorus cannot be fixed from the ambient air. It has to be mobilized from the soil material by weathering and/or by the plant itself. The low phosphorus concentration in most waste materials demands a very efficient uptake by the colonizing plant species. Some grass species possess genotypes which can perform quite well at a very low phosphorus level (Jowett 1959; Davies and Snaydon 1974). In addition to this efficiency, the vesicular-arbuscular mycorrhizal fungi can exploit the easily soluble phosphorus pools in waste material for the benefit of the plants (Gianinazzi-Pearson and Gianinazzi 1983). For a few years it has been known that VAM fungi occur not only in infertile, metal-poor tailings such as coal mine wastes (Daft and Nicolson 1974), but also in tailings rich in heavy metals (Gildon and Tinker 1981; Dueck et al. 1986), as already mentioned.

A further imbalance of major nutrients in mine tailings may be related to the high concentration of sulphur. Bacteria play an important role in the regulation of the availability of sulphur to higher plants by oxidation and reduction processes (Kelley and Tuovinen, this Vol.). The paucity of information about the adaptation of higher plants to various levels of sulphur (Antonovics et al. 1967) makes it uncertain, whether high concentrations of sulphur will selectively influence the revegetation process. At a high reduction of sulphur by anoxia there may be indirect effects by diminishing the availability of iron and manganese; at high oxidation of sulphur by microorganisms, problems of soil acidity and aluminium toxicity may arise (see Sect. 3.1).

3.3 Physical Structures

The water-holding capacity of soil materials depends on the physical structure of the material. On coarse material and on all materials in semi-arid and arid regions, drought may be an additional problem for revegetation, as demonstrated in the slow process of stabilizing the tailings of the gold mines in southern Africa (James 1966) and of copper-enriched sands in western Central Europe (Lohmeyer 1964). A biological solution will demand plant species with a high drought resistance, especially with deep rooting systems which have access to the deeper layers of the soil water, as long as these layers are aerobic. An extended root system, however, demands an increased amount of carbohydrates for the maintenance respiration and thus it affects the energy allocation between root and shoot. The price for this increased energy demand by deep and extended root systems is a diminished production of leaves and shoots, and a lowered vegetation cover. On coarse mine tailings in Europe only the perennial herb *Silene cucubalus* has succeeded in developing such a rooting strategy in addition to metal tolerance (Ernst 1974).

Moving materials in tailings and disposal sites can bury the developing vegetation. Buried plants are deprived of light, in addition to mechanical damage, thus reducing the growth potential of the plant. Naturally mobile environments such as sand dunes or screes indicate the impact of this stress factor: a very open vegetation. Therefore, the spatial arrangement of tailings and disposal sites has to prevent sliding of the material by avoiding the construction of steep, cone-shaped disposal sites and, if necessary, by application of physicochemical stabilizers (Dean et al. 1971; Morrey et al. 1983).

4 Submersed Disposal Sites

Disposal of mine wastes and dredged materials in surface waters has some additional impacts on ecosystems in comparison to upland disposal. Such submersed disposals may affect the dissolved oxygen budget of these waters. Furthermore, the daily and seasonal changes of pH and bicarbonate by photosynthetic activities, the excretion of organic compounds and the high turnover of biomass may increase the availability of metals and other contaminations to organisms, especially to aquatic animals (Förstner and Wittmann 1981).

4.1 Reduced Oxygen Conditions

Under fully waterlogged conditions there will be insufficient or even no externally dissolved oxygen supply to meet the respiratory demand of plant roots and submersed shoots. Therefore, oxygen-poor conditions will hamper the growth of these plants, unless they have specific morphological adaptations for an internal oxygen supply in the form of aerenchyma. These aerenchyma are air spaces in the ground tissue of roots, rhizomes and shoots and connect these plant organs with the atmosphere via the opening in the leaves, i.e. the stomata, or at least with the oxygen-producing systems of photosynthetically active leaves. Dredged materials and mine wastes with a high oxygen consumption due to bacterial activities will strongly select for aerenchymatous plant species.

The path of oxygen movement through the aerenchymatous roots does not stop at the root cortex but there is a radial oxygen loss from the roots to the root environment, the rhizophere (Armstrong 1975). This radial oxygen loss implies the formation of an oxygenated zone around the roots and may have an important function in the detoxification of reduced inorganic toxins such as sulphides (Havill et al. 1985; Hollis et al. 1975; Ingold 1982) and in lowering the availability of manganous and ferrous ions to plants (Singer and Havill 1985; Smirnoff 1981; Rozema et al. 1985a). Precipitation of iron and manganese oxides on the root surface can build up a barrier and an additional reactive zone for other heavy metals (Rozema et al. 1985b). These aspects may be very important for a reclamation of submersed disposal sites, but they have received nearly no attention.

A further consequence of oxygen depletion in the rooting zone of plants is the increased production of organic acids (Duss and Brändle 1982). These organic acids may stimulate the growth of bacteria and can cause an increased formation of metal-organic complexes, thus changing the availability of metals (Kelley and Tuovinen, this Vol.).

4.2 Metal Availability to Plants

Generally, terrestrial plants can take up dissolved substances in the soil solution only by roots; submerged and floating water plants have additional possibilities, i.e. shoots and leaves. Due to the intimate contact with the surrounding water, the morphological structure, i.e. the ratio between plant surface and plant volume, the physicochemical properties of the cell walls and the specific metabolic activity of the various plant organs can determine the adsorption and the absorption process. As demonstrated for the submerged *Elodea nuttallii* (Marquenie-van der Werff and Ernst 1979) the uptake of the heavy metals by roots was superior to that of leaves and shoots, in the case of cadmium by a factor of 1.6, in the case of zinc by even a factor of 5.0. Due to the photosynthetic activity of submerged leaves, there will be a strong change of the metal speciation near the leaf surface, where the pH of the water can increase to 10 to 12. At darkness the pH can drop down to 5.0. This daily fluctuation of the pH and thus of the speciation of metals may be the reason that the metal tolerance of submerged water plants is difficult to determine. An organ-specific uptake behaviour is the main reason for the difference in the pollutant concentration of water plants. Submerged or partially submerged plants such as species of *Elodea, Hydrilla, Potamogeton* and *Callitriche* are more effective in metal uptake than plants only floating on the water surface such as the *Lemna* species, *Wolffia* and *Spirodela polyrrhiza* (van der Werff and Pruyt 1982). Emerging water plants such as *Typha latifolia* or *Phragmites australis* are not very different from terrestrial plants, except their aerenchymatous roots which may be responsible for the diminished metal uptake (Taylor and Crowder 1983).

Excretions of organic compounds by roots in terrestrial ecosystems are known. The absorbing surfaces of humic substances and clay particles will at least camouflage or neutralize their effect on metal speciation. Excretion of complexing agents to the surrounding water, as demonstrated for certain populations of *Chlorella vulgaris* (Butler et al. 1980) and postulated for other algae (Fogg and Westlake 1955; Hall et al. 1979), will diminish the availability of metals to plants and thus reduce their toxicity. This aspect of speciation by biota, which may have a daily and a seasonal component, has been widely underestimated in analyzing the water quality of contaminated sites and the further impacts of metal turnover in submersed systems.

The intimate contact of water plants with the surrounding medium addresses another aspect, the evolution, i.e. the selection of resistance. As in terrestrial plants, population differentiation in water plants is a quite common reaction to differences in the abiotic environment. Most of the provided information is concerned with population-specific reaction to light quality (cf. *Hydrilla verticillata,* Verkleij et al. 1983), to thermal properties of the water body (cf. *Spirodela polyrrhiza,* Davidson and Simon 1981) or to major nutrients (cf. *Lemna gibba*, Landolt and Dann 1983). There is a paucity of reliable information on the population differentiation in aquatic angiosperms to heavy metals (Sarosiek et al. 1982), whereas this process has achieved more attention in algae (Harding and Whitton 1976; Say et al. 1977; Butler et al. 1980; Foster 1982; Bariaud and Mestre 1984). The mechanisms of metal resistance is based on a metal exclusion (Hall et al. 1979) or on an extracellular (Butler et al. 1980) or intracellular sequestration of the contaminants. Whether specific mechanisms for the regulation of the pH near the cell surface are involved in the evolution of tolerant populations demands further research. The introduction of metal-tolerant algae to submersed disposal sites is an unexplored area.

4.3 (Bio)turbation and Water Stratification

In the assessment of the impact of submerged disposal of dredged material and mine wastes two other aspects of the ecology of water bodies have to be considered: the bioturbation and the water stratification. Bioturbation, i.e. the disturbance of surface sediments by burrowing or surface-foraging organisms, will influence the transport of contaminants between the sediment and the overlying water and will change the redox potential in the highly differentiated zone of the upper sediment layer (Cappenberg 1975). It is well known that tubificid worms can alter the redox profiles in lake sediments (Davis 1974), thus the impact on the speciation of contaminants is poorly understood. Recent research in Lake Maarsseveen (Ringelberg, pers. commun.) has shown that the seasonal variation in the foraging behaviour of bream, a freshwater fish is responsible for a great part of the nutrient turnover in this freshwater lake.

Turbation of water bodies by dredging activities is principally an enlarged bioturbation. It thoroughly disturbs not only sediment profiles and oxidation processes, but can also affect the stratification in water bodies as far as lakes are concerned. These aspects of bioturbation and turbation of the stratification demand urgent research.

5 Concluding Remarks

The evolution of metal tolerance in terrestrial and aquatic plants gives a sound basis for revegetation of metal-contaminated mine tailings and disposal sites of dredged materials. Prior to every revegetation, a careful analysis has to be carried out to analyze the speciation of heavy metals, the amount of major plant nutrients, the water-holding capacity of the material in terrestrial disposal sites and the dissolved oxygen status in submersed dumps. All these environmental factors determine the selection of the appropriate plant material. The more the disposal sites can be stabilized, the better the areas will be revegetated. Reclamation of mine tailings and disposal sites of dredged materials does not allow the use of these areas for every purpose. It depends on the behaviour of possible toxic substances in the turnover and food web of the ecosystem. Healthy green may be dangerous green for herbivores (Ernst et al. 1983).

References

Andrews RD (1975) Tailings. Environmental consequences and a review of control strategies. Int Conf Heavy metals in the environment, Toronto, vol 2, pp 645–675

Antonovics J, Lovett J, Bradshaw AD (1967) The evolution of adaptation to nutritional factors in populations of herbage plants. In: IAEA (ed) Isotopes in plant nutrition and physiology. Int Atom Energ Ag, Vienna, pp 549–567

Armstrong W (1975) Water logged soils. In: Etherington JR (ed) Environment and plant ecology. Wiley & Son, New York London, pp 181–218

Ashenden TW, Stewart WS, Williams W (1975) Growth responses of sand dune populations of *Dactylis glomerata* L. to different levels of water stress. J Ecol 63:97–107

Bariaud A, Mestre JC (1984) Heavy metal tolerance in a cadmium-resistant population of *Euglena gracilis*. Bull Environ Contam Toxicol 32:597–601

Barner J (1978) Rekultivierung zerstörter Landschaften. Enke, Stuttgart

Baumeister W (1954) Über den Einfluss des Zinks bei *Silene inflata* Sm. Ber Dtsch Bot Ges 67:205–213

Bradshaw AD (1975) The evolution of metal tolerance and its significance for vegetation establishment of metal contaminated sites. Int Conf Heavy metals in the environment, Toronto, vol 2, pp 599–622

Bradshaw AD, Dancer WS, Handley JF, Shledon JC (1975) The biology of land revegetation and reclamation of China clay wastes of Cornwall. In: Chadwick MJ, Goodman GT (eds) Degraded environments and resource renewal. Blackwell, Oxford, pp 363–384

Butler M, Haskew AEJ, Young MM (1980) Copper tolerance in the green alga *Chlorella vulgaris*. Plant Cell Environ 3:113–126

Cappenberg TE (1975) Interrelations between sulfate-reducing and methane-producing bacteria in bottom deposits of a fresh-water lake. PhD thesis, Free Univ Amsterdam

Cox RM, Hutchinson TC (1980) Multiple metal tolerances in the grass *Deschampsia cespitosa* (L.) Beauv. from the Sudbury smelting area. New Phytol 84:631–647

Czuchajowska Z, Straczek T (1981) Germination of *Vaccinium vitis-idaea* seeds originating from natural stands with differing heavy metals pollution levels. Bull Acad Pol Sci Ser Sci Biol 29:135–140

Daft MJ, Nicolson TH (1974) Arbuscular mycorrhizas in plants colonizing coal wastes in Scotland. New Phytol 73:1129–1138

Dancer AS, Handley JF, Bradshaw AD (1977) Nitrogen accumulation in Kaolin mining wastes in Cornwall. Plant Soil 48:153–167, 303–314

Davé NK, Lim TP, Clantier NR (1985) Migration of metals and trace radionuclides from mine tailings to the environment. Int Conf Heavy metals in the environment, Athens, vol 2. CEP Consultants, Edinburgh, pp 24–26

Davidson D, Simon JP (1981) Thermal adaptation and acclimation of ecotypic populations of *Spirodela polyrrhiza* (Lemnaceae). Morphology and growth rate. J Therm Biol 6:121–128

Davies MS, Snaydon RW (1974) Physiological difference among populations of *Anthoxanthum odoratum*. III. Response to phosphate. J Appl Ecol 11:699–708

Davis RB (1974) Tubificids alter profiles of redox potential and pH in profundal lake sediment. Limnol Oceanogr 19:342–346

Dean KC, Havens R, Valdez EG (1971) USBM finds many routes to stabilizing mineral wastes. Min Eng 23:61–63

Dueck ThA, Ernst WHO, Faber J, Pasman F (1984) Heavy metal emission and genetic constitution of plant populations in the vicinity of two metal emission sources. Angew Bot 58:47–59

Dueck ThA, Visser P, Ernst WHO, Schat H (1986) Vesicular-arbuscular mycorrhiza decreases zinc toxicity in grasses in zinc-polluted soil. Soil Biol Biochem 18:331–333

Duss F, Brändle R (1982) Flooding tolerance in bulrush (*Schoenoplectus lacustris* L. Palla): Synthesis of different fermentation products and transport substances in the rhizome tissue under oxygen deficit. Flora 172:217–222

Duxbury T, Bicknell B (1983) Metal-tolerant bacterial populations from natural and metal-polluted soils. Soil Biol Biochem 15:243–250

Ernst WHO (1966) Ökologisch-soziologische Untersuchungen an Schwermetallpflanzengesellschaften Südfrankreichs und des östlichen Harzvorlandes: Flora B 156:301–318

Ernst WHO (1968) Der Einflups der Phosphatversorgung sowie die Wirkung von ionogenem und chelatisiertem Zink auf die Zink- und Phosphataufnahme einiger Schwermetallpflanzen. Physiol Plant 21:323–333

Ernst WHO (1972) Ecophysiological studies on heavy metal plants in South Central Africa. Kirkia 8:125–145

Ernst WHO (1974) Schwermetallvegetation der Erde. Fischer, Stuttgart

Ernst WHO (1975) Physiology of heavy metal resistance in plants. Int Conf Heavy metals in the environment, Toronto, vol 2, pp 121–136

Ernst WHO (1976) Okologische Grenze zwischen *Violetum calaminariae* und *Gentiano-Koelerietum*. Ber Dtsch Bot Ges 89:381–390

Ernst WHO (1981) Probleme bei der Begrünung und Aufforstung von Schwermetallhalden. In: Schwabe-Braun A (Red) Vegetation als anthropo-ökologischer Gegenstand. Cramer, Vaduz, pp 237–248

Ernst WHO (1982) Schwermetallpflanzen. In: Kinzel H (ed) Pflanzenökologie und Mineralstoffwechsel. Ulmer, Stuttgart, pp 472–506

Ernst WHO (1983) Okologische Anpassungsstrategien an Bodenfaktoren. Ber Dtsch Bot Ges 96:49–71

Ernst WHO, Weinert H (1972) Lokalisation von Zink in den Blättern von *Silene cucubalus* Wib. Z Pflanzenphysiol 66:258–264

Ernst WHO, Joosse-van Damme ENG (1983) Umweltbelastung durch Mineralstoffe. Fischer, Jena

Ernst WHO, Mathys W, Janiesch P (1975) Physiologische Grundlagen der Schwermetallresistenz – Enzymaktivitäten und organische Säuren. Westdeutscher Verlag, Opladen

Ernst WHO, Verkleij JAC, Vooijs R (1983) Bioindication of a surplus of heavy metals in terrestrial ecosystems. Environ Monit Assess 3:297–305

Fogg GE, Westlake DF (1955) The importance of extracellular products of algae in freshwater. Verhandl Int Ver Theor Angew Limnol 12:219–232

Förstner U, Wittmann GTW (1981) Metal pollution in the aquatic environment. Springer, Berlin Heidelberg New York

Foster PL (1982) Metal resistance of *Chlorophyta* from rivers polluted by heavy metals. Freshwater Biol 12:41–61

Gadgil RL (1969) Tolerance of heavy metals and the reclamation of industrial waste. J Appl Ecol 6:247–259

Gartside DW, McNeilly T (1974) The potential for evolution of heavy metal tolerance in plants. II. Copper tolerance in normal populations of different plant species. Heredity 32:335–348

Gemmell RP (1977) Colonization of industrial wasteland. Arnold, London

Gianinazzi-Pearson V, Gianinazzi S (1983) The physiology of vesicular-arbuscular mycorrhizal roots. Plant Soil 71:197–209

Gildon A, Tinker PB (1981) A heavy metal-tolerant strain of a mycorrhizal fungus. Trans Br Mycol Soc 77:648–649

Godbold DL, Horst WJ, Marschner H, Collins JC, Thurman DA (1983) Root growth and zinc uptake by two ecotypes of *Deschampsia caespitosa* as affected by high zinc concentrations. Z Pflanzenphysiol 112:315–324

Gregory RPG, Bradshaw AD (1965) Heavy metal tolerance in populations of *Agrostis tenuis* Sibth. and other grasses. New Phytol 64:131–143

Greszta J (1973) The recultivation of post-industrial territories. In: Protection of man's natural environment. Pol Acad Sci Warzawa, pp 396–417

Gries B (1966) Zellphysiologische Untersuchungen über die Zinkresistenz bei Galmeiformen und Normalformen von *Silene cucubalus* Wib. Flora B 156:271–290

Hall A, Fielding AH, Butler M (1979) Mechanism of copper tolerance in the marine fouling alga *Ectocarpus siliculosus*. Evidence for an exclusion mechanism. Mar Biol 54:195–199

Harding JPC, Whitton BH (1976) Resistance to zinc of *Stigeoclonium tenue* in the field and the laboratory. Br Phycol J 11:417–426

Havill DC, Ingold A, Pearson J (1985) Sulphide tolerance in coastal halophytes. Vegetatio 62:279–285

Hedrick PW (1984) Population biology. The evolution and ecology of populations. Jones & Bartlett, Boston

Hill JRC (1978) The use of range plants in the stabilization of phytotoxic mining wastes. Proc 1st Int Rangeland Congr, pp 707–711

Hogan GD, Rauser WE (1979) Tolerance and toxicity of cobalt, copper, nickel and zinc in clones of *Agrostis gigantea*. New Phytol 83:665–670

Hollis JP, Allan AJ, Pitts G, Joshi MM, Ibrahim JKA (1975) Sulphide diseases of rice on iron-excess soils. Acta Phytopathol Acad Sci Hung 10:329–341

Ingold A (1982) The effects of sulphide toxicity on the distribution of higher plant species in salt marshes. PhD Thesis, Univ London

James AL (1966) Stabilizing mine dumps with vegetation. Endeavour 25:154–157

Johnson MS, McNeilly T, Putwain PD (1977) Revegetation of metalliferous mine spoil contaminated by lead and zinc. Environ Pollut 12:261–277

Jordan MJ, Lechevalier MP (1975) Effects of zinc-smelter emissions on forest microflora. Can J Microbiol 21:1855–1865

Jowett D (1959) Adaptation of a lead tolerant population of *Agrostis tenuis* to low soil fertility. Nature (London) 184:43

Karataglis SS (1978) Studies on heavy metal tolerance in populations of *Anthoxanthum odoratum*. Ber Dtsch Bot Ges 91:205–216

Landolt E, Dann W (1983) Behaviour of ten clones of *Lemna gibba* at varied concentrations of nitrogen. Ber Geobot Inst ETH Rübel Zürich 50:86–96

Levitt J (1980) Response of plants to environmental stresses. Academic Press, London New York

Lohmeyer W (1964) Uber die künstliche Begrünung offener Quartzsandhalden im Bergbaugelände bei Mechernich. Angew Pflanzensoziol 20:61–71

Lolkema PC (1985) Copper resistance in higher plants. PhD Thesis, Free Univ Amsterdam

Lolkema PC, Donker MH, Schouten AJ, Ernst WHO (1984) The possible role of metallothioneins in copper tolerance of *Silene cucubalus*. Planta 162:174–179

Marquenie-van der Werff M, Ernst WHO (1979) Kinetics of copper and zinc uptake by leaves and roots of an aquatic plant, *Elodea nuttalii*. Z Pflanzenphysiol 92:1–10

Marquenie-van der Werff M, Out T (1981) The effect of humic acid as Zn complexing agent on water cultures of *Holcus lanatus*. Biochem Physiol Pflanzen 176:274–282

Mathys W (1977) The role of malate, oxalate, and mustard oils glucoside in the evolution of zinc-resistance in herbage plants. Physiol Plant 40:130–136

McCarvil J, Macham LPC (1985) Metal resistance and metal accumulation in bacterial populations from activated sludges. Int Conf Heavy metals in the environment, Athens, vol 1. CEP Consultants, Edinburgh, pp 66–68

McGrawth SP, Baker AJM, Morgan AN, Salmon WJ, Williams M (1980) The effects of interactions between cadmium and aluminium on the growth of two metal tolerant races of *Holcus lanatus* L. Environ Pollut A 23:267–277

Morrey DR, Cooke JA, Baker AJM (1983) The use of stabilizers and mulches in the revegetation of metalliferous wastes. Int Conf Heavy metals in the environment, Heidelberg, vol 2. CEP Consultants, Edinburgh, pp 996–1000

Nordgren A, Baath E, Soderstrom B (1985) Soil microfungi in an area polluted by heavy metals. Can J Bot 63:448–455

Palaniappan VM, Marss RH, Bradshaw AS (1979) The effect of *Lupinus arboreus* on the nitrogen status of China clay wastes. J Appl Ecol 16:825–831

Pulford JD, Kimber AJ, Duncan HJ (1983) Leaching of metals from acidic colliery spoil. Int Conf Heavy metals in the environment, Heidelberg, vol 2. CEP Consultants, Edinburgh, pp 1001–1004

Repp G (1963) Die Kupferresistenz des Protoplasmas höherer Pflanzen. Protoplasma 57:643–659

Rother JA, Millbank JW, Thornton I (1982) Seasonal fluctuations in nitrogen fixation (acetylene reduction) by free-living bacteria in soils contaminated with cadmium, lead and zinc. J Soil Sci 33:101–113

Rozema J, Luppes E, Broekman R (1985a) Differential response of salt-marsh species to variation of iron and manganese. Vegetatio 62:293–301

Rozema J, Otte R, Broekman R, Punte H (1985b) Accumulation of heavy metals in estuarine salt marsh sediment and uptake of heavy metals by salt marsh halophytes. Int Conf Heavy metals in the environment, Athens, vol 1. CEP Consultants, Edinburgh, pp 545–547

Rüther F (1967) Vergleichende physiologische Untersuchungen über die Resistenz von Schwermetallpflanzen. Protoplasma 64:400–425

Sarosiek J, Wozakowska-Natkanice H, Niemiec C (1982) The influence of heavy metals on dynamics of the *Lemna minor* population. Acta Univ Wratislaw 530, Prace Bot 25:11–39

Say PJ, Diaz BM, Whitton BA (1977) Influence of zinc on lotic plants. 1. Tolerance of *Hormidium* species to zinc. Freshwater Biol 7:357–376

Singer CE, Havill DC (1985) Manganese as an ecological factor in salt marshes. Vegetatio 62:287–292

Smilde KW, Driel W van (1977) Mangaangebrek in granen verbouwd op havenslib Nota 39. Inst Bodemvruchtbaarheid, Haren

Smirnoff N (1981) Iron tolerance and the role of aerenchyma in wetland plants. PhD Thesis, St. Andrews, Scotl

Smith RAH, Bradshaw AD (1979) The use of metal tolerant plant populations for the reclamation of metalliferous wastes. J Appl Ecol 16:261–277

Stewart WDP (1975) Nitrogen fixation by free-living microorganisms. University Press, Cambridge

Street HE, Goodman GT (1967) Techniques of revegetation in the lower Swansea Valley. In: Hilton KJ (ed) The Lowee Swansea Valley project. Longmans Green, London, pp 71–110

Symeonidis L, McNeilly T, Bradshaw AD (1985) Differential tolerance of three cultivars of *Agrostis capillaris* to cadmium, copper, lead, nickel and zinc. New Phytol 101:309–315

Taylor GJ, Crowder AA (1983) Uptake and accumulation of heavy metals by *Typha latifolia* in wetlands of the Sudbury, Ontario region. Can J Bot 61:63–73

Urquhart C (1971) Genetics of lead tolerance in *Festuca ovina*. Heredity 26:19–33

Van der Werff M (1981) Ecotoxicity of heavy metals in aquatic and terrestrial higher plants. PhD Thesis, Free Univ Amsterdam

Van der Werff M, Pruyt MJ (1982) Long-term effects of heavy metals on aquatic plants. Chemosphere 11:727–739

Verkleij JAC (1980) Isoenzyme variation in pioneer species from ecologically different habitats. PhD Thesis, Free Univ Amsterdam

Verkleij JAC, Bast-Cramer WB (1985) Cotolerance and multiple heavy metal tolerance in *Silene cucubalus* from different heavy metal sites. Int Conf Heavy metals in the environment, Athens, vol 2. CEP Consultants, Edinburgh, pp 174–176

Verkleij JAC, Pieterse AH, Horneman GJT, Torenbeek M (1983) A comparative study of the morphology and isoenzyme patterns of *Hydrilla verticillata* (L.f.) Rooyle. Aquat Bot 17:43–59

Wilkins DA (1957) A technique for the measurement of lead tolerance in plants. Nature (London) 180:37–38

Wilkins DA (1978) The measurement of tolerance to edaphic factors by means of root growth New Phytol 80:623–633

Williams ST, McNeilly T, Wellington EMH (1977) The decomposition of vegetation growing on metal mine waste. Soil Biol Biochem 9:271–275

Wong MH (1982) Metal cotolerance to copper, lead, and zinc in *Festuca rubra*. Environ Res 29:42–47

Wu L, Antonovics J (1975) Zinc and copper uptake by *Agrostis stolonifera* tolerant to both zinc and copper, New Phytol 75:231–237

Wu L, Bradshaw AD, Thurman DA (1975) The potential for evolution of heavy metal tolerance in plants. Heredity 34:165–187

Part II
Biological and Geochemical Assessment

Case Histories of Coastal and Marine Mines

D. V. ELLIS[1]

1 Introduction

Coastal mines have the option of disposing of their tailings and waste rock to the sea instead of on land. Marine mines, either nearshore or offshore, inevitably must dispose of their wastes to the sea. Such disposal can potentially contaminate or preclude marine fisheries, and affect tourism, recreation and public health. It is known that some marine discharging mines have avoided such impacts and others have not. Accordingly, it is important to document cases of actual experience so that the information is available for impact prediction and mitigation of new mine sites.

This chapter assembles relevant case material, provides (generally publicly available) references for more detailed information about each case, notes the patterns of impact and provides comments on how to minimise potential impact while planning is still at the conceptual design stage and how to quality control the minimised environmental impacts during operations. Some comments are also included on smelter and refinery discharges to the sea.

Mines and mining developments from which environmental information has been obtained are listed in Tables 1–3. Relevant information is scattered widely through mining and environmental literature, is hard to find in many cases, and undoubtedly, the material here can be extended. An important discussion on the general topic of marine tailings disposal took place at the XIV International Mineral Processing Congress in Toronto in 1982, and is reported in the Proceedings of the Congress (Poling 1982).

2 Case Histories

The cases described have been selected partly for their good documentation, but also for the types of environmental concerns that they illustrate. Other case histories noted, but left undescribed here are equally interesting, and future information from them should also be useful to other similar developments. In particular information should become available from Ok Tedi, PNG (Table 1) about the extent to which river discharged tailings affect estuarine resources, from

[1]Department of Biology, University of Victoria, Victoria, B.C. V8W 2Y2, Canada

Table 1. Coastal mines and their environmental impact assessment programs

Mine	Main product	Tailings disposal	Environmental monitoring[a]	Introductory reference
Canada				
Island Copper	Cu/Mo	Fiord 50 m depth (waste rock shoreline)	All listed components	Ellis 1982a Anon 1985a
Kitsault	Mo	Fiord 50 m depth	All listed components	Ellis 1982b
Polaris	Pb/Zn	Lake (relict marine inlet)	No information	Pullen 1982
Nanisivik	Pb/Zn	Lake (marine disposal assessed)	a,b,f,h	Thomas and Metikosh 1984
Westfrob	Fe/Cu	Fiord 12 m depth	a,b,f,h,j,m,n	Harding 1983
Yreka (closed)	Cu	River and shoreline, fiord	None	Harding 1983
Texada (closed)	Fe	Surface Open coast	No information	Harding 1983
Jordan River (closed)	Cu	10 m depth Open coast	a,b,d,f,h j,m,n	Ellis and Popham, 1983
Brynnor (closed)	Fe	Beach	Post-closure k	Levings 1975
Britannia (closed)	Cu	River and shoreline, fiord	n	Ellis and Popham 1983
USA				
Quartz Hill (under development)	Mo	Probably submarine, fiord	All listed components	Anon 1984b, 1985b Ellis 1982a
Greenland				
Black Angel	Pb/Zn	Fiord 42 m depth	a,d,f,g,h,j,n	Møller 1984
Norway				
Fosdalens Bergverk	Fe	Fiord 60 m depth Some barging	a	Down and Mill 1978
Repparfjord	Fe/Cu	Fiord 50 m depth	a	Down and Mill 1978
Stjernoy	Fe	Fiord	No information	Down and Mill 1978
UK				
Cleveland Potash	Potash	Submarine	a,f,j,l	Down and Mill 1978 Carter 1975
Hayle estuary (closed)	Sn/Cu	Estuary	f,n	Yim 1976
St. Austell (closed)	China clay	River and sea	j,k,l	Holme and Probert 1978 Wilson and Connor 1976
Mediterranean				
Pennaroya	Pb/Zn/Fe		No information	Down and Mill 1978
South America				
El Salvador (Chile)	Cu,Mo	Shoreline	a,f,j,k,l,m,n	Castilla 1983

Table 1 *(continued)*

Mine	Main product	Tailings disposal	Environmental monitoring[a]	Introductory reference
Potrerillos (closed)	Cu	Shoreline	No information	Castilla 1983
Toquepala Cuajone (Peru)	Cu	Shoreline	No information	No publications
East Asia/Oceania				
Ma On Shan (Hong Kong) (closed)	Fe	Shallow inlet	f,j,n	Wong et al. 1979
Marcopper (Philippines)	Cu	Surface Open coast	Most listed components	No publications
Atlas (Philippines)	Cu	Surface Open coast	Most listed components	Dira and Canete, 1980 Salazar and Gonzales, 1974 Alino 1984
Bougainville (P.N.G.)	Cu	River and shoreline	Most listed components	Chaps. 15, 26
Ok Tedi (P.N.G)	Cu/Au	River	Most listed components	Petr 1979 Pintz 1984
Australia				
Tamar River (Tasmania) (closed)	Au	River	Most listed components	Ayling 1974

[a]*Key to environmental monitoring components: a* Tailing discharge rates and chemical analyses; *b* acute bioassays (LD_{50}s, etc.); *c* other bioassays (long-term in situ, etc.); *d* turbidity (suspended solids, etc.), ambient levels; *e* density current (tailings plume); *f* tailing deposition (coring, bathymetry, chemical, mechanical, etc.); *g* physical oceanography (current measures, stratification, etc.); *h* chemical oceanography (trace metals, dissolved oxygen, nutrients, etc.); *i* biological oceanography (primary production, phytoplankton, chlorophyll, zooplankton, ichthyoplankton, etc.); *j* benthos (stocks and dynamics); *k* shellfish (stocks and dynamics); *l* fish and wildlife (stocks and dynamics); *m* shoreline (stocks and dynamics); *n* trace metal bioaccumulation and biomagnification.

Table 2. Marine mines and potential mining areas[a]

Area	Operations	Main product	Environmental monitoring[a]	Introductory reference
Continental shelf and shallow water				
Thailand-Phuket	Shallow dredging	Sn	d,i,j	Chansang et al. 1981 Brown and Holley 1982 Limpsaichol and Poopetch 1984
Malaysia	Shallow dredging	Sn	No information	Brown 1985
Indonesia	Shallow dredging	Sn	No information	Brown 1985
Alaska[b]	Exploration	Au	No information	–
Oregon	Exploration	Cr,Fe	No information	–
Offshore				
Red Sea Atlantis Deep	Suction dredge Pre-pilot operation completed	Zn,Cu,Ag, Co,Au	Most components of Table 1	Nawab 1984 Gideiri 1984
USA				
Gorda Ridge	Exploration	?	Will be substantial	Anon 1983
Hawaii	Exploration	Mn		Ozturgut et al. 1980
Several other possible sites	–	?		
Canada				
Explorer Ridge	Exploration	?	–	–
South Pacific Islands	Exploration	?	No information	–
Juan de Fuca Ridge[c]	Exploration	–	–	–

[a]See notes to Table 1.
[b]A letter to the author (1986) from the Alaska Department of Natural Resources states that shallow water prospecting and dredging for gold, platinum and other metals has occurred under lease since 1959. There is some known conflict with fishery resources.
[c]This deep ridge extends into international waters (Fig. 1) from within the Canadian 200-mile EEZ. It is now being explored by US interests, and presumably any development will come under the US National Environmental Protection Act.

Gorda Ridge (polymetallic sulphide beds) and Hawaii (manganese crusts), both of which developments under US EPA regulations must undertake substantial environmental impact assessment (Table 2). It is noticeable that very little environmental information is available about marine discharges from smelters and refineries (Table 3).

2.1 Continuing Assessments

Many operating mines maintain their own environmental staff and facilities, or retain consultants, to undertake continuing (or intermittent) assessments.

Table 3. Coastal smelters and refineries

Area	Main product	Waste disposal	Environmental monitoring	Introductory reference
Mediterranean				
Pechiney	Al	Outfall to sea	No information	–
Canada				
Kitimat	Al	To estuary	a,b,h, n (Fluoride)	Anon 1977 Hocking et al. 1980
UK				
Avonmouth	Pb,Zn	Outfall to estuary	No information	–
Australia				
Yabulu (Queensland)	Ni	Outfall	b,g,h,j,m,n	Knauer 1976 Knauer 1977 Reid 1978 Carey 1981
Risdon (Tasmania)	Zn	Barge to 200 m	Reported 'extensive'	No publications
SE Asia				
Thailand, Phuket	Sn	Not known	h,j,k	Brown and Holley 1982

2.1.1 Island Copper Mine, Canada (Fig. 1)

Island Copper mine is one of the massive open-pit, low-grade (0.52% copper) mines developed during the 1960s and 1970s. It processes approximately 40,000 t day^{-1} ore, most of which is discharged as waste tailings through a submarine outfall to an adjacent fiord. The submarine outfall with a seawater mixing chamber was a novel design in 1971, but has now been adopted elsewhere (see Sect. 2.3.1 and Pelletier 1982; Ellis 1982a). Much of the pit waste rock is dumped alongshore. The mine has been subject to environmental controls (Table 4) since 1970 (Pelletier 1982), when both Federal and Provincial (British Columbia) governments were starting to exercise their rights in the matter largely through fisheries and land resource legislation respectively. As an early case of control, during years of extreme public concern, the mine was subjected to a number of requirements; e.g. a monitoring program and an annual environmental report (Anon 1985a). It was also subjected to two public inquiries (Waldichuk and Buchanan 1980), and was featured as a reference case during two further public inquiries which produced mining industry pollution control objectives (Anon 1979).

The environmental monitoring program was required to be designed and overseen by a government-approved group of university engineers and scientists from 1970–1978, after which several members of the university group were retained by the mine as an environmental advisory group. The mine's environmental program has kept continuity of professional personnel, in both the university group and environmental staff. The unique oceanographic and engi-

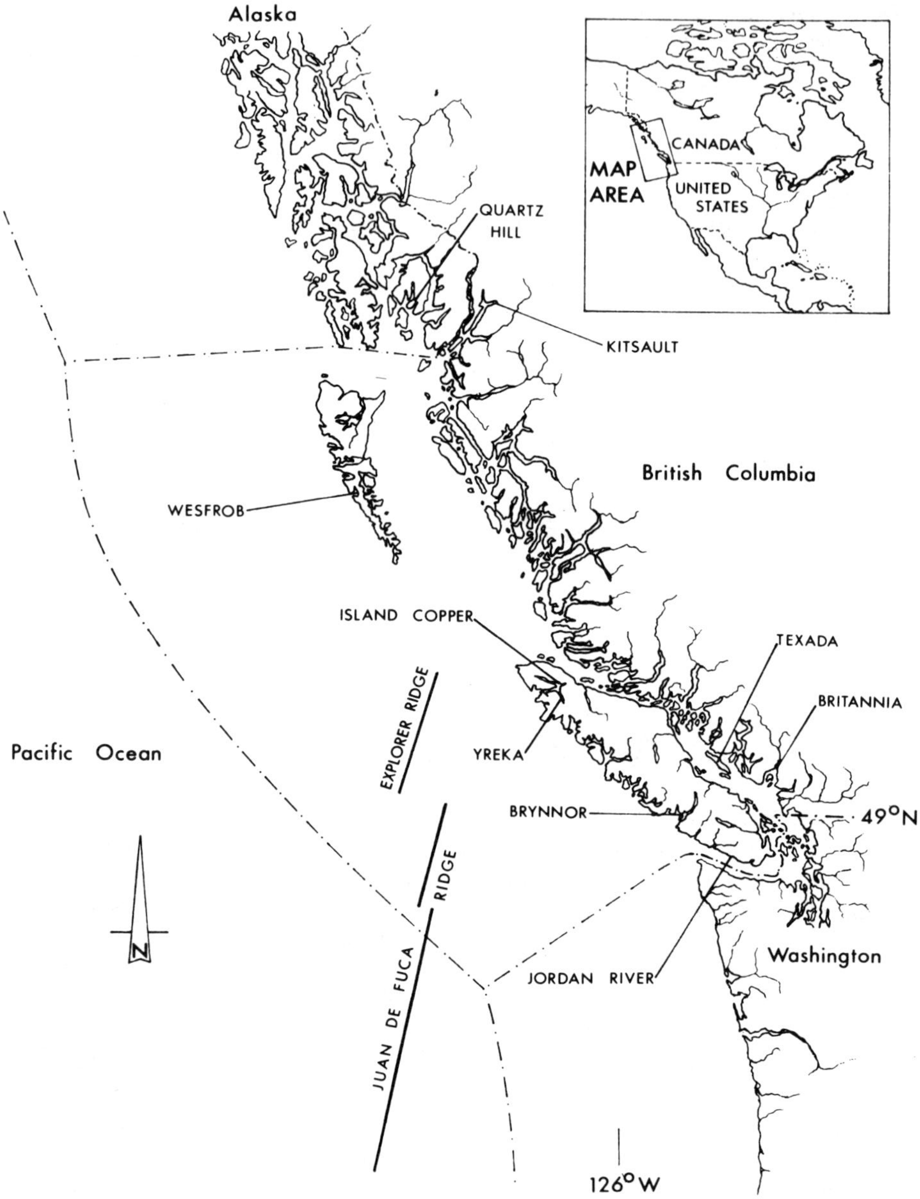

Fig. 1. The location of noted mine sites in western Canada and southeast Alaska. The 200-mile EEZ and ocean ridges now under exploration are shown

Table 4. Summary of environmental controls and monitoring at Island Copper Mine (for details see Pelletier, 1982)

Effluent	
Parameter	*Permitted levels*
Volume (Imp. gallons)	16×10^6 daily mean; 22×10^6 max.
% Solids (by wt.)	50
pH	7.5–11.5
Copper (dissolved)	50 $\mu g\ l^{-1}$
Molybdenum (dissolved)	500 $\mu g\ l^{-1}$
Cadmium (dissolved)	5 $\mu g\ l^{-1}$
Lead (dissolved)	50 $\mu g\ l^{-1}$
Arsenic (dissolved)	100 $\mu g\ l^{-1}$
Cyanide (dissolved)	250 $\mu g\ l^{-1}$
Mercury (dissolved)	1 $\mu g\ l^{-1}$
96-h Bioassays (monthly)	–

Receiving area

Controls
- Outfall design and placement

Monitoring (ongoing at various frequencies)
- Outfall performance
- Tailings surveys – physical/chemical
  - – seabed and water column
- Oceanography – current patterns
  - – ambient properties affecting biology
- Marine Biology – primary plant production
  - – artificial substrate monitoring
  - – zooplankton surveys
  - – benthos surveys
- Fishery species – crabs, shrimps, clams, fish
- Contaminant investigations – sediments
  - – crustacea, clams, fish, algae, marine grass
- River/lake – chemical parameters
- Weather – conventional parameters

neering design features of the case have attracted the attention of research scientists and engineers from university and government laboratories, which has ensured that much of the results have been published through refereed professional journals, hence are widely available (e.g. Hoff et al. 1982; Harding 1983; Pedersen 1984, 1985; Stucchi 1985). Social scientists have also taken an interest in the case. In summary, the consequences of the mine's control program are particularly well documented.

The mine has monitored all components identified in Table 4, mostly regularly on weekly, monthly, quarterly or annual schedules. The results obtained have been critically appraised (audited) for relevance at intervals, and the program has been changed (with government approval where necessary) as appropriate. In this way, the program has been updated, unproductive assessments have been eliminated and frequency of testing varied according to needs of the assessments.

There have been four, direct (primary) impacts identified: water turbidity, benthic smothering, trace metal bioaccumulation and shoreline habitat loss (from waste rock, not tailing). There are few secondary (derived) impacts consequent from these. The direct impacts are localised and determined by the pattern of tailings flow, i.e. by channelled density current over the sloping seabed (Hay 1982), only occasionally changing to widespread sheet flow. Eventually the tailings, which had raised the seabed on average 15 m (max. 35 m) by 1982 after 11 years of operation (Anon 1984a), deposit. Where deposition is rapid (not accurately measurable, but in general terms more than a few centimetres at a time), seabed organisms (benthos) are smothered. Biological sampling has shown that previously smothered areas can recover when quiescent such that within 1 year a community of small burrowing organisms will have re-appeared (Chap. 29, Vol. 2). The narrow channelling of the density current means that at any one time much of the seabed is quiescent and recovering. It should be noted that a commercial crab fishery is maintained in the receiving fiord in spite of the dynamic deposition-erosion seabed environment. These crabs must be finding their food from the seabed with its tailings.

The density current generates eddy systems from which a turbidity cloud has formed over the seabed (Ellis and Heim 1985). At the point where the tailings current reaches the fiord sill, an unusual tidally varying descending jet current hitting an opposing rock face (1 km distant) generates upwelling, bring varying amounts of resuspended tailings to the surface (Stucchi 1985).

Some of the seabed organisms show raised levels of trace metals on tissue analysis (Tables 5 and 6). These raised levels arise from particle-bound metals attached to, or passing through the guts of, the benthos, or they may be digested by the animals and incorporated into their tissues. This bioaccumulation does not appear to be giving rise to biomagnification up the marine food chain, as upper level herbivores and carnivores analysed do not show progressively raised levels of tissue metals.

The pit waste rock shoreline dump has progressively obliterated the original shoreline and shallow water habitat over 3 km. This shallow water habitat is used by juvenile salmon from nearby spawning streams, but there is no indication that the habitat loss has affected the salmon stocks.

The continued monitoring over 15 years to date has not detected kills of fish from surges of toxic effluent, although on one occasion effluent bioassay demonstrated temporary toxicity, which was rapidly controlled by replacement of a newly tried process chemical (Pelletier 1982). Nor has the monitoring indicated that more insidious biological resource losses are occurring through diminished primary and secondary biological production of finfish, shellfish and their food stocks. Although no specific social assessments have been made, there are no communities of residents dependent on the local biological resources whether for subsistence food, or as their economic base. There is, however, some local food gathering in the turbidity-affected area by native Indian and white residents, plus commercial crab fishing, some prawn fishing and occasional days of regulated commercial salmon fishing.

Table 5. List of species tested for tissue metal levels at various mine sites

Mine sites	Years of observations
Island copper (As, Cd, Cu, Hg, Mo, Zn, Pb)[a]	
Plants	
Fucus distichus, rockweed	1976 Annually
Zostera sp., eelgrass	1978 Annually
Zooplankton	
Total)	1972 Annually
Euphausids)	1975 Annually
Shrimp	1975 Annually
Cancer magister, Dungeness crab	1971 Annually
Bivalve molluscs	
Mya arenaria, soft-shelled clam	1971 Annually
Protothaca staminea, littleneck clam	1971 Annually
Macoma irus	1975 Annually
Mytilus edulis, blue mussel	1974 Annually
Humiliaria kennerlyii (a deep water clam growing on tailings beds)	1977 Annually
Saxidomus giganteus, butter clam	1971 Annually
Fish	
Many species	1971 Annually to 1976
Kitsault (Cu, Cd, Fe, Zn, As, Mo, Ni, Cr, Pb, Mg, Al)	
Fucus distichus, rockweed	1977, 1978, 1981
Shrimp	
Several species (analysed separately)	1978, 1980, 1981
Crab	
Lithodes aequispina	1978, 1980, 1981
Chionoecetes bairdi	1980, 1981
Bivalve molluscs	
Mytilus edulis, blue mussel	1977, 1978, 1981
Yoldia thraciaeformis/montereyenis	1978, 1981
Clinocardium, cockles – 2 species	
Fish	
Sole, several species (analysed separately)	1980, 1981
Kitimat (Fl)	
Fucus distichus	1976
Ectocarpus sp.	1976
Balanus glandulosa?	1976
Macoma inconspicua	1976
Nanisivik (As, Zn, Cd, Fe, Cu, Pb, Hg, Mn)	
Mya truncata	1975, 1979, 1982
Boreogadus saida	1974–1975 Baseline
Zooplankton	1974–1975 Baseline
Fucus vesiculosus	1974, 1976, 1979
Strongylocentrotus droebachiensis	1979
Laminaria solidungula	1976, 1979

Table 5 *(continued)*

Mine sites	Years of observations
Nanisivik (As, Zn, Cd, Fe, Cu, Pb, Hg, Mn)	
Agarum cribrosum	1979
Myoxocephalus quadricorni	1979
Myoxocephalus scorpioides	1979
Leptasterias polaris	1979
Serripes groenlandicus	1975, 1979
Cardium ciliatum	1975, 1979
Holothuria	1979
Palmaria palmata	1976
Astarte boreali	1975
Modiolaria nigra	1975
Hiatella arctica	1975, 1979
Buccinum – 3 spp.	1978
Sipho	1975
Bougainville (Cu, Pb, Zn, Cd, Hg, As, Mo)	
Many fish species (8 reported in Powell et al., 1981)	Several years
Tasmania (Zn, Cd, Cu, Cr, Pb)	
Crassostrea gigas, Pacific oyster	1973
Wesfrob (Fe, Pb, Zn, As, Cd, Ni, Hg)	
Bivalve molluscs	
Haliotis kamskatchana, abalone	1971–1974 ?
Mytilus edulis, mussel	1971–1974 ?
Hinnites, rock scallop	1971–1974 ?
+ Others intermittently	
Jordan River (Cu, Zn, Cr, Cd, Pb, Ag)	
Protothaca staminea, littleneck clam	1972–1978
+ Others initially	
Britannia Beach (Cu, Zn, by X-ray energy spectroscopy)	
Mytilus edulis, mussel	1980
Protothaca staminea, littleneck clam	1980
Macoma baltica	1980
Plants	
Fucus, rockweed	1980
Ulva, sea lettuce	1980
Laminaria, kelp	1980
Black Angel (Zn, Pb, Cu, Cd, Ni, Ag, Fe)	
Fish	
Anarrhicas minor, wolf fish	1973–1977
Platysomatichtys hippoglossoides, halibut	1973, 1974
Mytilus edulis, mussel	1972, 1973, 1976
Fucus, spp., rockweed	1972, 1973, 1976

Table 6. Mines and species with elevated levels of metals in a spatial or temporal pattern suggesting bioaccumulation from mine wastes (mg kg^{-1} dry w. unless otherwise shown)

Mine sites			Notes
Island Copper[a]			
Fucus, rockweed	Cu	50–60	May be attached particles
	Zn	30–100	
	Cd	4.0–4.5	
Mytilus edulis, mussel	Cu	2	Consistently higher at loading dock than reference docks (Cu ~ 1.5), but no trend to increase
	Zn	35	
Kitsault			
Yoldia	Pb	200–300	1981 increase from 10–50 in 1978
	Cd	30–35	1981 increase from 18–20 in 1978
			Deposit feeding small species
			Not always available
			Highly variable results
			No size-related data
Wesfrob			
	None		
Jordan River			
Protothaca staminea, littleneck clam	Cu	20–30	About double reference specimens
Britannia Beach			
Mytilus edulis,	Cu	645	Viscera
mussel	Zn	607	Gill
Protothaca staminea,	Cu	119	Viscera
littleneck clam	Zn	100	Viscera
Macoma baltica	Cu	160	Gill
	Zn	448	Viscera
Plants			
Fucus, rockweed	Cu	111	
Ulva, sea lettuce	Cu	345	
	Zn	218	
Black Angel (Pb is wet wt.)			
Fucus, rockweed	Zn	20–100 (1977)	Zn generally increased from 1973 to 1974 after discharge started
	Zn	100–500 (1974)	
	Pb	40–50 (1977)	Pb increased from ~ 10 in 1973
Mytilus edulis,	Zn	40–70 (1974)	Zn generally increased from 1973 to
mussel	Zn	100–550 (1977)	1974 after discharge started
	Pb	40–50 (1977)	Pb increased from ~ 10 in 1973
Wolf-fish liver	Pb	2.5	Predator
Kitimat			
Ectocarpus	F	317	Higher at 150 m from outfall than at 500 m
Amphipods	F	1168	Higher in amphipods than in barnacles or bivalves (*Macoma*)

Table 6 *(continued)*

Mine sites			Notes
Nanisivik			
Mya truncata	Pb	5.4	No biochemical impact
	Zn	584	No obvious pathologies (histological examination)
			Exceeds maximum levels for marine products
Fucus vesiculosus	Pb	×28.3 = 59	Increased
	Zn	× 5.6 = 493	
	Cd	× 2.6 = 1.66	
	As	× 1.4	
Strongylocentrotus droebachiensis	Pb	23.9	Increased
	Zn	× 4.3 = 287	
	As	× 1.1	
Serripes groenlandica	Pb	×19.7	
	Zn	× 2.3	
	Cd	× 1.9	
	As	× 1.7	
	Hg	× 4.7	
Cardium ciliatum	Pb	× 1.4	
	Zn	× 1.7	
	Cd	× 2.4	
	As	× 1.7	
	Hg	× 2.2	

[a]Range approximated from diagrams, 1983 data.

This level of impact minimisation to only increased water turbidity (mostly slight, but obviously visible in one area), limited benthic smothering with recovery, some trace metal bioaccumulation and habitat loss, correlates with two approaches to environmental controls at the site. Constraints were placed in 1970 on the rate of discharge, and on the chemical and physical composition of the effluent (Table 4). The location of the discharge was controlled, and the outfall was designed to maximise the density current effect, so as to minimise dispersion of tailings and surface re-appearance (Pelletier, 1982). The unusual hydrographic features of the fiord have not allowed 100% success in keeping tailings at depth.

With approximately 10 years of operations remaining, the mine is now attending to eventual recovery of the impacted receiving area. Temporary closures over the years have indicated that a widespread, but low level of increased turbidity (e.g. twice the ambient level of 0.4 NTU) will disappear within a few days as fine tailings particles settle. It is currently expected that final scouring of deposited tailings down to bedrock under the descending tidal jet will eliminate the present upwelling impact. The pit could be opened to the sea as an experimental fiord.

2.1.2 Black Angel Lead-Zinc Mine, Greenland (Fig. 2)

Black angel is a high-grade underground mine with a seawater-process concentrate mill whose history and impact is documented by several published papers (e.g. Asmund 1979; Nielsen and Hansen 1979; Møller 1984) and a series of government reports (usually printed in Danish). Tailings are discharged at depth to a silled basin at the mouth of a fiord (Agfardlikavsa). The mine has operated from 1973 and was opened following an industry assessment, Danish government assessment, review and controls.

One impact of consequence has been identified in the literature. It is that raised trace levels of Pb and Zn have been found in three of four species tested (Tables 4 and 5). Bioaccumulation is presumably occurring in a range of species. The hydrographic situation allows leaching from the tailings, followed by export of the metals to the other fiord, Qaumarujuk (Møller 1984). However, tissue analyses of the two carnivorous fish (Tables 5 and 6) do not indicate that biomagnification is occurring. Although one species does show some concentration of lead (2.5 mg kg^{-1} wet w.), it is less than lead found in mussels and rockweeds (40–50 mg kg^{-1} dry w., allowing for × 5 ratio dry wt./wet wt.). Local residents' villages are 25 km or more distant and various reports suggest that villagers are not subsistent on the two fiords for fishery resources.

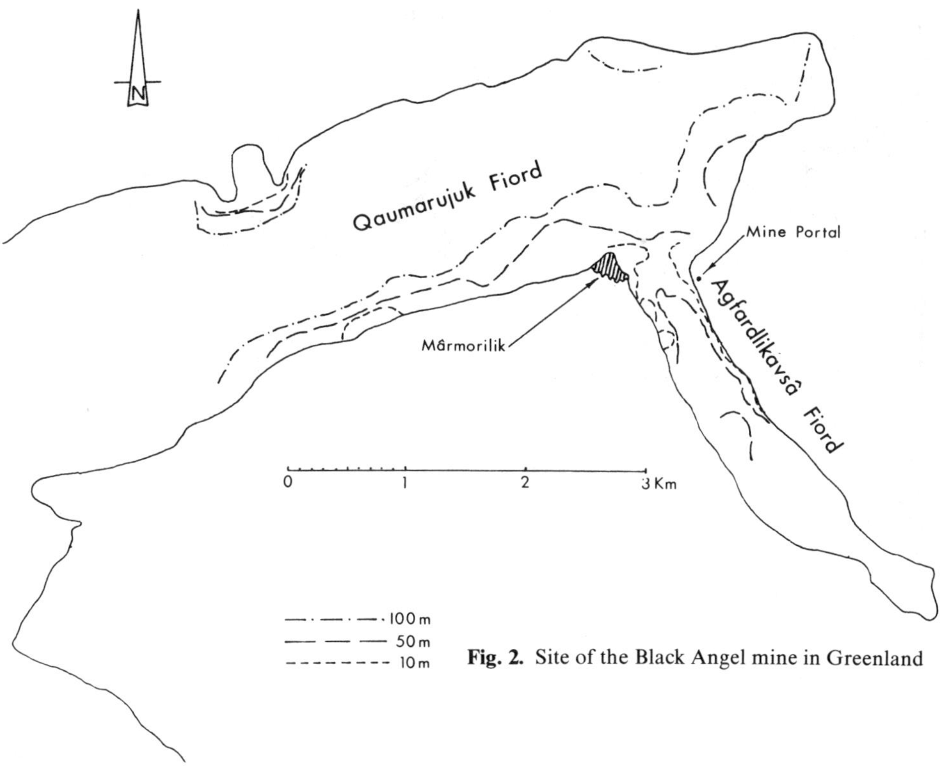

Fig. 2. Site of the Black Angel mine in Greenland

2.1.3 Bougainville Copper Mine, Papua New Guinea (Fig. 3)

Bougainville Copper Limited operates a low-grade, open-cut mine in the Crown Prince Range of Panguna, 670 m above sea level. Tailings are discharged into the Kawerong/Jaba river system 30 km from the sea. Some tailings settle out in the river system and the balance has formed a marine delta with accompanying dispersion and deposition nearshore (see also Chaps. 15 and 26).

The mine has had an environmental program since the start of the development and retains an environmental staff. To date a broadly based investigation of marine and land impacts has been developed, and in 1984 the company initiated a review of its environmental program, i.e. an audit of the environmental program by a team of senior scientists and engineers with appropriate expertise. The team comprised an environmental engineer as chairman, a chemist, a biologist (the author), a river engineer and a coastal engineer. The review of the program encompassed the scope, methodology, results and their significance. The environmental program is documented by an extensive series of in-house reports which

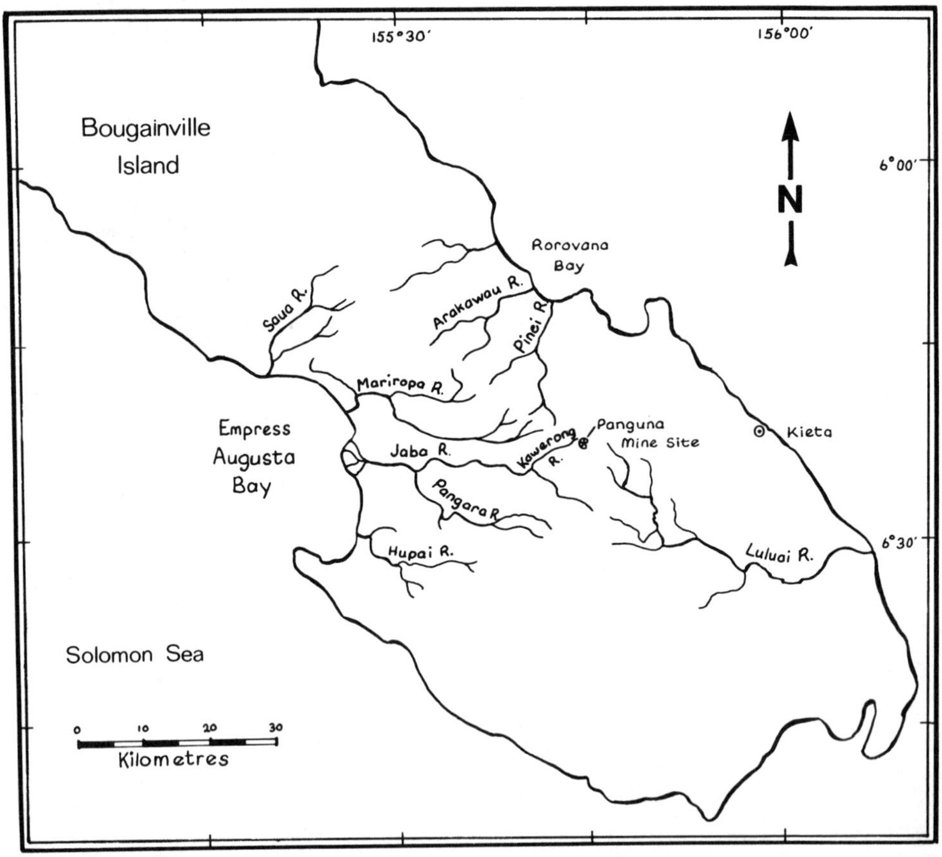

Fig. 3. Location of the copper mine at Panguna on the island of Bougainville, Papua New Guinea

were made freely available to the environmental auditors. The biological aspects of the program are summarised here (see also Chaps. 15 and 26).

The main results of the aquatic biology program are that, compared to reference sites, there are raised levels of soluble Cu in surface water at the mouth of the Jaba River and in seawater immediately overlying the tailings deposits. There is no elevation, however, of trace metals levels in edible fish from the bay (Powell, in-house data). There has, however, been loss of migratory fish from the Jaba River and its tributaries as a result of the tailings. The delta rapidly revegetates once areas become stable (Archer and Marshman 1984).

The conclusion of the environmental audit was that in general the environmental program at BCL had been professionally conducted, and the scope of the program covered the components of the environment known in other cases to be affected by tailings disposal. Points of specific detail were raised during the audit. Where company staff have agreed that changes in the program were justified they have been incorporated into the program. Particular mention was made by the auditors of post-mine tailings degradation, leaching of acid and copper from waste dumps and tailings and the difficulties faced by environmental staff working in remote locations, e.g. the need for continuing professional supervision to ensure adequate quality control. Throughout this environmental program, marine aspects could not be separated from considerations of land, freshwater and social aspects. Separate specialist environmental assessors must co-operate in their respective data gathering and reviews. Social information is available (e.g. Oliver, 1981), and compensation has been provided for local residents for resource losses. Interactions between physical and social compartments of the mining-environment system are diagrammed in Fig. 4.

Part of the significance for the mining industry of the environmental assessment program at Bougainville Copper is that it provides results from a tropical area where subsistence fishing drawing on the diverse marine fish species of warm water ecosystems is a way of life. Also Bougainville illustrates the use of a particular quality control procedure for environmental assessments, that of the environmental audit.

2.2 Pre-Operational Assessments

In recent years two mine sites under development have had the opportunity to draw on experience elsewhere and undertake relevant environmental studies to determine resources at risk and to indicate mitigative procedures.

2.2.1 Quartz Hill Molybdenum Mine, Alaska, USA (Figs. 1 and 5)

Quartz Hill is a proposed open-pit mine which will discharge 80,000 t day^{-1} tailings from its concentrator mill. Current expectation is that discharge will be to one or other of three suitable fiord basins (Anon 1984b; Anon 1985b). Pit overburden will be to land, and water supply may be from a salmon spawning river by means of gravel seepage extraction or other systems. The developers must comply with relevant legislation of both the US federal government and the State of Alaska.

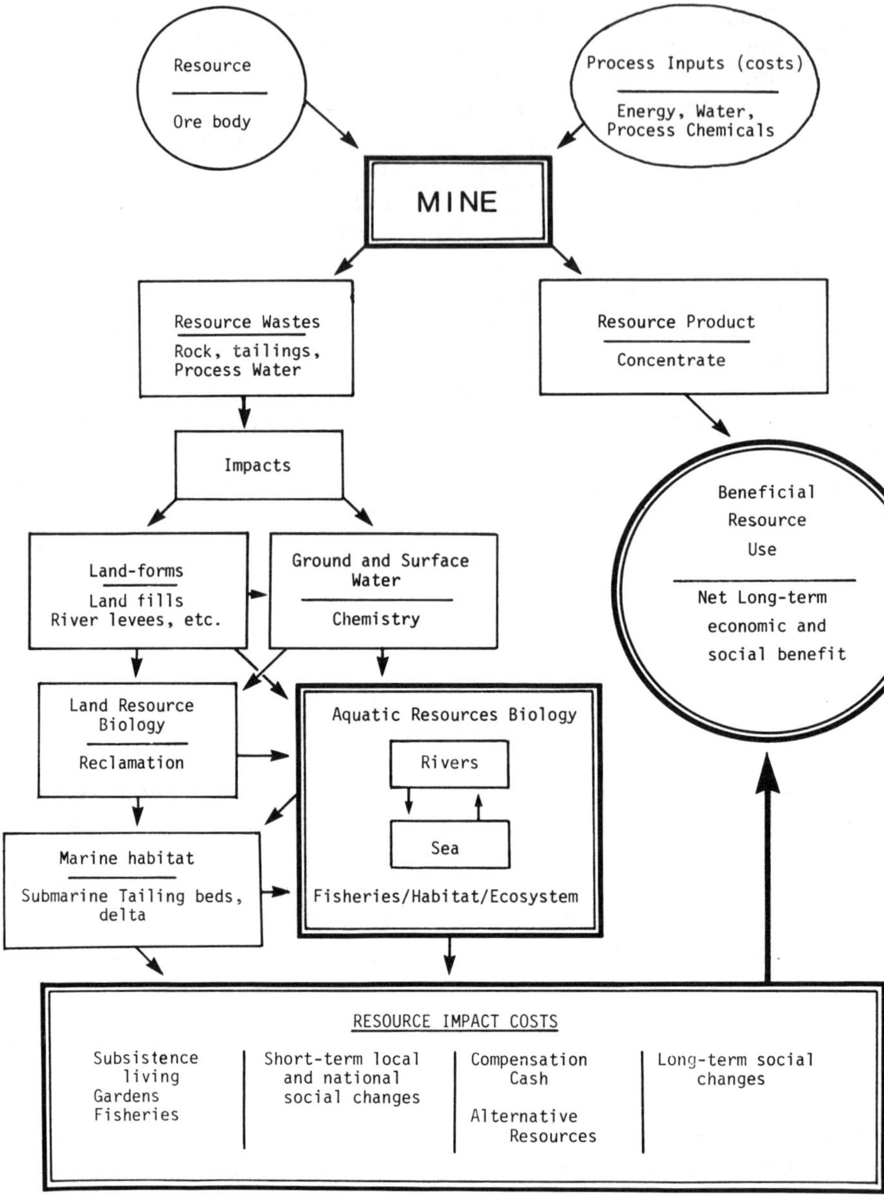

Fig. 4. Interactions between the mining development at Bougainville and the physical-social environment

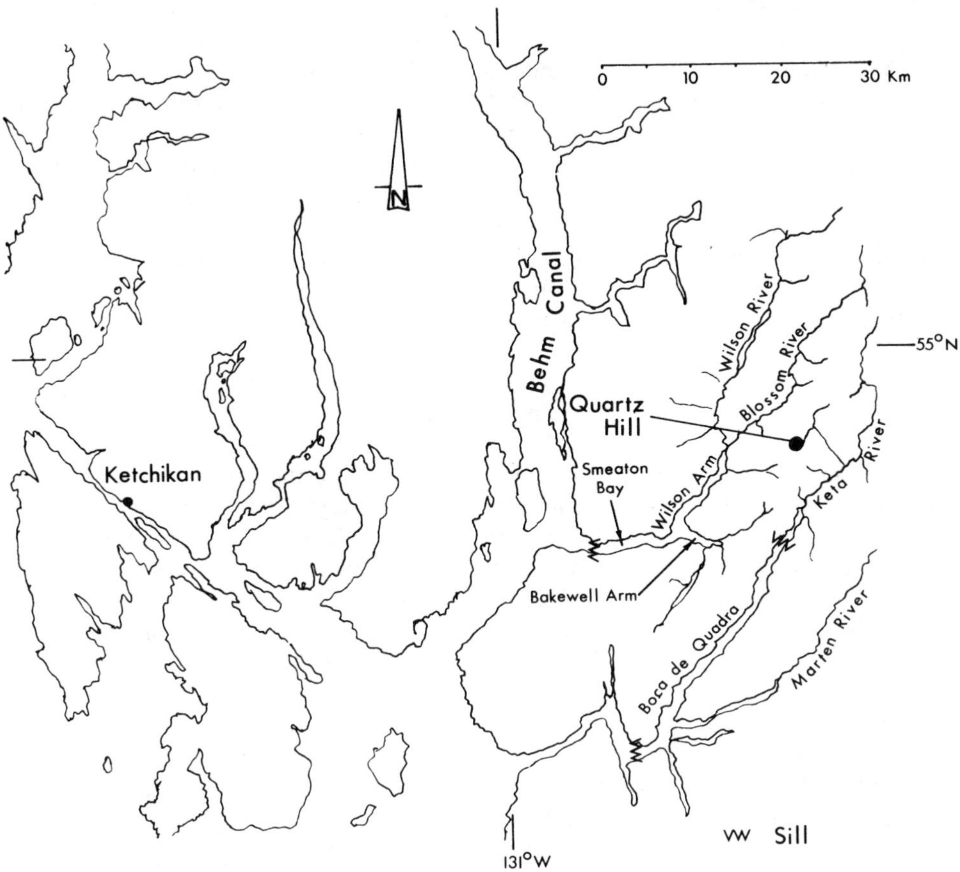

Fig. 5. The Quartz Hill mine site and proximity of deep fiord basins in Smeaton Bay (1 basin) and Boca de Quadra (2 basins)

This means that Quartz Hill is the first mine with intended marine discharge to fall under the National Environmental Policy Act and the concerns of the US Environmental Protection Agency.

The effect has been that the mine developers have undertaken very substantial baseline surveys to determine resources at risk (see Table 1), and to establish the state of the marine ecosystem, and its natural fluctuations, prior to action which could lead to impact (Ellis 1982a; Anon 1985b). In 1985 the development proceeded through formal channels, including meeting very tight specified deadlines, to a Draft Environmental Impact Statement (Anon 1984) and an Ocean Discharge Criteria Evaluation (Anon 1985b), both very large documents of several hundred pages each. These are subject to public review, which will lead in 1986 to a definitive EIS, on which a decision for approval or not (and, if approved, the environmental constraints) will be based. The major resources at risk are salmon and herring, with some potential losses of crustacean shellfish and aesthetics (of an attractive fiord with tourist and recreation potential). It is accepted in the draft

ODCE that established tailings line construction procedures and maintenance, and submarine outfall design, can prevent salmon impact.

The extent of the potential impacts on herring and shellfish are under dispute, due to differing opinions on the scale of turbidity elevations and other factors of physical oceanography and consequent biological production.

It has been noted that there is no subsistence dependence on fishing in the three possible tailings receiving basins, but that both herring (a roe industry) and salmon stocks are major contributors to the fishing industry of southeast Alaska, hence are sensitive resources. The case as it has developed has produced strong local responses from environmentally concerned citizens groups, including legal action seeking to prevent initial road construction and bulk ore sampling.

2.2.2 Atlantis Deep, Offshore Mining in the Red Sea (Fig. 6)

The Atlantis Deep is the site of what will probably be the first offshore mining development on metalliferous muds. A pre-pilot mining test was completed in 1979 (Nawab 1984; Gideiri 1984). The unconsolidated muds (19,550 t) were pumped up from a depth of 2,200 m, processed through a shipboard concentrator mill and tailings discharged through a pipeline at 400 m depth (Fletcher and Mustafa 1980).

This pilot project was monitored by an industry-retained team of scientists, whose measures encompassed the topics shown in Table 1. The tailings were detected as a descending and dispersing plume drifting away from the mining vessel, and remained below the discharge depth (as was to be expected from knowledge of the oceanographic conditions). It is concluded that tailings subsequently deposited at depth dispersed widely, hence did not cause impacts in the shallows where the majority of biological resources are found. It has been recommended that any further developments discharge tailings at greater than 1,100 m depth (Gideiri 1984).

It is to be noted that the Atlantis Deep lies astride an international boundary: Sudan and Saudi Arabia. Authority for proceeding with the mining development is held by a two-country agency, the Saudi-Sudanese Red Sea Commission.

2.3 Non-Industry Assessments (Government, University)

At a number of sites, assessments of impact have been made by government or university scientists rather than mine staff or retained consultants.

2.3.1 Kitsault Molybdenum Mine, Canada – Part Government (Fig. 1)

Kitsault has been in operation twice, from 1967 to 1972 and from 1981 to 1982. It is an open-pit mine, which on first operation discharged tailings without controls to a creek. In its second operation under pollution controls, the mine piped tailings to a fiord and discharged them at depth (Burling et al. 1981, 1983) through an

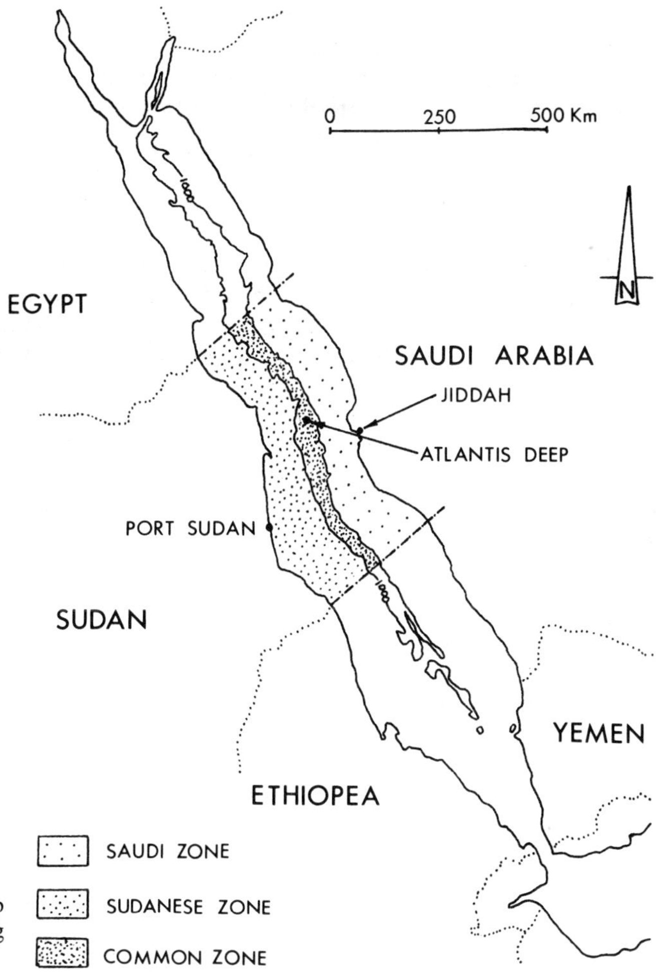

Fig. 6. The Atlantis Deep site in the Red Sea for mining offshore at depth (2,200 m)

outfall with a seawater mixing chamber (Ellis 1982). Prior to intended re-opening, the developers initiated specific assessments of relevant aspects of physical, chemical and biological oceanography, and sought permits for deep discharge of tailings.

Shortly before opening, government agencies focussed on a number of issues, and undertook associated environmental measures. The site also became a subject of media, political and religious group interest such that a public inquiry was called (Ellis 1982b). This met at the time of the second opening of the mine, when a number of novelties in EIA arose. In particular, oceanographic instruments brought into use for assessment both by government and industrial assessment groups were more sensitive than those used initially, and to slightly different parameters. This complicated assessment of the results, with the two groups developing different opinions whether the mine was in compliance with its permit. At no time during operations was it found possible to bring the working scientists together to resolve their technical and inferential differences. Closure of the mine due to market conditions postponed resolution of the problems.

2.3.2 El Salvador Copper Mine, Chile – An Independent Scientist (Fig. 7)

El Salvador is yet another major open-pit, low-grade copper mine discharging its tailings to the sea. It does so by means of an 80-km-long canal/river system discharging to a beach, initially near the city of Chanaral (1938–1974), but subsequently to another beach 8 km to the north (since 1974).

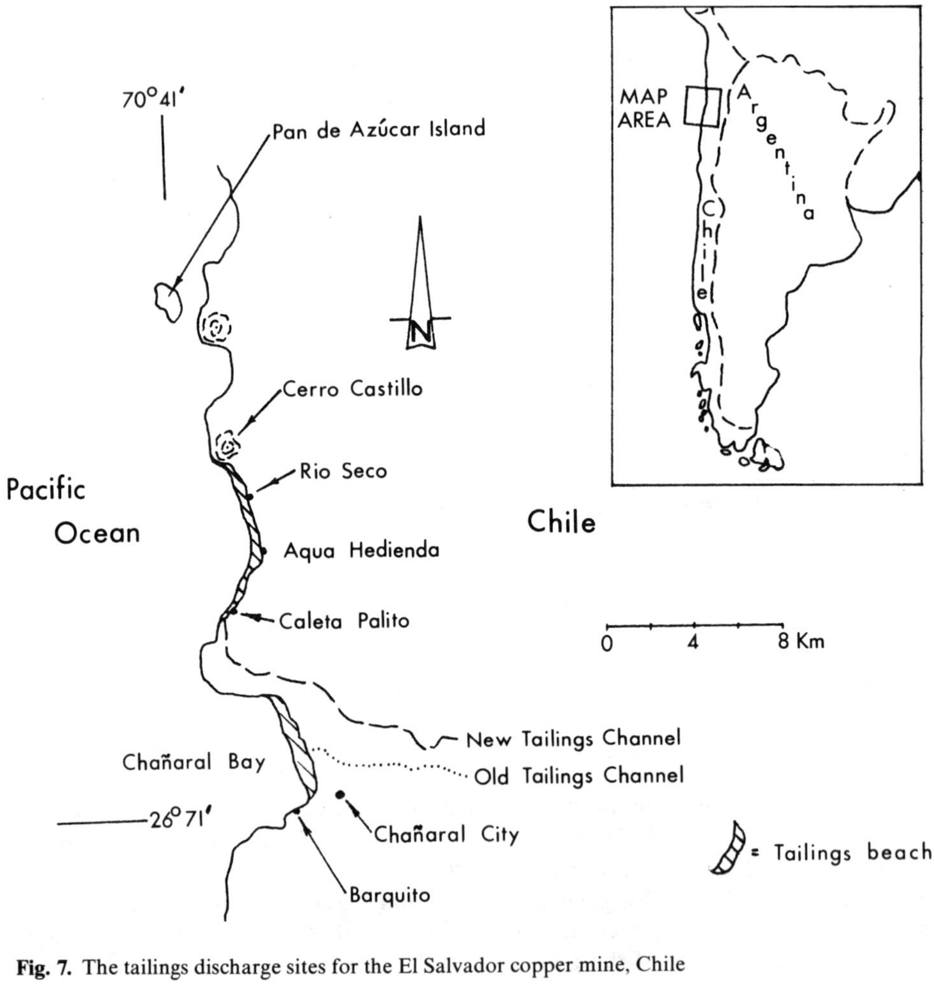

Fig. 7. The tailings discharge sites for the El Salvador copper mine, Chile

The impact has been partially assessed by university scientists (Castilla 1983; Castilla and Nealler 1978) who have stated that the discharges "hindered harbour activities, caused geomorphological coastal modification and affected seriously marine coastal ecosystems and recreational activities". Sand beach and sub-tidal beds were impacted to the point of becoming barren, on occasions by visible fish kills. Beach discolouration from pyrites, and water turbidity, were extensive. A local marine resort area became unusable after the shift in discharge point.

This case illustrates the potential for limited-scope impact measures by scientists skilled in appropriate techniques, but working within the very substantial constraints of funds from university-related sources. In this case some serious impacts of tailing discharge have been documented. It is reasonable to expect that other impacts such as trace metal bioaccumulation may be occurring. Since tissue metal testing and other appropriate oceanographic testing requires specialised (and expensive) facilities, they cannot always be available for a comprehensive assessment by scientists working under limited funding.

2.3.3 The Phuket Nearshore Tin Gravels, Thailand – An Independent Laboratory

The west coast of Thailand is part of a belt of tin-bearing onshore ore and nearshore wash-out gravel deposits extending from Burma to Indonesia. The gravel is mined at a number of nearshore locations, one of which near Phuket, although traditional, is a site of resource conflict as the Thai tourist industry develops and seeks expansion into this attractive area. In addition, the coast supports fishing, and has a number of coral reefs (Brown and Holley, 1982; Chansang et al. 1981) and other productive marine shallow and wetland habitats (Bussarawich et al. 1984; Nateewathana et al. 1984).

Assessments in this area have been initiated through an independent marine biological station catering to fundamental research needs of the country through a supporting network of universities and government agencies. Table 2 notes the nature of investigations from the station which relate to impacts from the tin-dredging operations. Research on mine-derived turbidity levels in the area provides the only direct measures of correlated reductions in primary production available to date in the published literature about marine mining impact (Limpsaichol and Poopetch et al. 1984).

This case is of interest in a number of ways. The most important is that it is the only case to date of assessments of regional impact from dispersed underwater mining sites in shallow water. The area is subject to not just the impact of tailings disposal, but also the impact of the extraction methods over a wide area of seabed. The results are of interest to other underwater mine sites in demonstrating potential to affect the resources of shallow water and what depths and distances are within impact. Undoubtedly more use could be made of the available expertise at this laboratory, providing information about the extent of the inevitable resource conflict between nearshore mining and fisheries.

3 Conclusions

Marine discharging mines have approached their problems of environmental impact in different ways, with some making substantial efforts to assess and reduce the impacts. In many cases, the government has required action and, in some cases, independent scientists have assessed what they could. From the accumulated results, some patterns of effective impact assessment and mitigation are now apparent.

3.1 Impacts

There are three impacts which have consistently occurred at marine mining sites and will probably occur to a limited extent at new sites even when controlling action is taken. The impacts are water turbidity, seabed smothering and trace metal bioaccumulation. Increased water turbidity potentially can reduce primary biological production, and hence may reduce fishery yields. The effect can be minimised by deep discharge, below the photosynthesising zone, preferably where upwelling will not occur. Seabed smothering can destroy benthic species of groundfish and shellfish, and their foodstocks, but colonisation occurs rapidly (see Chap. 29), and can be aided by discharge engineering which channels tailing dispersion to localised areas. Trace metals appear in marine organisms, especially seaweeds and bivalve molluscs (clams, oysters, etc.) as a result of bioaccumulation (Tables 5 and 6). This process has been constrained by process controls minimising trace metal discharge levels. These three kinds of impacts are liable to interact with complex social consequences, but the direct consequences can often be reduced by appropriate controls to levels which have little biological effect.

The direct impacts may induce some secondary effects. They are acute toxicity (fish kills), trace metal biomagnification and fishery losses. Acute toxicity can be virtually eliminated by process controls. Accidental surges of toxins can be monitored for by bioassays, so that process control action can be taken. Trace metal biomagnification potentially can occur if tailings discharge is to an embayment with little flushing and a closed food chain. Initial trace metal bioaccumulation can then escalate to biomagnification up the food chain of feeding shellfish and finfish. Where there are subsistent fishermen in the area, this is an important consideration which must be assessed, and if likely to occur must be avoided in some way. Fishery resources potentially can be lost as a consequence of water turbidity, seabed smothering, contamination and by cessation of fishing due to social concerns. In areas of subsistent fishing, loss of the resource is an important matter which should be assessed, and appropriate measures taken if needed.

It is noticeable that little is known about the impact of smelters and refineries (Table 3).

Controls to reduce the above impacts are generally implemented at two stages of the mining process: either in the process stage of milling, and by design, placement or operation of the outfall (Littlepage et al. 1984).

3.2 Assessment Procedures

A mine has a lifetime, and assessment procedures should change during the lifetime (Fig. 8). Initially, assessment by baseline resource surveys and pre-operational monitoring to check resource variability allows prediction of potential impacts, and prevention or reduction by good engineering design. Construction needs monitoring to control stream flow diversions, etc., and commissioning tests should be monitored for such impacts as controllable toxicities. During operations, initial short-term assessments monitor that immediate changes in the receiving

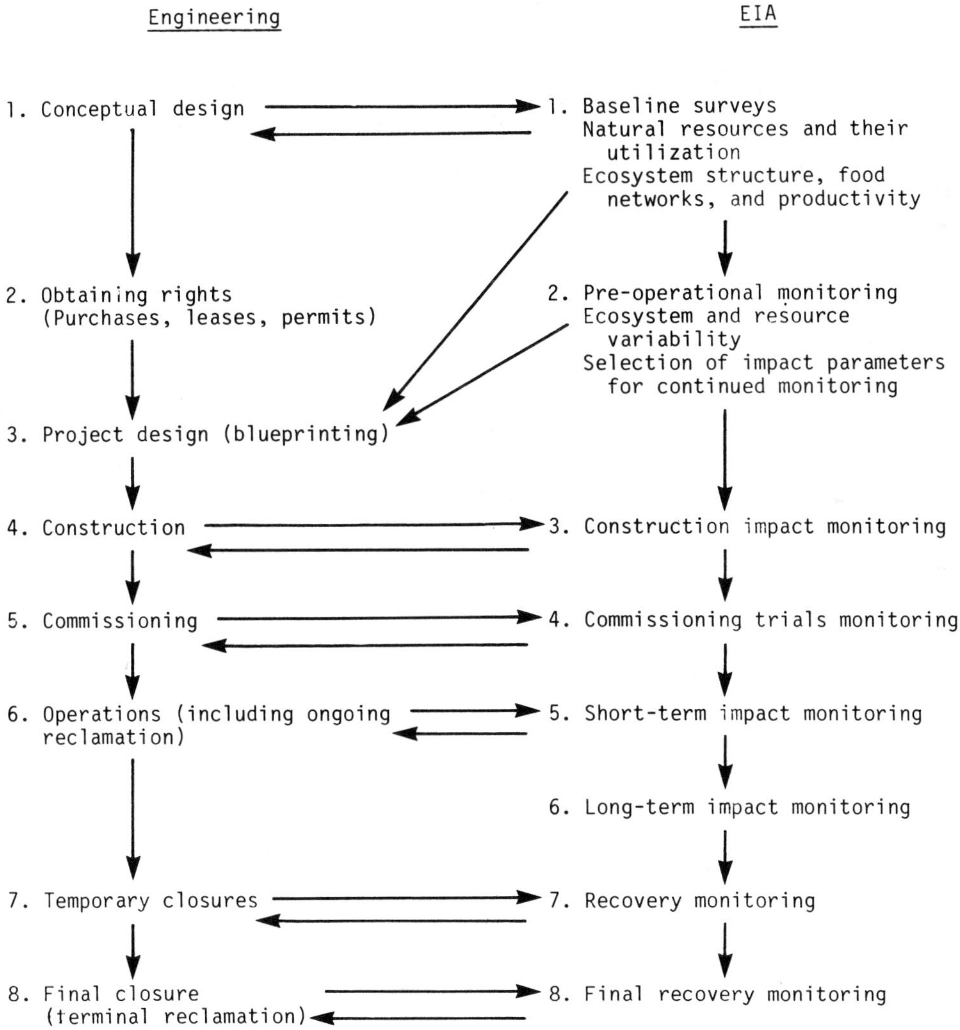

Fig. 8. The feedback of information from environmental impact assessment during the lifetime of a mine

area are as expected (and process control action should be taken, if not). They need to be followed by long-term assessments monitoring development of insidious slow changes such as trace metal biomagnification. Opportunities should be taken to monitor biological recovery during temporary mine closures, prior to monitoring and control of required reclamation (see Chap. 29). There are now many mines which have proceeded along this general path. Some of those that were brought into operation before this system developed, still require adequate hindsight assessments to check what resource costs have been externalised by the mines to be met by local residents or the taxpayer at large. Social consequences have been well documented at some mines, and could be assessed elsewhere.

Assessment should be considered as part of an informational feedback system to the mine developers and operators (Fig. 9). A mining development starts with a conceptual design. The design should incorporate a first round of information gathering about other resources, predictions of impact (externalised costs) and recommendations for mitigation. During operations, routine monitoring provides information for operating engineers to quality control their discharges, hence to quality control the receiving environment. Pollution control authorities can also assess monitoring results for surveillance and regulatory action if necessary. After closure, final reclamation can be monitored to check whether productive fishery resources have returned as intended. The continuing result of this informational feedback system is intended to be a minimised level of impact which is socially acceptable for the lifetime of the mine.

Assessments have been undertaken by in-house scientists, retained consultants (full-time consultants or university specialists), by government agencies and independent scientists. It is noticeable that comprehensive assessments have been achieved through mine initiatives. Program review is an important component of assessment, both regularly by an environmental supervisor, and periodically by a

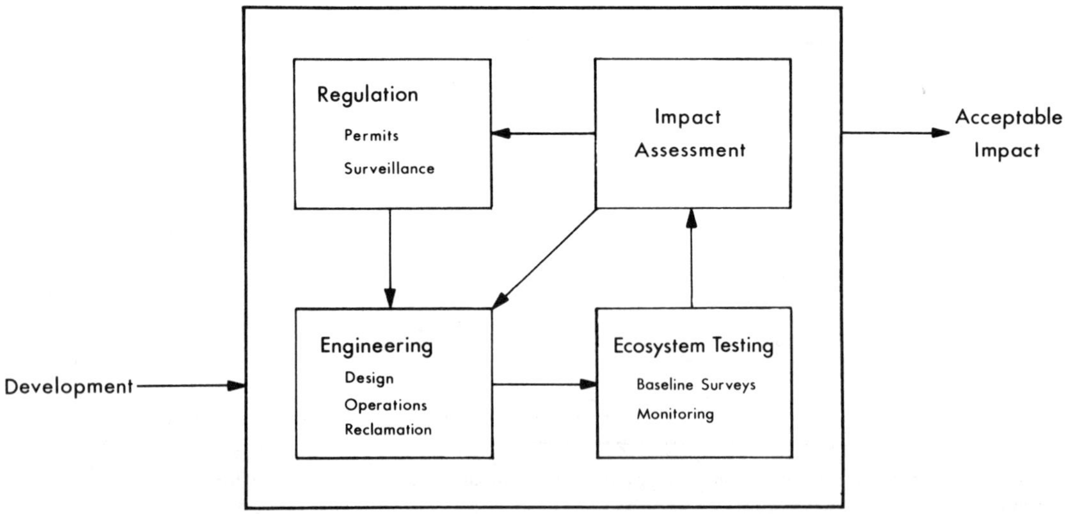

Fig. 9. The informational feedback system in environmental impact assessment

team of environmental auditors (Ellis 1985). Effective communication between mine and regulatory authorities is important. Program credibility has been aided by professional publication of results; which also ensures that assessing mines feedback provides a contribution to the common pool of knowledge on which they have drawn.

4 Marine Mines–Offshore

The International Deep-Sea Biology Symposium held in Hamburg, FRG, June 23–29, 1985 had a 1-day session on "Human Impact on the Deep Sea". Papers were presented on manganese nodule ecosystem and a controlled impact experiment, and the poster session described the Atlantis Deep project. A summary by H. Thiel "Man's Impact on the Deep Sea" was written for Deep-Sea Newsletter Vol. 11 (p.6), 1985. In the same issue T. Rice reported that SCOR Working Group 76 supports monitoring test operations, but acknowledges little likelihood of effects being detectable.

Addendum (Tables 1 and 3)

A recent paper by Rygg (1986) on esoteric benthic analytical methods goes a long way to clarifying the extent of monitoring in Norway on mine pollution. Areas are: Nordrana (unstated mining), Orkdalsfjorden (copper mine), Sofjorden (zinc smelter), Kristiansandfjorden (nickel). References are: Haug et al. (1974), Rygg (1984, 1985a,b, 1986), Rygg and Skei (1984), Skei (1981), Skei et al. (1982).

Acknowledgements. The author greatly appreciates the help from individuals in industrial and government agencies and elsewhere who provided the opportunity to collate the information presented here. In particular co-operation from the management and environmental staff of Island Copper mine, US Borax, Amax of Canada, Bougainville Copper Mine, Marcopper Mining Corporation, the Waste Management Branch of British Columbia, the Environmental Protection Service of Canada, various agencies in Thailand (Mining and General Management Co. Ltd., Offshore Mining Organisation, Department of Mineral Resources, the Phuket Marine Biological Centre), in Australia (Federal and Queensland Governments) and the US Minerals Management Service. Some financial support was obtained from the University of Victoria Faculty Research Fund. Leonard Craig and Elizabeth Gilbert drafted the figures and Katharine Ellis typed the manuscript.

References

Alino PM (1984) The effect of mine tailings on the structure of coral communities at Toledo, Cebu. 3rd Symp Our environment, Singapore

Anon (1977) Brief of Alcan Smelters and Chemicals Ltd. submission to the public inquiry into pollution control objectives for the mine, mine-milling and smelting industries of British Columbia, Can

Anon (1979) Pollution control objectives for the mining, smelting and related industries of British Columbia 1979. BC Minist Environ, Victoria, BC, Can, 15 pp

Anon (1983) Proposed polymetallic sulfide minerals lease offering Gorda Ridge area. US Min Manag Serv, Reston, VA, 580 pp
Anon (1984a) Annual environmental assessment report. Island Copper Mine, BC, Can, 2 vols
Anon (1984b) Draft environmental impact statement. Quartz Hill molybdenum project mine development. US Dep Agric, Forest Serv, Admin Doc 133
Anon (1985a) Annual environmental assessment report. Island Copper Mine, BC, 2 vols
Anon (1985b) Ocean discharge criteria evaluation. Quartz Hill mill tailings dispersal. US Environ Protect Ag, Region 10
Archer IM, Marshman NA (1984) Revegetation at Bougainville Copper. Min Mag 115:307–313
Asmund G (1979) Water movements traced by metals dissolved from mine tailings deposited in a fjord in north-west Greenland. In: Freeland HJ, Farmer DM, Levings CD (eds) (1979) Fjord oceanography. Plenum, New York, 347–354
Ayling GM (1974) Uptake of cadmium, zinc, copper, lead and chromium in the Pacific oyster *Crassostrea gigas*, grown in the Tamar River, Tasmania. Water Res 8:729–738
Brown BE (1985) Marine pollution research in south-east Asia. Mar Poll Bull 16(3):95–97
Brown BE, Holley MC (1982) Metal levels associated with tin dredging and smelting and their effect upon intertidal reef flats at Ko Phuket, Thailand. Coral Reefs 1:131–137
Burling RW, McInerney JE, Oldham WK (1981) A technical assessment of the Amax/Kitsault molybdenum mine tailings discharge to Alice Arm, British Columbia. Rep Minist Fisheries and oceans, Ottawa, 154 pp
Burling RW, McInerney JE, Oldham WK (1983) A continuing technical assessment of the Amax/Kitsault molybdenum mine tailings discharge to Alice Arm, British Columbia. Rep Minist Fisheries and ocean, Ottawa, 65 pp
Bussarawich S, Nateewathana A, Hylleberg J (1984) Distribution of marine benthic amphipods off Phuket Island, with emphasis on tin mining and a model of species-individual relationships. Phuket Marine Biol Center. Res Bull 32:21 pp
Carey J (1981) Nickel mining and refinery wastes in coral reef environs. In: Gomez ED, Birkeland CE, Buddemeier RW, Johannes RE, Marchs JA, Jr, Tsuda RT (eds) Proc 4th Int Coral reef Symp, vol 1. Univ Philippines, pp 138–146
Carter L (1975) Disposal of wastes to tidal waters. In: Jones MJ (ed) Minerals and the environment. Inst Min Metall, England, pp 149–157
Castilla JC (1983) Environmental impact in sandy beaches of copper mine tailings at Chanaral, Chile. Mar Poll Bull 14(12):459–464
Castilla JC, Nealler E (1978) Marine environmental impact due to mining activities of El Salvador copper mine, Chile. Mar Poll Bull 9(3):67–70
Chansang H, Boonyanate P, Charuchinda M (1981) Effect of sedimentation from coastal mining on coral reefs on the northwestern coast of Phuket Island, Thailand. In: Gomez ED, Birkeland CE, Buddemeier RW, Johannes RE, Marsh JA, Jr, Rsuda RT (eds) Proc 4th Int Coral reef Symp, vol 1. Univ Philippines, pp 129–136
Dira DC, Canete AV (1980) ACMDC's tailings disposal today. In: Sixth symposium on mineral resources development and the 27th Annu Mine Safety Conf, November 27–29, Baguio City
Down CG, Mill AJB (1978) Marine disposal of mine and mill tailings. In: Jones MJ (ed) Proc 11th Commonwealth mining and metallurgical Congr, Hong Kong, pp 421–431
Ellis DV (ed) (1982a) Marine tailings disposal. Ann Arbor Sci Press, Ann Arbor, Mich, 368 pp
Ellis DV (1982b) Kitsault, B.C. – technical communication in a non-technical world. Mar Poll Bull 13(3):78–80
Ellis DV (1985) Environmental audits. Mar Poll Bull 16(5):171–172
Ellis DV, Heim C (1985) Submersible surveys of benthos near a turbidity cloud. Mar Poll Bull 16(5):197–203
Ellis DV, Popham JD (1983) Accidental formation and subsequent disappearance of a contaminated beach. In: McLachlan A, Erasmus T (eds) Sandy beaches as ecosystems. Junk, The Hague, pp 719–726
Fletcher AW, Mustafa Z (1980) Progress on the exploitation of metalliferous sediments of the Red Sea. In: National and international management of mineral resources, London, May. Inst Min Metall, pp 3–16
Gideiri YBA (1984) Impacts of mining on central Red Sea environment. Deep-Sea Res 6–8A:823–828
Harding L (1983) Use of fjords for disposal of marine tailing. Coastal zone 83. In: Proc 3rd Symp Coastal and ocean management, vol 1, San Diego, pp 32–51

Haug A, Melsom S, Omang S (1974) Estimation of heavy metal pollution in two Norwegian fjord areas by analysis of the brown algae *Ascophyllum nodosum*. Environ Pollut 7:179–192

Hay AE (1982) The effects of submarine channels on mine tailing disposal on Rupert Inlet, B.C. In: Ellis DV (ed) Marine tailing disposal. Ann Arbor Sci Press, Ann Arbor, Mich, pp 139–182

Hocking MD, Hocking D, Smyth TA (1980) Fluoride distribution and dispersion processes about an industrial point source in a forested coastal zone. Water Air Soil Poll 14:133–157

Hoff JT, Thompson JAJ, Wong CS (1982) Heavy metal release from mine tailings into seawater – a laboratory study. Mar Poll Bull 13(8):283–286

Holme NA, Probert PK (1978) Disposal of solid waste in the marine environment with particular reference to the China clay industry. In: Goodman GT, Chadwick MJ (eds) Environmental management of mineral wastes. Sigthoff & Noordhoff, Alphen aan den Rijn, Netherlands pp 239–253

Knauer GA (1976) Immediate industrial effects on sediment mercury concentrations in a clean coastal environment. Mar Poll Bull 7(6) 112–115

Knauer GA (1977) Immediate industrial effects on sediment metals in a clean coastal environment. Mar Poll Bull 8(11) 249–254

Levings CD (1975) Biological reconnaissance of an abandoned mine tailings disposal area in Toquart Bay, Barkley Sound, B.C. Fish Res Board Can Mss Rep Ser 1362, 9 pp

Limpsaichol P, Poopetch T (1984) The effect of dumping suspended solids at sea on chlorophyll A concentrations and primary production values. Phuket Mar Biol Center, Res Bull 33, Phuket, Thailand

Littlepage JL, Ellis DV, McInerney J (1984) Marine disposal of mine tailing. Mar Poll Bull 15(7) 242–244

Møller JS (1984) Hydrodynamics of an arctic fjord. Inst Hydrodyn Hydr Eng, Tech Univ, Denmark, Ser Pap 34, 197 pp

Nateewathana A, Tantichodok P (1984) Species composition, density and biomass of macrofauna at Ko Yao Yai, Southern Thailand. Proc Asian Symp Mangrove environment-research and management, pp 258–285

Nawab ZA (1984) Red Sea mining; a new era. Deep Sea Res 31:6–8A 813–822

Nielsen TK, Hansen N-EO (1979) Mixing and exchange processes in a small Greenland sill-fjord. In: Freeland HJ, Farmer DM, Levings CD (eds) Fjord oceanography. Plenum, New York, pp 219–226

Oliver D (1981) Aspects of modernisation in Bougainville. Papua New Guinea. Work Pap Ser, Pacific Islands Stud, Univ Hawaii, 91 pp

Ozturgut E, Lavelle JW, Steffin O, Swift SA (1980) Environmental investigations during manganese nodule mining tests in the north equatorial Pacific in November 1978. US Natl Ocean Atmos Admin Tech Mem ERL MESA-48, 50 pp

Pedersen TF (1984) Interstitial water metabolite chemistry in a marine mine tailings. Rupert Inlet, B.C. Can J Earth Sci 21(1):1–9

Pedersen TF (1985) Early diagenesis of copper and molybdenum in mine tailings and natural sediments in Rupert and Holberg inlets, British Columbia. Can J Earth Sci 22(10):1474–1484

Pelletier CA (1982) Environmental data handling and long-term trend monitoring at Island Copper mine. In: Ellis DV (ed) (1982) Marine tailings disposal, chap 6. Ann Arbor Science Press, Ann Arbor, Mich, pp 197–238

Petr T (1979) Possible environmental impact on inland waters of two planned major engineering projects in Papua New Guinea. Environ Conserv 6(4):281–286

Pintz WS (1984) Ok Tedi. Evolution of a Third World mining project. Mining journal books, London, 206 pp

Poling G (1982) Session 10. Round Table Sem Submarine and lake disposal of mill tailing. Proc 14th Int Mineral processing Congr, Toronto

Pullen TC (1982) A 3,000-mile arctic towing odyssey. Can Geog Dec 81/Jan 82

Reid JG (1978) Metallurgical operations at the Yabulu nickel refinery of Queensland Nickel Pts Ltd. Aust Inst Mining metallurgy Conf

Rice T (1985) SCOR working group 7b. (Ecology of the Deep Sea Floor) Hamburg Symp.Dead Sea Newslett 11: 6–7

Rygg B (1984) Bløtbunnfauna undersøkelser, Et godt verktøy ved marine resipientvurderinger. Rep Norw Inst Water Res (Oslo)

Rygg B (1985a) Effect of sediment copper on benthic fauna. Mar Ecol Prog Ser 25:83–89

Rygg B (1985b) Distribution of species along pollution-induced diversity gradients in Norwegian fjords. Mar Poll Bull 16:469–474

Rygg B (1986) Heavy-metal pollution and log-normal distribution of individuals among species in benthic communities. Mar Poll Bull 17:31–36

Rygg B, Skei J (1984) Correlation between pollutant load and the diversity of marine soft-bottom fauna communities. In: Proc Int Worksh Biological testing of effluents (and related receiving waters). OECD/US EPA/Environ Can, pp 153–183

Salazar RC, Gonzales RI (1974) Design, construction and operation of the tailings pipelines and underwater tailings disposal system of Atlas Consolidated mining and development corporation in the Philippines In: Aplin CL, Argal GO, Jr (eds) Tailings disposal today, Chap 20. Miller Freeman, San Francisco, pp 477–511

Skei JM (1981) Dispersal and retention of pollutants in Norwegian fjords. Rapp P-v Reun Cons Perm Int Exp Mer 181:78–86

Skei JM, Price NB, Calvert SE, Holthedahl H (1982) The distribution of heavy metals in sediments of Sorfjord, West Norway. Water Air Soil Pollut 1:452–461

Stucchi DJ (1985) The tidal jet in Rupert-Holberg inlet, B.C. Atmosphere Ocean 23:(2)118–136

Thiel H (1985) Man's impact on the deep-sea. Deap Sea Newslett 11:6

Thomas DJ, Metikosh S (1984) The effect of a lead-zinc mining operation on the marine environment of Strathcona Sound, NWT, Canada. Proc Conf Environmental contamination, London

Waldichuk M, Buchanan RJ (1980) Significance of environmental changes due to mine waste disposal into Rupert Inlet. Fisheries and oceans Canada and B.C. Minist Environ, Victoria.

Wilson KW, Connor PM (1976) The effect of China clay on the fish of St. Austell and Mevagissey Bays. J Mar Biol Assoc UK 56:769–780

Wong MH, Chan KY, Kwan SH, Mo CF (1979) Metal content of the two marine algae found on iron ore tailings. Mar Poll Bull 10:56–59

Yim WWS (1976) Heavy metal accumulation in estuarine sediments in a historical mining of Cornwall. Mar Poll Bull 7(8) 147–150

Development of Dredged Material Disposal Sites: Implications for Soil, Flora and Food Quality

W. VAN DRIEL[1] and J.P.J. NIJSSEN[2]

1 Introduction

In order to keep waterways and ports accessible to shipping, regular dredging of the sedimented materials is necessary. In densely populated and industrial countries nearly all sediments are considerably contaminated with inorganic and organic pollutants. Disposal of the huge quantities of dredge spoils with minimum effect on the quality of the environment is a major problem. International conventions forbid disposal of severely contaminated sediments in the open sea. Confined disposal of contaminated dredged materials on land raises questions about groundwater quality, and about adverse effects on biota and ecosystems.

In this chapter attention will be paid to the effects of large-scale disposal of dredged materials on land with respect to groundwater quality and accumulation of contaminants in the food web and to risks for human consumption. Possibilities of capping contaminated disposal sites with uncontaminated substrates in order to diminish uptake of contaminants in food crops will be discussed. Most examples in this survey are derived from the authors' experiences with respect to handling and cropping of dredged materials. These materials, originating from the rivers Meuse and Rhine, sediment in the ports of Rotterdam, are mixed with marine sediments from the North Sea, and are dredged and partly disposed in layers of 5–8 m in adjacent polders.

2 Method of Filling

In the Rotterdam area a suction dredger is almost always used to fill inland disposal sites. A pipeline runs from the suction dredger which lies in a suitable position in the river or harbour. The disposal site is divided by small dykes into a number of sections of varying shape and the size of these may exceed 10 ha. A seepage ditch is excavated round the site and the water level in this ditch is usually kept lower than that in the surrounding area. The water from the seepage is pumped to the river via a settling basin and in this way a separate drainage system is created.

[1]Institute for Soil Fertility, P.O. Box 30003, NL 9750 RA Haren (Gr.), The Netherlands
[2]Department of Public Works, Environmental Advisory Section, Municipality of Rotterdam, P.O. Box 6633, NL 3002 AP Rotterdam, The Netherlands

The dredged material is admitted to one of the sections of the site via one or more discharge pipes and in each section there is an equivalent number of so-called discharge weirs. On admission the dredged material has a density of 1,120–1,160 kg m^{-3}. After admission the density quickly increases to 1,200 to 1,300 kg m^{-3}. By this means the water under overpressure is driven upwards. This water is drained off via the discharge weir.

The dredged material is distributed in layers, the thickness of which varies between 0.60 and 1.5 m. Next, the dredged material is given a chance to ripen over a period of 1 to 2 years. Due to shrinkage (consolidation) and ripening of the dredged material, the thickness of the layers decreases. After some time a layer of, e.g., 1.0 m is reduced to 0.60 m. This method is favoured in order to facilitate the use of the former disposal site for agriculture, housing etc.

To encourage the ripening processes extra measures, such as the construction of a close network of drainage trenches, are often taken. The number of layers which are placed in this way varies.

The dredged material originates from various harbours and fairways and the proportion of sand and silt therefore varies. During the introduction of the dredged material, a vertical build-up of layers is formed (various layers of silt and sand are encountered). Some mixing also occurs in the horizontal plane; relatively coarse sandy material is found close to the discharge opening, while near the discharge weirs the material is relatively fine. Research carried out by the Public Works department of Rotterdam (Public Works of Rotterdam, 1986) has indicated that most of the mixing in fact occurs close to the discharge openings (see distribution of organic matter and fraction <16 μm in Fig. 1) and that over the rest of the discharge site the silt is distributed relatively homogeneously in the horizontal plane (uniform composition).

2.1 Ripening of the Dredged Material

2.1.1 Physical Ripening

After the final discharge the area is trenched. Because the underlying layers must be considered completely impermeable all the water which is released must be drained from the upper surface.

Drying out therefore begins at the upper surface. As a result of drying out the properties of the newly formed ground change rapidly; this process is termed physical ripening.

Physical ripening is defined as: "A soil forming process, whereby a sediment which is exposed to dryer hydrological conditions than those in which it is in equilibrium, is changed more or less irreversibly into a compacter, aerated and permeable material, of which the physical properties are dependent upon the new hydrological conditions and that can be termed soil." (Rijnierse 1983).

Rijnierse considered the physical ripening process as an initial soil-forming process. It is differentiated from other soil-forming processes by the rapidity with which the process takes place. Processes during which the rinsing in and out of

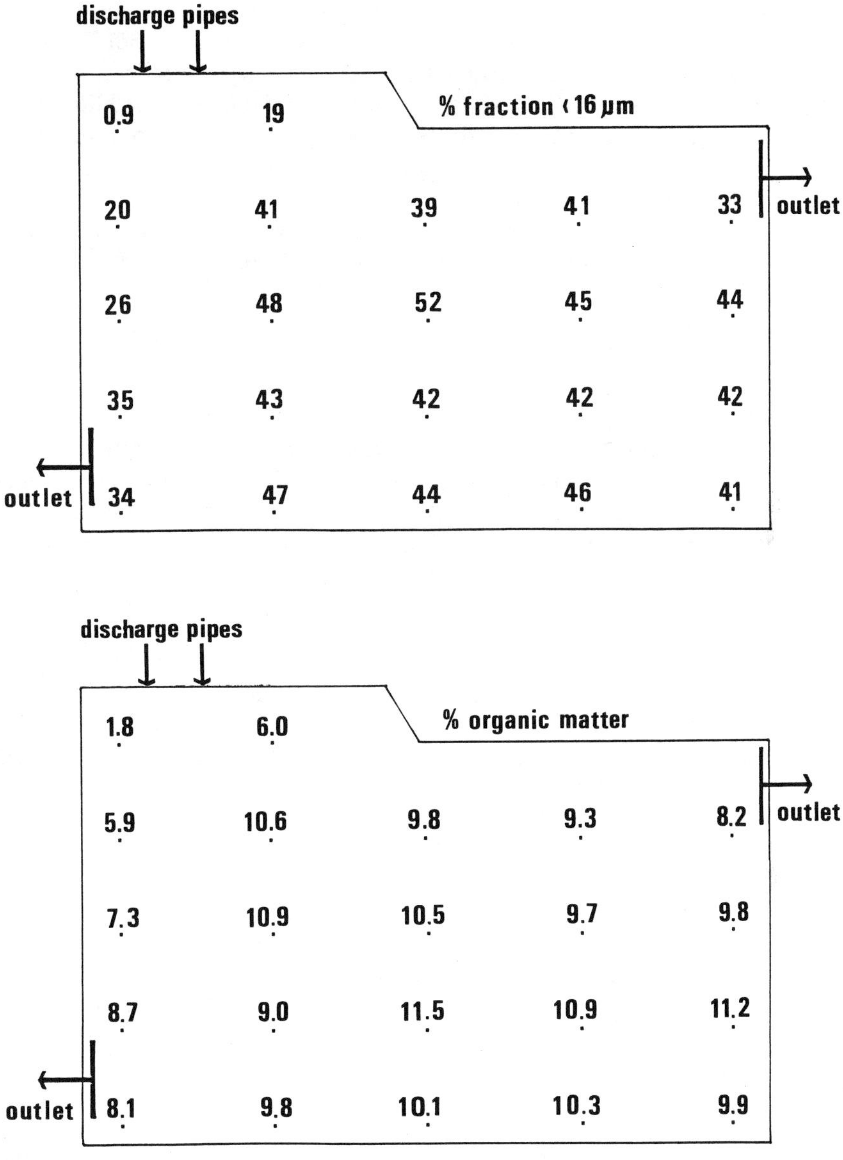

Fig. 1. Typical distribution of clay and organic matter over a section of a disposal site

organic matter and lime occur require considerably more time before observable changes in the profile occur. These slow occurring processes are also observed in the exposed sediments. The physical ripening is thus the first stage in a series of soil-forming processes (Pons and Zonneveld 1963).

In 1983 Rijnierse developed a mathematical model with which the physical ripening process could be simulated. The model was developed for the simulation of the physical ripening process of soil in the IJsselmeerpolders. On the basis of this model description it has been concluded that the model is also applicable to dredged material on land fill sites. From the calculations for dredged material disposal sites it appeared that the lutum and seepage loss had a great influence on the rate of the ripening process. Ripening proceeds more quickly if one or more years with a great evaporation excess occur at the beginning of the process (Municipality of Rotterdam et al. 1984).

The rate of ripening of the ground can be expressed in the form of the ripening factor and the n-value. The value is calculated as follows:

$$n = \frac{A - p(100 - L - H)}{L + b\,H}$$

in which:

A = the number of grammes of water per 100 g dry matter;
b = the relation between the water-binding capacity of a specific weight of organic matter and that for the same amount of lutum;
p = the number of grammes of moisture, bound by 1 g non-colloidal material, i.e. dry soil minus lutum and organic matter;
L = the number of grammes of lutum per 100 g dry matter;
H = the number of grammes of organic matter per 100 g dry matter.

The constants b and p differ for each sediment. For dredged material it appears that the combination b = 3 and p = -3 gives the best fit (Willet 1970). As the value decreases below 1.0 and approaches 0.5 the soil can be considered ripe. In this way at a site in the Broekpolder the ripening was followed from year to year in relation to a planting experiment (see Fig. 2). In the planting test it was investigated whether a commercial forest (mixed deciduous trees) could be supported on dredged material.

On the basis of Fig. 2, it can be concluded that the ripening is limited to the direct surface and that the rate of ripening decreases with depth.

As was shown by the experimental planting, ripening is influenced by vegetation. Investigation into the influence of vegetation on the ripening of dredged material used as fill indicated that vegetation has a great effect (Adviesbureau Arnhem 1980). The results are given in Table 1.

From Table 1 it can be concluded that deep-rooting vegetation (trees etc.) produces a greater effect. In the research, plots in both the Broekpolder and on the Eem island which were comparable with regard to soil mechanics parameters were used. A greater degree of drying out was reached in the Broekpolder.

2.1.2 Chemical Ripening

In contrast to the physical ripening very little research has been carried out on the chemical ripening process. From the literature it is known that in cases of exposed brackish or saline clay soils the composition of the cation complex of the clay

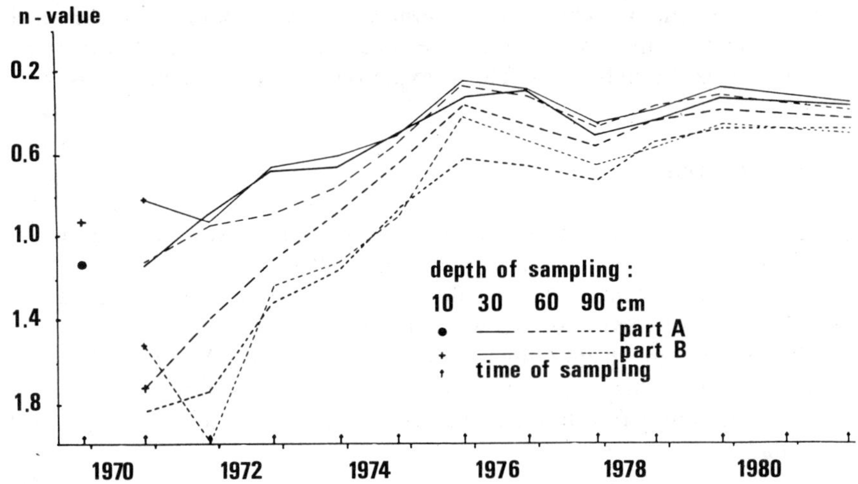

Fig. 2. Development of the n-value during the ripening of dredged material

Table 1. Overview of the results of field research on the effect of various types of vegetation upon the ripening process (Adviesbureau Arnhem, 1980)

	Eem island			Broekpolder		
	Ungrown	Weeds	Forest	Agricul-tural area	Sporting area	Forest
Depth of roots (m)	–	0.65	0.75	1.40	1.00	1.60
Ripening (m)	0.50	0.65	0.75	1.40	1.00	1.60
n-Value	0.8–0.9	0.8–1	0.7–0.9	0.6–0.8	0.7–0.8	0.6–0.8
Hydraulic resistance (m/24 h)	0.3–0.5	>10	>10	0.8–1.5	0.6–1.2	1–2
Broekpolder (1.20 m below the surface)						
Eem island (0.5 m below the surface)						
Organic matter (0.4–0.5 below the surface)	12–14%	10–12%	10–12%	8%	8%	5–6%

fraction changes (Na clay as compared to Ca clay). Moreover, initially aerobic or anaerobic degradation of organic material occurs in the dredged material.

In fact very shortly after the first appearance of vegetation there is an increase in the fraction of organic matter in the top-soil and a certain steady state condition can be observed. This does not occur in cultivated soil, in which there is a gradual net decrease in the organic matter until a stable condition is reached.

The appearance of oxygen and the associated mineralogical changes/transformation processes are particularly important with regard to the fate of the contaminants present in the dredged material. The dredged material which is introduced is anoxic. The anoxic character has a primary influence on the contaminant content in the water phase, i.e. the contaminant contents are usually lower than in the oxidic sediments. This is explained further in Section 5.4.

2.2 Desalination

In many cases harbour and fairways are located in an estuarine area as a result of which the dredged material is saline or brackish.

Initial salt contents of 400 to 6,000 mg $Cl^{-}l^{-1}$ are not unusual. With a view to the establishment of vegetation a lower salinity is desirable.

To obtain some insight into the nature and degree of the desalination of former dredged material disposal sites, the process has been studied by means of mathematical modelling (modified by Adviesbureau Arnhem 1978) using the following formula:

$$C = \frac{(\alpha O + A)\ (C_b) - k\ C_3}{(\alpha O + A)\ (e\alpha o/\varphi + At)} + k(t)\ C_3/\alpha O + A$$

and

$$k(t) = at + b$$

in which:

A = the groundwater drainage (m);
O = the surface water drainage (m);
K = the seepage (m s^{-1});
φ = the total volume of pore water in the discharge site per unit of surface area (l);
C_b = initial salt concentration (mg l^{-1});
C_3 = salt concentration in the seepage water (mg l^{-1});
t = time (s);
α = relation between the concentration and the area of the drainage of the pore water of the discharge site (–);
q,b = constants.

Calculations with this formula, using the values given in the figure, give the results shown in Fig. 3.

The calculation is valid for the same silt series. The desalination in the unsaturated zone cannot be described in a model, although measurements can be made (see Fig. 4). It may be concluded that desalination is a slow process. In the unsatured zone the desalination has progressed furthest as depth decreases in relation to the groundwater level and the influence of capillary transport of salt from the deeper layers is least.

In the unsaturated zone desalination increases in the upper layers where the influence of the capillary transport of salt in the deeper layers diminishes considerably.

The velocity of the desalination in the unsaturated zone can be accelerated by the application of a greater drainage depth.

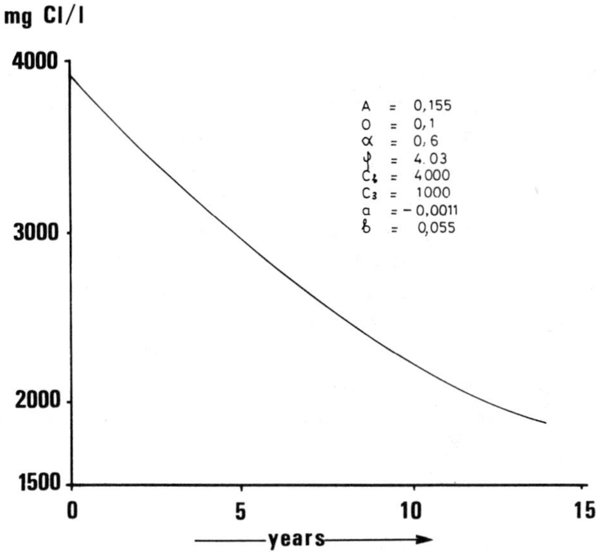

Fig. 3. Calculated desalination of dredged material in the saturated zone (in terms of chloride concentration)

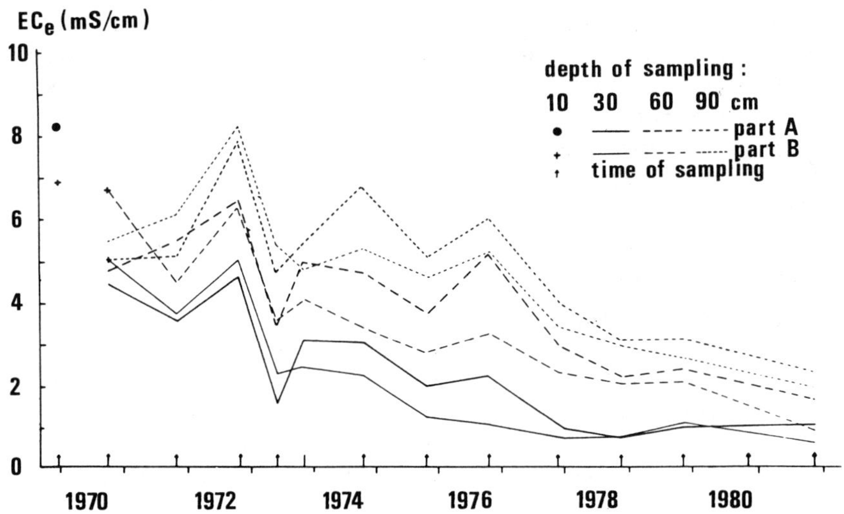

Fig. 4. Measured desalination of dredged material in the unsaturated zone (in terms of electrical conductivity)

2.3 Contaminants in Dredged Materials

Waterways in densely populated areas are mostly polluted with nutrient elements, heavy metals and organic contaminants. The pollutants originate from waste water discharges into the waterway, and they are mainly associated with the fine-grained sediment fractions. The processes in surface water include the breakdown of easily decomposable organic materials and the association of the pollutants with the suspended particles, leading to a residue that with respect to the contaminant pattern corresponds with that of sewage sludge, the residue of the waste water purification process. Local industrial discharges of specific pollutants can alter this pattern. In harbours situated near the sea the heavily polluted fluvial sediments are diluted with less contaminated marine sediments that move upstream. Thus, in the bottom sediments a gradual decrease in contaminant concentrations is observed. The diversity of dredged materials with respect to contaminant composition is very large.

2.3.1 Inorganic Contaminants

To establish the degree of pollution of dredged materials it is essential first to define the natural, uncontaminated levels of the components in the relevant sediments and soils. For sediments the composition of fossil shales, as compiled by Turekian and Wedepohl (1961) are often referred to. For some elements, e.g. Co, Cr, Cu and Ni, these values are considered too high as compared with recent, unpolluted sediments of fluvial and marine origin in The Netherlands (Salomons and Förstner 1984). Recent observations of heavy metal levels in uncontaminated clay soils are in good agreement with the background concentrations proposed by Salomons and Förstner (1984) (van Driel and Smilde 1981) (Table 2). Comparison of contaminant levels in sediments is complicated by the association of contaminants to the finest particles. In sediments of the same origin a linear relation is found between the concentration of the fine-grained fraction and the metal concentration. This can be accounted for in various ways:

1. Müller (1979) isolated the fraction $<2\ \mu m$, and determined in that fraction the contaminant concentration.
2. De Groot et al. (1982) investigated 10–20 subsamples and extrapolated the contaminant concentration to 50% $<16\ \mu m$.
3. Various other procedures have been proposed, all with the advantage that sediment contaminant levels can be compared irrespective of their particle size distribution.

The relative enrichment of heavy metals in dredged materials can be expressed as the quotient of the corrected contaminant concentration and its background concentration. Müller (1979) developed an expression based on the isolated particle size fraction $< 2\ \mu m$: the Geoaccumulation Index:

$$I_{geo} = \log_2 \frac{C_n}{1.5 * B_n}$$

Table 2. Heavy metal concentrations in upland dredged material disposal sites (mg kg^{-1} dry matter)

Element origin	As	Cd	Cr	Cu	Hg	Ni	Pb	Zn
Rotterdam-NL-1 (1964)[a,d]	135	13	499	242	8.2	53	399	1644
Rotterdam-NL-2 (1967)[a,d]	64	6	213	141	4.5	42	213	782
Rotterdam-NL-3 (1969)[a,d]	60	9	253	158	5.3	45	239	837
Rotterdam-NL-4 (1975)[b,e]	38	21	496	211	5.9	58	341	1296
Hamburg-FRG[b,f]	122	9	90	237	8.7	45	268	1238
Neckar-FRG (1979)[b,e]	nd[c]	41	611	269	nd[c]	75	210	726
Bridgeport-USA[b,g]	5	42	3360	4280	nd[c]	412	599	3260
Dutch background clay soils[b,h]	13	0.34	73	20	0.2	26	30	76

[a]Metal concentrations extrapolated to 50% $< 16 \mu m$.
[b]Metal concentrations as observed in situ.
[c]nd: Not determined.
[d]Van Driel (1980).
[e]Van Driel et al. (1985).
[f]Herms et al. (1984).
[g]Wolf et al. (1978).
[h]Van Driel and Smilde (1981).

where C_n is the concentration of the element n in the isolated fraction $< 2 \mu m$, and B_n is the geochemical background concentration in average fossil shale according to Wedepohl and Turekian (1961). The factor 1.5 accounts for variations in the geochemical background concentration. Most observations on metal contents, however, are based on the analysis of the entire, unfractionated sample, rather than on its $< 2 \mu m$ fraction. The association of heavy metals to the finest mineral and organic fractions has consequences for their mobility and bioavailability. In the USA the maximum cumulative applications of some heavy metals to arable soils are related to the cation exchange capacity (CEC) of that soil. At CEC of 0–5 mEq/100 g soil the tolerable maximum concentrations are: 1.5 mg kg^{-1} Cd, 40 mg kg^{-1} Cu, 17 mg kg^{-1} Ni, 170 mg kg^{-1} Pb and 80 mg kg^{-1} Zn. For CEC ranging from 0–15 mEq/100 g soil these values are double, for CEC over 15 mEq/100 g soil four times higher (Chaney 1983). In the Netherlands guidelines for contaminants in the soil are also being developed on the basis of CEC.

2.3.2 Organic Contaminants

Dredged materials contain numerous synthetic organic components. Most information is available on the chlorinated hydrocarbons, including pesticide residues and PCBs and the polycyclic aromatic hydrocarbons (PAH) (Table 3). Their concentrations are in the $\mu g\ kg^{-1}$ or in the mg kg^{-1} ranges.

Table 3. Concentrations of chlorinated organic contaminants in upland dredged material disposal sites (mg kg^{-1} dry matter) (Van Driel 1980)

	HCB	Aldrin	Dieldrin	Endrin	Isodrin	Telodrin	PCB[a]
Rotterdam-NL-2 (1967)	0.40	5.4	19.0	12.4	27.4	1.9	1.19
Rotterdam-NL-3 (1969)	0.27	4.4	4.9	1.3	4.4	0.8	1.97
Rotterdam-NL-5 (1969)	0.02	0.07	0.1	0.02	0.3	0.06	0.62
Rotterdam-NL-6 (1975)	0.10	0.01	0.05	0.01	0.01	0.01	2.05

[a]Sum of PCBs, calculated as Aroclor 1260.

More abundant is the composite group of oil and grease, comprising hydrocarbons, fatty acids, soaps, fats, waxes and mineral oils. Detailed information on concentrations and significance of the separate components is lacking. Concentrations of oil and grease range from 100 mg kg^{-1} to 4% (Disalvo et al. 1977, White et al. 1977). Because of large differences in biodegradability of the separate components, the composition of the oil and grease fraction gradually changes. Dramatic effects on plant growth have not been observed.

3 Transfer of Contaminants from Dredged Materials to Plants

The elevated contaminant levels observed in most dredged materials give rise to concern about their potential effects on biota. Occurrence of these effects depends on the bioavailability of the contaminants in the soil medium.

Factors controlling plant availability are the nature and properties of the contaminant, of the soil and of the plant species. Predominant soil factors are texture, concentration and nature of organic matter and pH. High pH, high clay and organic matter contents reduce the plant availability of most metals.

The organic micro-pollutants are mainly associated with the soil organic matter fraction. Transfer to the plant is also dependent on the physico-chemical properties of the organic contaminant molecule.

Dredged materials often have higher silt and organic matter contents than the corresponding soils. This results in lower bioavailabilities of the contaminants. In the long run the organic matter will be partly decomposed with a corresponding increase in bioavailability.

Soil pH is a very important factor in controlling mobility and availability of heavy metals. Dredged materials with high sulphide contents and with a relatively low calcium carbonate level turn acid during conversion of the wet soil from anaerobic to aerobic conditions. Dredged materials with high calcium carbonate levels remain neutral. In all soils losses of soluble calcium salts gradually lead to decalcification of the soils, and thus to lower pH. These processes may be accelerated under wet conditions.

Other soil factors such as soil salinity and redox conditions may affect heavy metal bioavailability in dredged materials. These conditions are subject to large variations in sediments dredged from brackish or saline waterways, and in partly dewatered soil profiles respectively.

Plants differ in their ability to accumulate heavy metals. Some metals (Cu, Fe, Mn, Mo, Zn) are essential to plant life, but the non-essential elements can also be accumulated and can interact with the essential elements in their biochemical functioning.

Monocotyledonous plant species accumulate less metals than dicotyledonous species (Cottenie 1981), but differences between species, varieties and subspecies may mask these differences. Moreover, growth conditions, nutrient status, temperature and transpiration rate determine growth rate and element uptake. The organs of the plant also differ in their ability to accumulate metals: for most plant species seeds and fruits accumulate less metal than leaves and roots. It has been attempted to simulate the processes in the rooting medium responsible for the flow rate of metals into the roots by chemical extraction procedures. Promising results have been obtained with the chelating extractants EDTA (Lakanen and Erviö 1971) and DTPA (Lindsey and Norvell 1978), and with neutral salt extractants based on $CaCl_2$ (Sauerbeck et al. 1985) and $NaNO_3$ (Häny and Gupta 1980), but the soil-plant system apparently is too complicated to be simulated by a chemical extractant. The neutral salt extractants mentioned give better results than the total metal contents of the soil in indicating differences in mobility and bioavailability of heavy metals.

The best soil extractant is the plant itself. Plant bioassays have been developed at the US Army Waterways Experiment Station, Vicksburg, Mississippi, USA. The objective of the plant bioassay is to verify the results of laboratory simulations and to indicate phytotoxicity and bioaccumulation of heavy metals in plants growing on dredged material disposal sites (Folsom, Lee, Bates 1981). The plant bioassay is conducted mostly under fresh water conditions with the marsh plant *Cyperus esculentus*, which may be cultivated in an aerobic or anaerobic rooting environment, under upland and flooded growing conditions respectively. For salt water marsh conditions *Spartina alterniflora* and some other salt water marsh plant species are used (Folsom et al. 1981). These plant bioassays give additional information to the soil chemical and physical data of the dredged materials, and supply a first indication of the effects of the contaminant into the food web. Transfer of contaminants from dredged materials to lower animals is being studied in animal bioassays (Simmers et al. 1985a, 1985b; Stafford 1986).

Lee et al. (1982) and van Driel et al. (1985) conducted plant bioassays with dredged materials from US and European waterways and rivers respectively. All dredged materials in these studies contained large amounts of calcium carbonate and showed pH values of 7.0 to 7.5. *Cyperus esculentus* accumulated under aerobic soil conditions more cadmium, copper, nickel and zinc than under anaerobic conditions (van Driel et al. 1985). In both studies *Cyperus* appears to be an excellent indicator for the cadmium and zinc status of the dredged materials. Response to the other metals was poor. Van Driel et al. (1985) compared the results of the bioassay of *Cyperus* with those of pot experiments containing the same substrates with lettuce, spring wheat, radish and red fescue grass. On a dry weight basis cadmium accumulation in lettuce and radish leaves is higher and it is lower in radish tuber, wheat straw, grass leaves and wheat grain than the accumulation in *Cyperus* leaves. For the elements chromium, copper, nickel, lead and zinc the correlation with *Cyperus* is poor.

The results of pot experiments with food crops cultivated on calcareous dredged materials indicate that in spite of the very high levels of heavy metals growth of most plant species is not impaired; the observed differences in yield cannot be attributed to the levels of the heavy metals in the substrate. Only for cereals (wheat, oats) and sugarbeets have growth reduction and manganese deficiency symptoms been observed, associated with low manganese concentrations in the leaves. This reduced manganese uptake is associated with soil factors such as high organic matter contents combined with high calcium carbonate content and possibly with high zinc concentrations in the substrate. The deficiency can easily be controlled by foliar sprays with manganous sulphate. The treatment results in increased yield and manganese concentrations (Smilde and van Driel 1982).

All food and feed crops cultivated on calcareous dredged materials contain more arsenic and heavy metals than crops on corresponding uncontaminated calcareous fluvial clay soils (Smilde and van Driel 1982). Accumulation varied among elements, descending in the order Cd, Zn, Cu, As, Hg, Pb, Ni.

Results from pot experiments have been verified in field experiments and in field observations. Tables 4, 5 and 6 give a survey of some results, and of the normal concentrations of the same crops grown on uncontaminated soils obtained in an inventory in The Netherlands (Wiersma et al. 1985).

Table 4. Composition of food crops cultivated at dredged material disposal sites in The Netherlands

	As		Cd			Hg		Pb		Zn	
	Av.[a]	Ref.[b]	Av.	Max[c]	Ref.	Av.	Ref.	Av.	Ref.	Av.	Ref.
Substrate[d] (mg kg^{-1} dry matter)	56	11	6.5	19.0	0.43	4.66	0.11	193	23	692	76
Crop (μg kg^{-1} fresh weight)											
Potato tuber	10	12	110	242	30	1.08	3.0	83	28	7820	2300
Carrot tuber	21	22	90	149	40	4.65	2.0	34	40	3500	1400
Onion	11	10	32	40	6	0.60	0.2	14	20	3590	2520
French beans	10	4	9	19	3	0.88	0.45	110	5	4650	2550
Brussels sprouts	4	–	53	92	–	1.78	–	92	–	10360	–
Spinach	33	11	191	232	60	3.72	5.00	182	90	9120	–
Lettuce	14	11	130	376	50	1.57	2.00	41	90	5130	2990
Endive	–	–	–	665	20	–	–	63	80	–	–

[a]Av. Average composition of crops grown under field conditions at contaminated dredged material disposal sites (Van Driel et al. 1977; Smilde, Van Driel 1982; Van Driel pers.commun. 1986).
[b]Ref. Average composition of crops cultivated in non-contaminated fields in the Netherlands (Wiersma et al. 1985; Smilde, Van Driel 1982, Van Driel pers.commun. 1986).
[c]Max. maximum cadmium concentrations observed in crops and dredged material.
[d]Substrate: average (maximum) concentrations of the substrates used for cultivation of the relevant crops.

Table 5. Maximum concentrations of five elements observed in some vegetables cultivated at contaminated dredged material disposal sites in The Netherlands (dry weight basis)

Crop	As ($\mu g\ kg^{-1}$)	Cd ($mg\ kg^{-1}$)	Hg ($\mu g\ kg^{-1}$)	Pb ($mg\ kg^{-1}$)	Zn ($mg\ kg^{-1}$)
French beans	186	0.19	68.0	2.73	13
Spinach	617	3.86	98.3	3.14	180
Lettuce	245	6.26	26.0	1.95	133
Endive	106	13.3	38.0	1.26	311
Brussels sprouts	61	0.66	70.0	6.09	71

Table 6. Composition of feed crops cultivated at dredged material disposal sites in The Netherlands

	As		Cd			Hg		Pb		Zn	
	Av.[a]	Ref.[b]	Av.	Max[c]	Ref.	Av.	Ref.	Av.	Ref.	Av.	Ref.
Substrate[d] ($mg\ kg^{-1}$ dry matter)	56	11	6.5	19.0	0.43	4.66	0.11	193	23	692	76
Crop ($\mu g\ kg^{-1}$ dry weight)											
Wheat	45	37	820	1640	90	1.9	4.0	230	140	96000	23000
Oats	121	190	510	650	90	15.0	8.0	2120	300	75000	15000
Barley	85	67	130	300	150	2.1	6.0	250	270	77000	26000
Green maize	182	170	1180	2550	430	27.6	16.0	2780	2200	156000	–

[a,b,c,d]: see Table 4.

It appears that the concentrations of As, Hg and Pb in the crops cultivated on the dredged material disposal sites are of the same order as the corresponding reference concentrations in The Netherlands. This illustrates the relatively low availability of these elements in this type of dredged material. Greve and Wegman (1977) studied accumulation of chlorinated hydrocarbons in four crops on the same type of contaminated dredged material. The 'drins' accumulated were in the order endrin, dieldrin, telodrin and isodrin and aldrin. The highest concentrations have been found in carrot tubers, lower concentrations in potatoes and onion bulbs, and no accumulation took place in brussels sprouts.

4 Quantifying of Contaminant Intake

4.1 General

First, it should be noted that very little is known about many of the contaminants present in dredged material. However, when assaying the potential risk associated with the redevelopment of dredged material disposal sites, due to the following factors, cadmium is considered to be the critical element, because:

1. Of the contaminants investigated, cadmium is the one which most readily transferred to the vegetation grown on dredged material disposal sites, as compared to crops cultivated in control soils (Smilde et al. 1982);
2. The margin between the permitted intake is given by the WHO standard and the normal average intake via the sources water, air and food is relatively small. Under certain conditions (heavy smoking), the WHO values may be exceeded;
3. Due to the increasing background loading of food, the average intake will tend to increase rather than decrease.

This means that when the intake of cadmium in edible vegetable matter appears to lead to possible health effects it is not necessary to repeat the calculations for other contaminants. When the intake of cadmium is below the guidelines which have been set, it is recommended that an estimate should be made of the risks which may arise from the simultaneous intake of more contaminants.

4.2 Norms

At present, soil norms have not been sufficiently developed and specified to be able to show a direct relation between a value in the soil and the possible occurrence of health effects. The assessment of whether crops grown on dredged material (in gardens or allotments) are suitable for human consumption could be based on crop norms.

In this case crop norms are derived norms aimed at protecting the human consumer. The crop norms are legally binding, i.e. products with an excessive contaminant content which are offered for sale can be withdrawn from the market.

The permitted total intake of contaminants is indicated in guidelines such as those which, amongst others, the World Health Organization has formulated. Part of the total intake is contained in food, including edible vegetable matter. The WHO norms aim to prevent the occurrence of harmful effects on the health of man. Such health norms already exist for a number of contaminants.

Existing health norms are most applicable to the risk-carrying intakes resulting from the consumption of crops grown on former dredged material disposal sites. On the basis of the occurrence or non-occurrence of a risk-carrying intake and, in the case of such an occurrence, including the circumstances of its occurrence, a statement can be made on the possible health effects.

4.3 Methodology

The quantification proceeds as follows:

1. The calculation of the margin in the weekly intake (MI) as the difference between the WHO value (PWI, provisional tolerable weekly intake) and the average weekly intake in food in the Netherlands (NI);
2. The calculation of the extra intake (EI) resulting from the intake of cadmium with edible vegetable matter grown on dredged material, as compared with the normal intake from edible vegetable matter;

3. The calculation of the risk-carrying intake as a result of the intake of cadmium with edible vegetable matter grown on dredged material (RI), as the difference between the extra intake and the average margin which is valid for the population of the Netherlands, in the weekly intake.

4.4 Basis

The quantification of intake is based on the following assumptions:

1. The WHO value expressed as PWI is 400 to 500 μg per week per head; the lowest value is used in the calculations;
2. The normal weekly intake is assumed to be equal to the average intake of cadmium with food per head. The contribution of drinking water is included, while in contrast, the contribution of smoking is omitted;
3. The amount and composition of edible vegetable matter consumed is assumed to be equal to the average amount per head of the Dutch population;
4. Consideration is limited to home cultivation of potatoes, pulses and other vegetables;
5. No potatoes, pulses or other vegetables other than those grown for home consumption are eaten;
6. In the calculations, the average cadmium content shown in crops grown on dredged material with an average Cd content is taken as the average for the intake of Cd with crops grown directly upon dredged material;
7. The maximum cadmium content found in crops grown directly upon dredged material with an average Cd content is taken as the upper threshold for the intake of Cd with crops grown directly upon dredged material;
8. The average for the intake of Cd with crops not grown directly upon dredged material is based on factors in relation to crops grown directly upon dredged material.

4.5 Values of the Parameters

In Table 7 the values for the parameters such as those used to quantify the risk-carrying cadmium content for crops grown directly upon dredged material are shown.

4.6 Quantification

The results of the quantification are shown in Table 8. The risk-carrying intake (RI) is determined as being the difference between the extra intake (EI) and the margin in the weekly intake (MI). The margin in the weekly intake is estimated to be 225 μg per week per capita. The intake of cadmium with crops grown directly upon dredged material is, by a factor of 10, higher than the intake via the average vegetable consumption; for maximum cadmium content the intake is higher by a factor of 20.

Table 7. Values of the parameters used in quantifying the cadmium intake

Parameter	Value	Unit
PWI cadmium	400[a]	μg
Normal intake of cadmium (NI)	175[b]	μg per week per head
Potatoes	1.28[c]	kg per week per head
Pulses	0.03[c]	kg per week per head
Other vegetables	1.20[c]	kg per week per head
Cd normal potatoes	33[c]	$\mu g\ kg^{-1}$ = ppb (fresh weight)
Cd normal pulses	3[c]	$\mu g\ kg^{-1}$ = (fresh weight)
Cd other normal vegetables	12[c]	$\mu g\ kg^{-1}$ = (fresh weight)
Cd dredged material average, potatoes	110[d]	$\mu g\ kg^{-1}$ = (fresh weight)
Cd dredged material average, pulses	8[d]	$\mu g\ kg^{-1}$ = (fresh weight)
Cd dredged material average, other vegetables	390[d]	$\mu g\ kg^{-1}$ = (fresh weight)
Cd dredged material maximum, potatoes	242[d]	$\mu g\ kg^{-1}$ = (fresh weight)
Cd dredged material maximum, pulses	19[d]	$\mu g\ kg^{-1}$ = (fresh weight)
Cd dredged material maximum, other vegetables	692[d]	$\mu g\ kg^{-1}$ = (fresh weight)

[a]WHO (1974).
[b]CCRX (1983).
[c]Ellen (1977).
[d]IB (1977).

Table 8. Intake of cadmium with vegetables grown directly on dredged material (μg per head per week)

Cd intake edible vegetable matter	Crops grown directly on dredged material	Average crop composition	Extra intake in dredged material (EI)	Risk-carrying intake (RI)
For average Cd content in edible vegetable matter	610	60	470	245
For maximum Cd content in edible vegetable matter	1,140	60	1,080	855

The data presented indicate that the total estimated cadmium intake for the consumption of edible vegetable matter, and also that grown on a dredged material disposal site, exceed the WHO value for the average consumer (factor 0.5). Under certain conditions the WHO value can be exceeded twice as determined from the calculation of the risk-carrying intake (the measure for excess above the WHO value) by maximum cadmium contents present in crops grown for home consumption.

In addition, factors which are not related to the use of soil developed on dredged material can also lead to the higher WHO values. In this context smoking and a high consumption of certain, heavily contaminated foodstuffs, including offal are considered (CCRX 1983).

5 Transfer Reduction

5.1 Introduction

In the previous sections it was shown that transfer of contaminants from the soil to organisms occurs and that there is a possibility of increased risks for mankind.

The question arises as to how the transfer of contaminants and the risks accompanying this transfer can be minimized.

The transfer of contaminants can be reduced by the addition of lime, by immobilization of contaminants and by covering the contaminated soil.

Kuntze et al. (1984) demonstrated that liming of acid dredged material disposal sites leads for some crops to better growth, and for most heavy metals to reductions of 10–90% of their concentrations in seven vegetable crops. They recommend to control the pH of the disposal sites by periodic liming. The success of liming is restricted to acid-reacting soils, to less sensitive crops and to moderately contaminated disposal sites.

Immobilization of metal-binding, ion-exchange resins does not seem to be a practical possibility. Covering the contaminated substrate with a layer of uncontaminated topsoil proved to be more effective. The thickness of this layer has to be accomodated to the rooting depth of the relevant crops (Kuntze et al. 1984). On the whole, at present long-term experience of the effects of these measures to reduce transfer to plants is relatively sparse. Covering layers are already in use at waste disposal tips and at disposal sites for slag, fly ash and other waste products, and in the case of dredged material, on land intended for residential use.

5.2 Covering Layers

When consideration is given to the covering of a dredged material disposal site, the choice of the structure of this layer will be based upon the requirements to be formulated with regard to the future use of the site.

In general, the following demands apply to the insulation of a soil covering layer (Laboratorium voor Grondmechanica 1986):

1. As much prevention of direct contact with the contaminated soil as possible;
2. As much prevention as possible of percolation of the precipitation excess into the contaminated soil (with a view to preventing further dispersion);
3. As much prevention as possible of upward-directed transport of contaminants from the contaminated soil to the soil surface.

The maximum compliance with these functional demands can lead to the application of a cover composed of layers, which from the surface layer consists of:

1. Usable layer: the uppermost layer of the covering, which must comply with the demands relating to the use of the site;
2. Drainage layer: for the drainage of the percolation through the usable layer insofar as this has not infiltrated into the underlying layer;
3. Underlying layer: to prevent the percolation of precipitation into the contaminated soil (6) and the upward seepage of contaminated water;
4. Supporting layer: to balance irregularities in the subsoil, especially in the case of a thin layer and a fragile, impermeable layer (3);
5. Gas drainage: to expel gas if there is a possibility of considerable gas production in the soil to be isolated or if there is significant transport of contaminants in the gas bubbles;
6. Contaminated soil.

At sites for the disposal of dredged material the dredged material itself can fulfill the function of the impermeable layer (3). A well-consolidated soil derived from dredged material has a permeability (10^{-9} to 10^{-10} ms^{-1}) which can be reached with an upper covering of natural material (Ministerie van Volksgezondheid en Milieuhygiene 1982; Municipality of Rotterdam et al. 1984 and Laboratorium voor Grondmechanica 1981).

After this both the supporting layer and the gas-draining layers (gas-draining layers are especially required in the case of household waste sites) can be omitted.

The nature of the usable layer is closely related to the intended function of the site. In this chapter agricultural use is considered.

The usable layer must also be sufficiently thick and humus-rich soil, clay or peat can be considered. It must be possible to place a drainage layer between the usable layer and the contaminated soil.

The layer also has the function of (1) draining the precipitation excess; (2) interrupting capillary transport; (3) insofar as possible, preventing root penetration.

In this way a layered structure is built up with a sand drainage layer over the contaminated soil and a usable layer above the sand, both layers being of sufficient thickness.

In order to obtain an effective system the drainage depth must lie at the interface between the drainage layer and the contaminated soil.

5.3 Upward Transport of Contaminants

For the determination of the effectivity of the covering layer it is especially important to ascertain to what degree the combination of downward (percolation) and upward (capillary) transport in the covering layer may lead to increased concentrations in the covering layer. Moreover, diffusion of contaminants from the dredged material may also be involved.

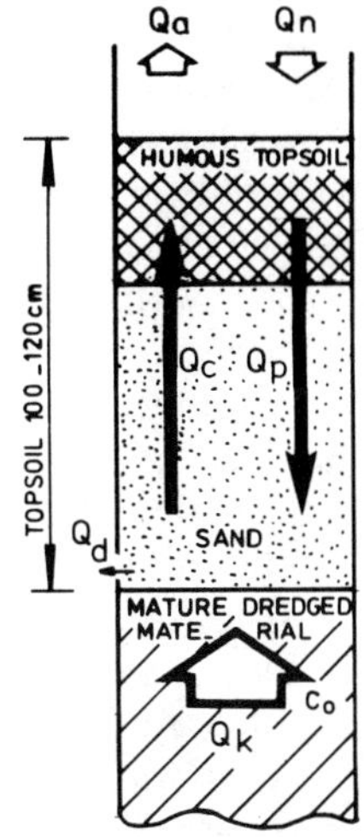

Fig. 5. Diagram representing flows of water and solute in dredged material covered with a sand layer and a humus topsoil

To obtain some insight into this process, numerical and analytical calculations are carried out taking into account diffusion, dispersion and processes of adsorption and desorption (Heidemij Advies 1983).

A schematic representation as shown in Fig. 5 is derived from these calculations. No consideration was given to the processes of solution or precipitation, which might be expected to occur around the level of the water table. The calculations are made for an 'average' situation which is as carefully chosen as possible. The situation which can be called average has:

1. A specific drainage resistance of 100 days and a seepage of 0.25 mm day^{-1};
2. A dry density of 1,800 kg m^{-3} for sand;
3. A dispersion distance in the sand layer of 0.025 m;
4. Diffusion coefficient of 10^{-11} m^2 s^{-1} in the sand layer;
5. A soil/contaminant system in which completely reversible, linear adsorption occurs. It is hereby assumed that the adsorption interaction between the contaminants in the water phase can be reflected by a K value of 1 for the sandy material, 10 for the humus-containing top layer and 1,000 for the silt. Thus, the K value is equal to the CEC value of the solid matter times the density of this phase divided by the electrolytic concentrations of the solids (mEq m^{-1}) in the dissolved phase (m/m);
6. No degradation of the contaminants.

The calculations are carried out analytically for a stationary evaporation situation in a dry summer with a total capillary flow of 0.12 m. The results of the analytical calculation are compared with a mathematical/numerical calculation with the model ONZAT (Van Drecht 1982) over a period of 10 years, in which varying meteorological conditions have occurred.

Comparison of the results of the analytical sensitivity calculation and the mathematical calculation for a number of years showed a good correlation.

The results, produced by the calculations carried out and translated to the possible practical situations, led to the following conclusions:

1. The combination of the upward transport of contaminants resulting from moisture transport and diffusion and the downward transport resulting from washing out by clean precipitation produces a quasi-stationary contamination profile in the groundwater of the covering layer (see Fig. 6).
2. The concentration of contaminants in the groundwater (taken as 100% in the pore water in the dredged material) is reduced by a factor of 10 for each 20-cm increase, under the condition that the adsorptive behaviour of the contaminants is limited and there is a minimal adsorption capacity for the covering layer. This is valid, for example, for benzene in normal filling sand. For contaminants with a greater adsorption capacity (such as more complex organic micro-pollutants), this reduction in concentration will be quicker, i.e. it will take place over a shorter distance.
3. Due to meteorological variations, the concentration profile outlined above can fluctuate by up to about 30 cm depending on the adsorption behaviour of the contaminant in question (see Fig. 6). This means that a concentration of 1% may occasionally occur in the groundwater about 70 cm above the dredged material, while under average conditions this concentration is found approximately 40 cm above the dredged material.
4. With regards to the hardened surfaces it can be supposed that the concentrations are of the same magnitude as those in the 'green' profile under consideration. The reduction in the precipitation is almost compensated by the reduction in the evaporation. Moreover, due to shallow horizontal groundwater movements during wet periods, a certain degree of washing out occurs under roads and buildings.

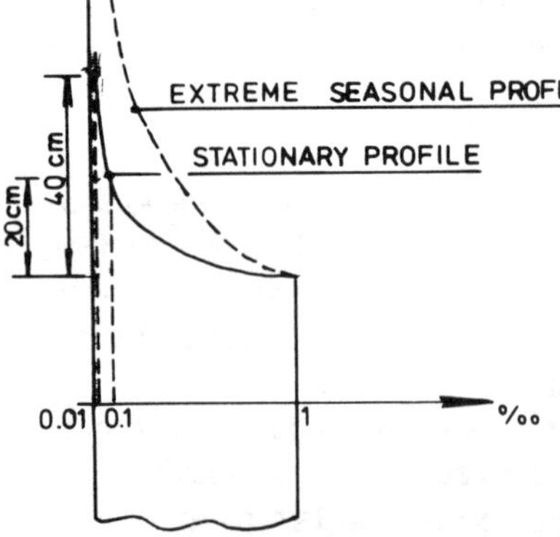

Fig. 6. Calculated profiles of contaminants in groundwater in the situation of Fig. 5 under normal and extreme seasonal conditions

5.4 Quality of the Groundwater

Three zones can be distinguished in the uppermost layer of the soil:

1. The unsaturated oxidized zone (soil water);
2. The transition zone (gley zone);
3. The saturated reducing zone (groundwater).

To a large extent the concentrations of contaminants in the groundwater determine the intake of contaminants by plants.

The deeper groundwater is anaerobic (oxygen-free) unlike the groundwater around the water table, which is aerobic (oxygen-rich). This difference in the macro-chemistry (in addition to the oxygen content, the redox potential, and pH also differ) can lead to significant changes in the concentration of contaminants in the water phase. This is valid, in particular, for the concentration of heavy metals, which is largely determined by the macro-chemical conditions.

In the framework of the Project Note/EIR for the large-scale site for the disposal of dredged material (Municipality of Rotterdam et al. 1984), an extensive investigation was carried out on the concentrations in the deeper groundwater at the site. Following this an investigation was carried out on the concentrations in the aerobic, unsatured zone (groundwater). The metal concentrations in the groundwater are calculated with the chemical model CHARON. In making the calculations the points of departure are as follows:

1. The movements of the groundwater are determined on the basis of a generalized schematization combining the water movements and meteorological data over a long period (1951–1980) and during an extremely dry year (1976). For the investigative study it was not deemed necessary to make a more detailed schematization;
2. The composition of the dredged material is given in Table 9 for a limited number of parameters which were considered most relevant. To determine how much influence a 'clean cover layer' of filling sand has on the quality of the ground- and

Table 9. Composition of dredged material (dry weight basis)

Parameter	Concentration of dredged material (mg kg^{-1})	Concentration[a] cover layer (mg kg^{-1})
Cd	13	1
As	22	2
Zn	794	30
Cu	118	5
Cr	179	20
Pb	216	14
oil	1,796	25
EOCl	8.67	0.2
PCB's	1.36	–

[a]Quality of sandy material derived from maintenance dredging offshore (at sea) (Municipality of Rotterdam et al., 1984).

Table 10. Calculated concentrations in the anaerobic groundwater under the phreatic surface with and without a covering layer ($\mu g\ l^{-1}$)

Chemical System	Metal	Harbour silt profile depth m.v. (cm)	Concentration	Covering sand depth-m.v. (cm)	Concentration
Unsaturated,		10– 30		10– 30	
oxidized	Cd		2 - 4		0.4– 0.7
	Zn		70 -350		2 - 4
	Cu		3 - 4		0.3– 0.5
	Cr		100 -200		5 - 7
	As		10 - 15		3 - 5
Gley zone		30– 60		30– 80	
	Cd		0.2– 0.5		0.1– 1
	Zn		10 -400		1 - 1.5
	Cu		0.2– 4		0.3– 1
	Cr		10 - 50		4 -40
	As		3 - 8		2 - 4
Reducing shallow		60– 70		80–100	
	Cd		0.4– 0.5		0.4– 0.5
	Zn		5 - 15		5 -15
	Cu		1 - 2		0.5– 1.5
	Cr		25 - 50		25 -50
	As		20 - 40		7 - 1
Deep		90–100		100–120	
	Cd		0.4– 0.5		0.4– 0.5
	Zn		5 - 15		5 -15
	Cu		1 - 2		0.5–15
	Cr		25 - 50		25 -50
	As		10 - 50		20 -50

surface waters, a covering layer of the quality shown in Table 9 was also taken into consideration in the model.

3. For the anaerobic zone under the phreatic surface the concentrations as given in Table 10 for the deep reducing zone are used;
4. The surface water is schematized as a basin into which there is a movement from the aerobic and anaerobic zones in a relation which has yet to be determined.

The outcome of the calculation is shown in Table 10. The application of a sandy, ‘clean’ top layer does, however, lead to significantly lower concentrations. Field measurements to verify the calculated values are lacking, so a calibration of the model results is not possible. Due to other technical complications nothing can be done about this in the short term.

5.5 Covering Layer on Disposal Intended for Agricultural Use

Practical and experimental experience on the application of covering layers on disposal sites intended for agricultural use are equally limited.

In general, if a former dredged material disposal site is covered, the contaminant levels in the crops are initially reduced. Indicative full-scale experimental data show that with a clay covering layer of 70 cm, a reduction in the cadmium content by a factor of 10 can be achieved in carrots and potatoes (Van Driel 1985). The reduction occurs because, amongst other factors, the concentration of contaminants in the moisture content in the rooting zone is relatively low or even 0 in the case of organic contaminants. It must, however, be noted that this is valid only for covering layers composed of moderately fine sand. Clayey covering layers have a better capillary transport capacity, but also a greater adsorption capacity, so that certainly in the long term the effectivity of the covering layers may decline. The use of a so-called sandwich construction, whereby for example, a thick capillary breaking sand layer is applied between the contaminated soil and the covering layer, may possibly extend the effectivity over a longer period.

Such a construction requires that the groundwater table be continuously maintained at or below the boundary between the coarse sand and the contaminated soil. Further experimental research is certainly required.

In Table 11 the risk-carrying intake resulting from the consumption of homegrown produce cultivated both directly and indirectly on dredged material is shown. In the second case a covering layer consisting of 70 cm of clay is used.

Table 11. The effect of cultivation of foodcrops on dredged material, with and without a covering layer, on the cadmium intake (in μg per head per week)

Covering layer		Cd intake in edible vegetable matter	Extra intake (EI) from dredged material	Risk-carrying intake
None	average	610	470	245
	maximum	1.140	1.080	855
70 cm clay	average	25	-35	-190
	maximum	120	60	-105

For the sake of completeness Table 11 also shows the Cd intake for edible vegetable matter in the situations given.

According to the assumptions made, the consumption of edible vegetable matter not grown directly upon dredged material, but upon a covering layer of 85 cm of clay, will most probably not lead to the occurrence of detrimental effects on human health. In this case maximum intake of Cd with edible vegetable matter results in its doubling as compared to the intake with the average vegetable consumption; with an average content this intake is halved. It must, however, be emphasized that the conclusions stated above are based upon incidental measurements.

6 Evaluation

As shown upland disposal of dredged material yields an excellent substrate for plant growth. This is caused by the relatively high silt and organic matter contents, the presence of nutrients and the favourable texture of the dredged material. However, due to the presence of contaminants in the dredged materials agricultural use is limited. By transfer of contaminants from soil to plant accumulation occurs in the edible parts of the plants. The accumulation in the plant differs from plant species to plant species and from contaminant to contaminant. Green plant leaves accumulate more heavy metals than other plant organs. Most organic contaminants are less accumulated than the heavy metals.

The driving forces behind the transfer processes from soil to plant are insufficiently understood. Soil factors such as high pH, high organic matter and silt contents tend to decrease transfer of metals to the plant, but a simple regression between contaminant concentration in the soil and in the plant is not available.

For this reason official guidelines for contaminant concentrations in soil have not yet been published. On the basis of plant experiments, field observations, and the maximum permissible levels of contaminants in food crops, provisional guidelines have been formulated. These provisional guidelines depend both on soil type and on the type of agricultural use. If these provisional guidelines are exceeded a more detailed investigation has to be performed. This investigation leads to a risk analysis. On the basis of the risk analysis, one can decide whether the contamination of the soil is within allowable limits or not. The risk analysis performed in this study shows that the WHO norms for cadmium may be exceeded. This means that in some cases agricultural use of upland disposed dredged material is not allowable and that remedial actions have to be undertaken. Soil cleanup does not appear to be feasible on economic grounds. In this chapter two measures in relation to agricultural use of the soil have been mentioned: liming and covering of the contaminated soil.

Liming of contaminated soil may be effective if the transfer of metals is sufficiently depressed. In calcareous river sediments, e.g. dredged from the Rotterdam harbour basins, liming apparently has no effect.

The use of a soil cover layer system seems to be more effective. Model calculations show that in the case of a soil cover of at least 1.0 m there is no net transfer of dissolved contaminants from the contaminated soil to the clean soil cover. A prerequisite is that the groundwater level is fixed at 1 m below the surface. Experiments with plants show a decrease in the accumulation of metals by the plants, but roots of deep-rooting plants can still reach the contaminated soil layer and contribute to the metal status of the above-ground plant organs.

Long-term experience with soil cover layers is still not available. A repeated risk analysis with data of crops grown on a soil cover layer showed that the exceeding of the WHO norms for cadmium may be prevented.

If soil contamination is to be controlled, the decision will depend on many factors, e.g. the outcome of risk analysis, the cost of cleanup operations, the future destination of the soil etc. Applying soil cover systems seems to be a promising option, but more study is needed.

References

Adviesbureau Arnhem (1978) Hydrological research Broekpolder (in Dutch). Arnhem, NL

Adviesbureau Arnhem (1980) Ripening of dredged material in relation to flora elements (in Dutch). Arnhem, NL

CCRX (1983) Metingen van radioactiviteit en xenobiotische stoffen in het biologisch milieu in Nederland. VROM, Leidschendam, NL

Chaney RL (1983) Potential effects of sludge-borne heavy metals and toxic organics on soils, plants and animals, and related regulatory guidelines. Beltsville Agricultural Research Centre, Beltsville, USA

Combs SM, Dowdy RH, Gupta SC, Larson WE, Gast RG (1982) Chemical properties and mineralogical characteristics of selected dredged materials. J Environ Qual 11:255–261

Cottenie A (1981) Trace metals in agriculture and in the environment. IOWNL, Bruxelles

Disalvo LH, Guard HE, Hirsch ND, Ng J (1977) Assessment and significance of sediment associated oil and grease in aquatic environments. Technical Rep D-77-26. US Army Eng Waterways Exp Stn, Vicksburg Miss, USA

Drecht G van (1982) Model van het niet stationaire transport van water en een opgeloste stof in een kolom grond (model "ONZAT"). RID, Leidschendam, NL

Driel W van (1980) Contaminanten in havenslib Samenstelling van de bovenlaag van 22 loswallen. Inst Bodemvruchtbaarheid, Haren, NL

Driel W van, Smilde KW (1981) Heavy-metal contents of Dutch arable soils. Landwirtsch Forsch Sonderh 38:305–313

Driel W van, Smilde KW, Luit B van (1985) Comparison of the heavy metal uptake of *Cyperus esculentus* and of agronomic plants grown on contaminated Dutch sediments. Misc paper D-83-1. US Army Eng Waterways Exp Stn, Vicksburg Miss, USA

Duvoort van Engers LE (1983) Zuiveringsslib, produktie, bestemming en kwaliteit. H_2O 16:586–587

Ellen G (1977) Het voorkomen van zware metalen in voedsel. Voeding 38:443–460

Folsom jr BL, Lee CR, Bates DJ (1981) Influence of disposal environment on availability and plant uptake of heavy metals in dredged material. Technical Rep EL-81-12. US Army Eng Waterways Exp Stn, Vicksburg Miss, USA

Gemeente Rotterdam, Rijkswaterstaat, Openbaar Lichaam Rijnmond (1984) Fysische processen, consolidatie, ontwikkeling, draagkracht en waterhuishouding depot. Rotterdam, NL

Gemeentewerken Rotterdam (1986) De verdeling van de minerale fractie en organische stof op een loswal. Rotterdam, NL

Greve PA, Wegman RCC (1977) Gechloreerde koolwaterstoffen en extraheerbaar organisch gebonden chloor in opgespoten Rotterdams havenslib en daarop verbouwde gewassen. RIV Rapport 25/77 Tox-ROB. Bilthoven, NL

Groot AJ de, Zschuppe KH, Salomons W (1982) Standardization of methods of analysis for heavy metals in sediments. Hydrobiologica 92:689–695

Gupta SC, Arson WE, Gast RG, Combs SM, Dowdy RH (1978) The agricultural value of dredged material. Technical Rep D-78-36. US Army Eng Waterways Exp Stn, Vicksburg Miss, USA

Gupta SC, Larson WE, Gast RG, Combs SM (1980) Potential use of dredged material to improve marginal agricultural soils. Physical properties and plant growth. J Environ Qual 9:653–660

Häny H, Gupta SK (1980) Ein Vergleich verschiedener methodischer Ansätze zur Bestimmung mobiler Schwermetallfraktionen im Boden. Landwirtsch Forsch Sonderh 37:267–274

Heidemij advies (1983) Verticaal transport van chemische stoffen in schone afdekgrond op gerijpte baggerspecie nr 630/3156.1. Arnhem, NL

Herms U, Scheffer B, Bartels R (1984) Schwermetallgehalte in Boden und Pflanzen von Hafenschlickspulfelder. Statusbericht dreijähriger Untersuchungen. Fachseminar Baggergut Hamburg: 143–166

Kuntze H, Herms U, Pluquet E (1984) Bodentechnologische Masznahmen zur Sanierung schwermetallbelasteter Spulfelder. Statusbericht dreijähriger Untersuchungen. Fachseminar Baggergut Hamburg: 287–307

Laboratorium voor Grondmechanica (1981) Toepassing gerijpt havenslib in de dijkbouw. Delft, NL

Laboratorium voor Grondmechanica (1986) Risicoanalyse en isolatietechnieken bij bodemsanering. Delft, NL

Lakanen E, Ervio RA (1971) A comparison of eight extractants for the determination of plant available micronutrients in soils. Acta Agr Fenn 123:223–232

Lee CR, Folsom BL, Engler RM (1982) Availability and plant uptake of heavy metals from contaminated dredged material placed in flooded and upland disposal environments. Environmental International 7:65–71

Lee CR, Smart RM, Sturgis TC, Gordon RN, Landin MC (1978) Prediction of heavy metal uptake by marsh plants based on chemical extraction of heavy metals from dredged material. Technical Rep D-78-6. US Army Eng Waterways Exp Stn, Vicksburg Miss, USA

Lindsay WL, Norvell WB (1978) Development of a DTPA soil test for zinc, iron, manganese and copper. Soil Sci Soc Am J 42:421–428

Ministerie van Volksgezondheid en Milieuhygiene (1979) Definitief Milieu-effectrapport berging baggerspecie. Leidschendam, NL

Ministerie van Volkgezondheid en Milieuhygiene (1982) Bodemafdichting met de in de natuur voorkomende materialen. Staatsuitgeverij, 's-Gravenhage

Muller G (1979) Schwermetalle in den Sedimenten des Rheins. Umschau 79:778–783

Municipality of Rotterdam, Ministry of Transport and Waterways (1984) Project note/EIR for the large scale site for the disposal of dredged material (in Dutch). Rotterdam, NL

Pons LJ, Zonneveld IS (1963) Soil ripening and soil classification. ILRI Publ el3, Wageningen

Rijnierse K (1983) A model simulation of the physical ripening of soils in the IJsselmeerpolders (in Dutch). Rijksdienst voor de IJsselmeerpolders, publicatie nr 52. Staatsuitgeverij, 's-Gravenhage

Salomons W, Förstner U (1984) Metals in the hydrocycle. Springer, Berlin Heidelberg New-York Tokio: 77–78

Sauerbeck DR, Stypereck P (1985) Significance of ion species and concentration for the determination of "available" cadmium in soils. Newsletter FAO European Cooperative Network on Trace Elements: 21–25. Ghent, Belgium

Simmers JW, Rhett RG, Crawley DK (1985) Prediction of the bioavailability of heavy metals to animals colonizing an intertidal wetland created with contaminated dredged material. Heavy metals in the environment Athens: 208–210

Smilde KW, Driel W Van, Luit B van (1982) Constraints in cropping heavy-metal contaminated fluvial sediments. Sci total environ 25:225–244

Stuurman FJ, Peeters JP (1986) Proefveldresultaten over de periode najaar 1982 – najaar 1985. Rijksinstituut voor onderzoek in de bos- en landschapsbouw "De Dorschkamp" rapport nr 6. Wageningen, NL

Turekian KK, Wedepohl KH (1961) Distribution of the elements in some major units of the earth's crust. Bull Geol Soc Am 72:175–192

White CT (1978) An evaluation of oil and grease contamination associated with dredged material containment areas. Technical Rep D-77-25. US Army Eng Waterways Exp Stn, Vicksburg Miss, USA

Wiersma D, Goor BJ van, Veen NG van der (1985) Survey of cadmium, lead, mercury and arsenic in Dutch crops and their corresponding soils. Rapport 8-85. Inst Bodemvruchtbaarheid, Haren, NL

Willet JR (1970) Eindrapport rijpingsproef Broekpolder. Heidemij Arnhem, nr 70.020

Wolf PL, Gallagher JL, Pennington CH (1978) Field bioassay test for detecting contaminant uptake from dredged material by marsh plants. Misc Paper D-78-6. US Army Eng Waterways Exp Stn, Vicksburg Miss, USA

World Health Organization (1974) Toxicological evaluation of certain food additives with a review of general principles and of specifications. Technical report Series no 539

Biological Assessment of Environmental Impact of Dredged Material

W. AHLF[1] and M. MUNAWAR[2]

1 Introduction

Aquatic ecosystems contain four potential reservoirs for contaminants: water, sediment, particulates and biota. In particular, sediments represent a very concentrated pool of toxic substances and may themselves represent an environmental problem. The dredging and/or disposal of sediments in dumping areas may adversely affect water quality and the health of aquatic organisms.

Since dredging operations and subsequent disposal cause resuspension of sediments, an increased load of suspended matter affects the water column and the organisms residing in it. The greatest potential for impact with solid-phase material is expected on benthic or epibenthic organisms because these organisms are in close contact with deposited sediment for long periods of time. Thus, to assess the environmental impact of contaminated sediments, information is needed on sediment quality, sediment-water interactions and the resulting impact on organisms inhabiting water columns and sediment.

Chemicals interact in various ways with organisms, and these interactions cannot be predicted with precision using chemical analyses alone (Cairns and van der Schalie 1980). Although bulk chemical characterization alone provides no indication of biological damage, such data are needed to determine the origin, degree and nature of contamination. To allow the prediction of potential environmental effects on aquatic organisms during and after dredged material disposal, different bioassays are required. A bioassay is defined as "any test in which organisms are used to detect or measure the presence or effect of one or more substances or conditions" (EPA 1973). Two general types of dredged material bioassays are of interest: those addressing liquid-phase or water-column effects, and those concerned with solid-phase effects, which are ingested by many aquatic organisms (Engler 1980).

It is important to note that there are different vectors of contaminant uptake for aquatic organisms (Luoma 1983). Therefore, predicting the effect of a pollutant in an ecosystem requires a selection of suitable test organisms. From the viewpoint

[1]Technische Universität Hamburg-Harburg, Arbeitsbereich Umweltschutztechnik, Eißendorfer Straße 40, 2100 Hamburg 90, FRG

[2]Great Lakes Laboratory for Fisheries & Aquatic Sciences, Fisheries & Oceans Canada, Canada Centre for Inland Waters, Burlington, Ontario, Canada L7R 4A6

of an ecosystem, it may be preferable to use a wide variety of organisms belonging to different trophic levels, including the primary producer.

This chapter proposes a strategy for determining the potential environmental impact of dredged material and comparisons with present bioassessment procedures are undertaken.

2 Strategy

In Fig. 1, the proposed strategy for risk assessment of dredged material is shown. The flowchart is based on tests used in the assessment of the ecotoxicological effects on aquatic life (Calamari et al. 1979). The scheme is refined in order to minimize the environmental impact of dredged material. The solid lines in the flowchart indicate the direction of increasing difficulty and specificity of each level. The dotted lines show that at each level, a risk assessment is possible when the results of the test represent either a toxic or hazardous bioconcentration of contaminants. In relation to aquatic disposal of dredged material, the assessment strategy can be described as an order of increasing effectiveness in the evaluation of the environmental impact, but unfortunately with increasing monitoring costs.

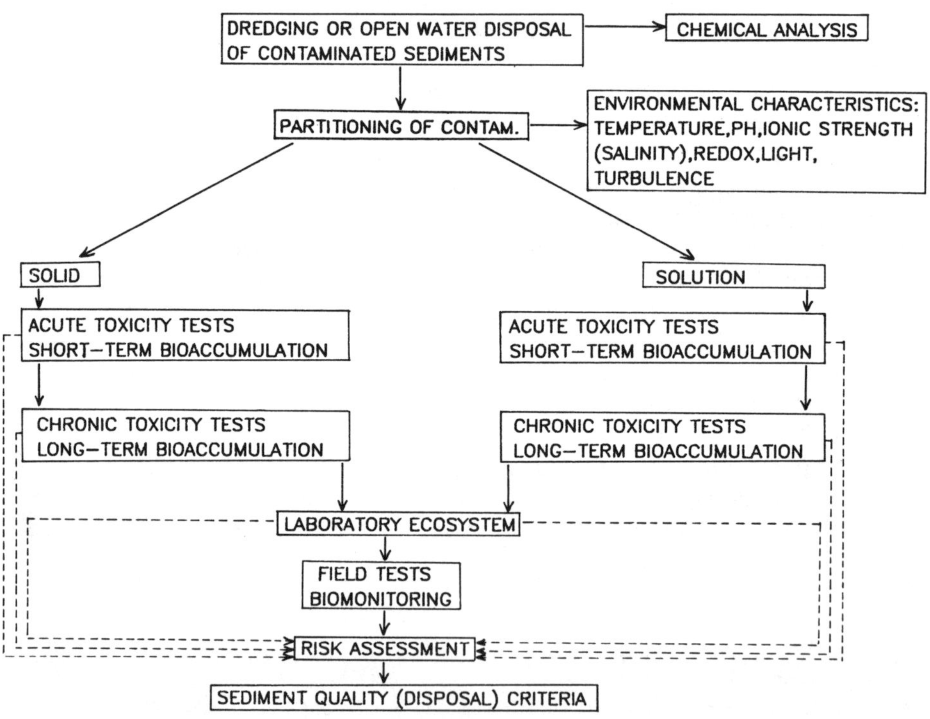

Fig. 1. Flow Chart for the Assessment of Biological Impact of Dredged Material

Contaminant bioavailability and bioaccumulation should be tested under various conditions at each level in order to determine the influence of environmental variables. Brief descriptions of biological tests useful for monitoring the effects of dredged material are presented in the following.

2.1 Acute Toxicity and Short-Term Bioaccumulation

Studies of acute toxicity measure the lethal response after 24 or 96 h of exposure in various bodies of water. Test species should be carefully chosen from amongst the most commonly used organisms and standardized procedures should be applied. The tests should include at least three different trophic levels, namely primary producers, primary consumers and secondary consumers. Normally, organisms such as green algae, daphnids and fish are utilized. Bioconcentration is a measure of the potential of a chemical to accumulate in the tissues of aquatic organisms. For several contaminants (e.g. heavy metals), bioaccumulation occurs a short time after the organisms come into contact with the contaminant. The rapid uptake could be the result of surface adsorption and may be often estimated by determining the amount of toxic substances in the test organisms used in the toxicity assays.

2.2 Chronic Toxicity and Long-Term Bioaccumulation

The long-term sublethal effects of dredged material disposal are measured after at least 96 h of exposure (usually some weeks) using flow-through water. Chronic toxicity tests should be done on the species discussed earlier. The influence of the main environmental variables (pH, temperature, etc.) on acute toxicity should also be determined.

Central to the integrity of a long-term assay is the problem of precisely measuring a parameter that accurately reflects sublethal damage. Tests on the reproduction, development and growth of aquatic organisms are commonly used as well as experiments on physiological parameters, taking into consideration levels of organization lower than the whole organism (i.e. organs, enzymes). Particularly important due to the high degree of sensitivity are the experiments on the whole life cycles or on critical stages of development of the test organism (Calamari and Marchetti 1978).

Because long-term experiments with consumers require feeding the organisms, studies of bioaccumulation potential must separate the relative importance of two vectors of uptake: water and food. In addition, the effects of metabolic activity on bioaccumulation of contaminants should also be determined.

2.3 Laboratory Ecosystem

Possible ecosystem level effects are to be studied in multispecies laboratory microcosms. Since the biota may also act on the contaminants of dredged material by changing its nature, concentration and partitioning, the use of more environmentally realistic laboratory test systems would seem appropriate. However, the sediment-water distribution observed in a closed laboratory experiment may not be the same as the distribution in open systems in nature.

2.4 Field Tests and Biomonitoring

In contrast to microcosms designed to experimentally simulate more natural environmental conditions in the laboratory, in situ test methods verify the findings of laboratory tests in field situations. A major weakness is that this type of evidence has not proven adequate for extrapolation to other ecosystems (Cairns 1981), but relating field measurements to the changes occurring in laboratory tests may contribute to the meaningful interpretation of the bioassay results.

The use of biological indicator organisms to monitor pollution of aquatic ecosystems is also suggested as a tool for corroborating laboratory bioassay data. The indicator organisms should either be living in or exposed to the area of interest. However, in such studies, test organisms accumulate contaminants from different sources, including suspended matter and water. Therefore, interpretation of bioaccumulation data should be carried out in conjunction with laboratory studies.

2.5 Risk Assessment and Sediment Quality Criteria

Risk assessment does not necessarily represent an ordered sequence of elements involving the recognition of hazards, the measurement of impact and the comparison of the measurements. Rather, all possible combinations of elements exist in practice. The flowchart of tests is, in itself, a structured form of continuing risk assessment. This procedure allows the evaluation of contaminants, significant undesirable effects and hazardous bioaccumulation, while also predicting the nature and magnitude of environmental impact. Development of standard sediment quality criteria requires the knowledge of the highest concentration of contaminants in dreged material to which organisms can be exposed without observing a toxic effect under various biological and environmental conditions and for different times.

3 Sediment Bioassay Techniques

Most aquatic bioassays have focussed on water-column effects, while relatively few bioassays have been conducted to directly monitor the effects of contaminated sediment, including dredged material. Hakanson (1984) emphasized the importance of using sediments to establish and define the relationship between aquatic taxicology and risk. This overview is intended to provide state of the art bioassessment procedures used in the dredging and disposal of sediment.

3.1 Short-Term-Tests

Water-column toxicity has been monitored using both sediment elutriates and sediment suspensions. The short-term impacts on water quality can be evaluated by the elutriate test (Shuba et al. 1977a), but the evaluation of sediment quality must consider the amount of on-site mixing and dilution occurring in the water column at the disposal site. The bioassays should then be interpreted in light of the amount of dispersion and dilution occurring in the field.

Most existing short-term bioassays are static procedures that simulate a range of dose-response relationships. Since dredging operations and subsequent disposal results in the resuspension of sediments in the water column, it is important to simulate such effects during experiments and to investigate their influence on planktonic organisms. It is difficult to measure the interaction of sediment and micro-organisms while retaining the ability to analyze each separately. Hegeman and Keenan (1983) proposed three general techniques for algae: (1) direct contact between algae and sediment, (2) separation of sediment and algae by a semi-permeable membrane and (3) encapsulation of sediment within agar. Bioassay studies conducted using sediments and bacteria have been limited in number and scope. Hendricks (1971) used a 0.3 M phosphate buffer at pH 7.0 to elute loosely bound chemical constituents from river sediments. The buffer elutriates contained protein, ammonia and hexose sugars in concentrations of four to six times greater than that found in river water or in equivalent elutriates using river water instead of buffer. He then measured the respiration rate of pathogenic and non-pathogenic enteric bacteria in these elutriates and demonstrated that the phosphate elutriate stimulated respiration rates in comparison with respiration measured using river water.

Samolloff et al. (1983) developed a protocol for the chemical fractionation of sediments and biological testing of these fractions. Fractions obtained were directly tested for toxicity using *Salmonella typhimurium*. The major toxic constituents of the sediments were neutral compounds that were eluated from Florisil columns by 1:1 hexane-dichloromethane.

Increasing metal concentrations generally had a marked inhibitory effect on the synthesis of amylase, cellulase and urease, and on numbers of substrate-hydrolyzing bacteria in all sediments studied by Wainwright and Duddridge (1982). Titus and Pfister (1982) described the competition between bacteria and sediment for cadmium under various incubation conditions. The bacteria were placed in the sediment in separate dialysis bags. The amount of cadmium taken up by bacteria was about ten times higher than cadmium adsorbed by the sediment when based on dry weight.

Multimedia/multitrophic level bioassays have been carried out to determine the extent and severity of environmental contamination at hazardous waste sites. Comparative toxicological profiles were presented for different organisms. Algae and *Daphnia* were the most sensitive test organisms to heavy metals and insecticides (Miller et al. 1985).

US Environmental Protection Agency guidelines for conducting liquid-phase phytoplankton communities were outlined by Shuba et al., (1977b). In a more recent study, Hannan and Patoillet (1979) monitored the growth of *Phaeodactylum tricornutum* in elutriates of sediments collected from a diversity of marine and estuarine sites. Cultures were maintained in combinations of growth media and elutriates and growth rate was monitored over 20 to 44 h by fluorescence. Although few cases of acute toxicity were found, elutriates often induced a prolongation of the lag phase of growth.

Laube et al. (1979) studied the mobilization and accumulation of sediment-bound cadmium and copper by freshwater algae. Algae were placed in dialysis sacs and suspended in a system of circulating water and contaminated sediment. The distribution of metals in water, sediment and algae was measured at selected intervals for up to 72 h. The authors demonstrated indirect biological uptake from the sediments and concluded that algae may play an important role in the biomobilization of sediment-bound heavy metals in natural systems.

The influence of environmental variables on bioaccumulation and toxicity of algae was examined using an apparatus in which a 0.45-μm pore diameter membrane separated the algae and sediments. The amounts of metals accumulated in the test algae were compared to chemical associations of metals in sediments but there were no clear relationships with geochemical phases. Bioaccumulation was affected by decreasing pH and marked differences were apparent among the six elements studied. Toxicity decreased with increasing pH (Ahlf, 1985). Metal accumulation in *Enteromorpha intestinalis* was lowered by increasing salinity of the media, but cadmium uptake was highest at intermediate levels of salinity (Ahlf et al. 1986).

An experimental design to study the impact of sediment-associated contaminants on phytoplankton and zooplankton was described in detail by Munawar et al. (1984). In these experiments, indigenous natural phytoplankton populations or communities collected from oligotrophic/eutrophic environments were exposed to increasing dosages of sediment elutriates and their photosynthetic production was measured by means of a carbon-14 uptake technique (Munawar et al. 1978, 1983). At the end of the test, the natural phytoplankton was size fractionated using Nitex screens or filters and productivity was analyzed separately. Details of this algal fractionation bioassay (AFB) procedure are given in Munawar and Munawar (1982, 1987) and Munawar et al. (1985). This new technique is used to assess the differential response of algal size assemblages to contaminant stress. In some of the research Munawar et al. (1987a,b,c) deals with bioassay techniques for the assessment of small quantities of bottom or suspended sediments and picoplankton as test organisms respectively. Severn et al. (1987) also tested sediment elutriates with picoplankton.

US Environmental Protection Agency guidelines for performing liquid-phase and suspended particulate phase animal bioassays were outlined by Shuba and Tatem (1977). These bioassays were intended for use in the evaluation of the impact of suspended particulates and dissolved chemicals released from sediment during ocean disposal operations.

Prater and Anderson (1977) conducted sediment bioassays on eight sediments collected from Duluth and Superior Harbours, Lake Superior using the burrowing mayfly, *Hexagenia limbata*, the isopod *Assellus communis*, the water flea *Daphnia magna* and the fathead minnow, *Pimephales promelas.* The use of recirculating water over test sediment was described in detail by Prater and Hoke (1980).

Mortality was determined after a 96-h exposure period. *Daphnia magna* and *H. limbata* were the most sensitive species to sediment contamination, while *P. promelas* was relatively insensitive.

Ray et al. (1981) studied the bioaccumulation of heavy metals from two sediments by three marine invertebrates (*Nereis virens, Macoma balthica, Crangon septemspinosa*). It was apparent that the bioavailability of the metals in the sediments to the animals was generally related to the EDTA-extraction values. A high correlation between the uptake of copper and the amount of copper present in the manganese oxide phase was demonstrated by Diks and Allen (1983). The authors suggested that the redox potential and pH in the gut of the tubificid worms is such that manganese oxide coatings were dissolved. Copper, solubilized by the dissolution of this coating was then available for uptake by the organism. Many statistically significant relationships between trace metal partitioning and their accumulation in molluscs were also found by Tessier et al. (1983).

Malueg et al. (1984) conducted acute toxicity tests on *Daphnia* and *Hexagenia* to determine their ability to simulate the toxicity of natural freshwater benthic communities. In general, the number of organisms, biomass, species diversity and dominant types of organisms correlated with metal content and acute laboratory toxicity.

A bioassay technique for estuarine sediments was developed with two species of free-living nematodes (Tietjen and Lee 1984). Both species grew better in the control and other sediments containing low concentrations of PCBs, PAHs and heavy metrals. In sediments having PCB and PAH concentrations (μm kg^{-1}) of more than 270 and 8700 respectively, population growth of both species tended to be less than half that found at lower concentrations.

3.2 Long-Term Tests

The long-term sublethal effects of dredged material on microorganisms are not well-known. Non-acute toxicity tests with animals produced various responses and are subject to continuing study.

The proposed algal fractionation bioassay includes long-term studies with phytoplankton (Munawar and Munawar 1987). Gannon and Beeton (1969, 1971) studied the selection of contaminated Great Lakes sediments by benthic organisms. The experimental animals used were the amphipods *Pontoporeia* and *Gammarus* and larvae of the midge *Chironomus*. Animals were placed in test chambers containing contaminated and control sediments and sediment preferences were monitored. These experiments showed that test organisms preferred some sediments over others. However, there were no clear relationships between sediment contamination and selection by the animals.

Wentsel et al. (1977) performed chronic toxicity tests to determine the effects of heavy metal contamination in sediment on growth and survival of the freshwater chironomid, *Chironomus tentans*. Early instar larvae were maintained for 17 days in sediments having various levels of heavy metal contamination. Growth was inhibited by heavy metal contamination and a linear relationship was observed between the square root of larval length and metal levels in the sediment.

Ray et al. (1980) measured tissue cadmium levels in the polychaete *Nereis virens* maintained in contaminated sediment-seawater systems over 25 days and then measured subsequent depuration rates over 72 days. Bioaccumulation from the sediment and seawater were assessed by maintaining control animals in contaminated seawater alone.

Rosssi (1977) exposed the polychaete *Neanthes arenaceodentata* to seawater and sediment contaminated with water-soluble petroleum hydrocarbons and radiolabelled naphthalenes and observed uptake over 4 weeks. The experiments suggested that hydrocarbons bound to sediment and to particulate organic matter are less available to the worms than those in solution.

A marine polychaete *Capitella capitata* was exposed during its complete life cycle to Puget Sound sediments and although high initial mortality occurred, the life cycle was completed in all sediments tested (Chapman and Fink 1984).

Halter and Johnson (1977) studied PCB desorption from contaminated sediment and subsequent uptake by fathead minnows (*Pimephales promelas*). Acetone solutions of PCBs were applied to sediment in various concentrations. Fish were exposed to various PCB levels in sediments and were either allowed access to the substrate or were separated from the bottom by screens. Tissue PCB loads were measured over 32 days. The rate of dissolution of PCBs from sediment was also measured. Although fish having direct access to the sediments accumulated PCB residues at six times the rate of those denied access, chromatograms of Aroclor 1254 residues in water, sediment and in the two groups of fish indicated that PCBs were taken up primarily from water in all cases. The authors attributed higher uptake in fish having access to the bottom to grazing at the sediment surface. This resulted in elevated levels of dissolved PCBs in the buccal cavity and subsequent uptake via the gills.

McCain (1978) exposed the English sole *Paraphyrus vetulus* to experimentally-oiled sediments over 4 months in the laboratory and showed that alkanes and aromatic hydrocarbons were readily accumulated. Petroleum hydrocarbon distribution among tissues and pathological consequences of uptake were also examined. Malins et al. (1984) tested Puget Sound sediments revealing hepatic lessons in English sole and other demersal fish species which were attributed to elevated levels of metals and polycyclic aromatic hydrocarbons.

Representative samples of partially-drained sediment dredged from Hamburg Harbour were placed in relatively uncontaminated water of various salinities, and the chemical changes were monitored for 2 months. Net accumulation of some heavy metals and chlorinated hydrocarbons by exposed plaice and brown shrimp was demonstrated (Berghahn et al. 1986).

3.3 Laboratory Ecosystem

Multispecies laboratory microcosms are suggested as tools for demonstrating ecological effects both biologically and statistically. Some examples of microcosms, the statistical properties of the data and design criteria for microcosms were presented by Crow and Taub (1979).

A method for assessing the biogeochemical fate and effects of contaminants was developed using a three-phase aquatic microcosm (TPMA) having sediment, water and gas phases to represent aquatic environments. TPMA benthic dissolved oxygen consumption was typical of natural environments. Trace additions of heavy metals depressed the productivity of TPAMs using Lake Powell sediments (Porcella et al. 1982).

The fate of kepone and mirex were determined in a model aquatic environment employing sediment, bluegill sun-fish *Lepomis macroshirus*, and *Daphnia magna* (Skaar et al. 1981). Bluegills accumulated kepone and mirex 10600 and 12300 times respectively, and mirex tended to persist in fish tissue but kepone did not. Retention of mirex introduced to bluegills by daphnids was three times greater than that for kepone.

A microcosm system for testing interactive effects of cadmium and nutrients was described and evaluated by Hendrix (1981). Community metabolism, community composition by trophic groups and output/input ratios for nitrates, manganese and iron were the best response indicators for cadmium. A hierarchy of microcosm experiments was recommended to include static microcosms with and without sediment, flow-through microcosms with and without sediments and microcosm subsamples from specific natural ecosystems.

A mesocosm experiment was carried out to determine the rate at which water columns in estuarine areas grossly polluted by sewage, heavy metals and hydrocarbons can recover and the rates at which pollutant concentrations within sediments decrease when the input of pollutants is terminated (Oviatt et al. 1984). Grossly polluted sediment from Providence River (Rhode Island), measurably polluted sediment from mid-Narragansett Bay, and relatively unpolluted sediment from Rhode Island Sound were placed in mesocosms containing clean, flowing seawater. After 5 months, the behaviour of the grossly polluted system was not dramatically different from the other two situations and seasonal patterns of phytoplankton, zooplankton and benthic fauna were similar, although pollutant material remained detectable in the sediments.

3.4 Field Studies

In contrast to microcosms designed to experimentally simulate more natural environmental conditions, field studies were conducted to monitor the effects of contaminated sediments in situ. The monitoring of bioaccumulation potential and toxicity of contaminants in situ has been undertaken by several workers. Munawar and Munawar (1987) have recently provided a detailed account of phytoplankton bioassays with emphasis on field techniques and site-specific data collected in the North American Great Lakes. They conducted simulated dredging impact experiments in a contaminated pond near Toronto, Ontario (Munawar et al. 1986). Sediments were mixed with the water column and the impact was measured regularly during a 3-day period using algal fractionation bioassays in conjunction with carbon-14 uptake techniques. The results indicated both inhibition and enhancement of the relative assimilation activity or production/biomass quotients.

In situ bioassays were also conducted to evaluate the impact of dredging, disposal and navigational activities in Toronto Harbour (Munawar and Munawar 1987; Munawar and Norwood 1987). The experimental work consisted of monitoring dredging and disposal sites during three phases of activity (before, during and after). The navigational impact study was comprised of monitoring the chemical, biological and physical water parameters before and after the passage of a ship in the harbour. When compared with the resulting impact of dredging and disposal, it was obvious that navigation significantly inhibited P/B quotients of algal size assemblages.

Munawar and Munawar (1987) have also developed in situ plankton cages (IPCs) to evaluate the effects of contaminants on caged algae. They deployed these cages at various sites in the North American Great Lakes including sediment-water interfaces, mixing zones, mouth of tributaries and effluent areas. These cages provided a rapid, less expensive tool for evaluating both bioavailability and bioaccumulation under in situ conditions with natural size assemblages of phytoplankton. Preliminary data collected in Toronto Harbour showed a differential response when exposed to contaminants/nutrients. The ultraplankton ($< 20\ \mu m$) showed significant enhancement of carbon-14 uptake, whereas microplankton/net plankton ($> 20\ \mu m$) showed significant inhibition over the control. Therefore, it seems that physiological response data collected by means of cages along with chemical data has great potential as an on-site sensitive and rapid procedure for toxicological screening of contaminants including those found during dredging/disposal activities.

A field method for the rapid estimation of chemical and microbial activities in sediments is based on the quatitative reduction of the dye resazurin by chemical-reducing substances and by dehydrogenase in microorganisms (Liu and Strachan, 1981). The chemical and biological activities were differentiated by using m-cresol to inhibit the dehydrogenase.

Mayes et al. (1977) grew the aquatic macrophyte *Elodea canadensis* in wooden containers placed in a metal-contaminated lake and an uncontaminated lake. Plants were rooted in sediments from each lake in each of the two environments. Since the plants were exposed to all combinations of contaminated and uncontaminated sediment and water, the experimental design permitted differentiation between uptake from sediment and water. Therefore, the relative importance of shoots and roots in bioaccumulation were demonstrated.

A simple monitoring technique for determining potential heavy metal loads of algae was developed, based on algal exposure in chambers immersed in the water body of interest (Ahlf and Weber 1981). Standardized, uncontaminated algae grown in laboratory cultures were exposed in the River Elbe at regular intervals over 3 years. Monitored levels of heavy metals in the tissues of algae were compared to values of metals in sediments of longitudinal sections and demonstrated a clear relationship with contaminated areas (Ahlf 1983). Luoma et al. (1982) found significant correlations between estuarine sediments and the tissues of benthic algae, and suggested that sediment-bound metals could provide an important source of uptake by algae.

The effects of dredging in St. Jean, Quebec, temporarily decreased plankton populations as a result of increased turbidity, and laboratory experiments indicated that the turbidity could eventually kill the fish (Lalancette 1984).

Mussels are becoming increasingly popular as monitoring tools for the assessment of environmental contamination (Phillips 1976) and are also used to evaluate the impact on biota of contaminated dredged materials (Marquenie et al. 1985). Reproduction development was not impaired in deployed mussels (*Mytilus* and *Modiolus*) at several New England dump sites; however, epithelial necrosis, cell sloughing, abnormal sperm follicles and small ova were identified during one sampling event at one site (Arimoto and Feng, 1983). A number of surveys and experiments were carried out by Langston (1984), designed to investigate As availability in selected estuarine and marine organisms. Seasonal variations in dissolved As inputs are reflected in the As content of the seaweed *Fucus vesiculosus* and also in two gastropod species, *Littorina littorea* and *L. littoralis*, which derive significant amounts of As by grazing on *F. vesiculosus*. By contrast, no seasonal variation in As concentration was observed in tissues of the bivalve *Scrobicularia plana*. Transplant experiments and radiotracer studies confirm that sediments are the major source of As to this deposit feeder.

Laboratory recirculating bioassay tests with *Daphnia* and *Hexagenia* showed good correlation between acute toxicity of natural sediments in the laboratory and in situ community composition from three metal-contaminated areas (Malueg et al. 1984).

Moore (1979) studied benthic densities and diversities in a contaminated bay of Great Slave Lake (Northwest Territories, Canada) where sediments contained elevated concentrations of heavy metals from mining waste. Benthic collections were made in areas of heavy, moderate and negligible contamination. Species diversity was an ineffective indicator of contamination, but a strong negative correlation between the concentrations of sediment-bound metals and total abundance of benthic macro-invertebrates was shown. Moore et al. (1979) attempted to separate the effects of sediment and water contamination by heavy metals on benthic invertebrate communities in two Northwest Territories lakes. Since the concentrations of heavy metals in two adjoining lakes were statistically similar, the authors suggested that the lower density and diversity in one lake were attributable to higher levels of dissolved contaminants (primarily copper and cyanide) observed in this lake.

Tagatz et al. (1983) studied the effect of creosote on field- and laboratory-colonized estuarine benthic communities. Species diversity and dominance were altered by an increase in creosote concentration; molluscs were more tolerant than annelids, arthropods and echinoderms.

Stainken (1984) conducted a study on the xenobiotic sediment-benthic relationships in Raritan Bay and Lower New York Bay. Macro-invertebrate community structure was influenced primarily by substrate characteristics and local currents. A trend of decreasing species richness was observed as the total concentration of sediment hydrocarbons and PCBs exceeded 300 to 400 mg l^{-1}.

No long-term changes in benthic communities were observed from dredging and disposal within the Dawho and North Edisto estuaries of South Carolina (VanDolah et al. 1984) and recovery from dredging occurred within 3 months. The authors suggested that the absence of a major long-term disruption to the benthos in the disposal area was probably due to (1) strong tidal currents, which rapidly dispersed the moderate amount of mud sediments released; (2) surface disposal, permitting wider dispersal; and (3) disposal during late autumn, a period of low faunal recruitment.

Sodergren (1984) evaluated a dredging operation in Lake Trummen, Sweden, and determined that no significant biological uptake of chlorinated hydrocarbons resulted in the fish sampled.

4 Summary

Relatively simple and implementable liquid, suspended particulate and solid-phase bioassays have been carried out for assessing the short-term impact of dredging and disposal operations on aquatic organisms. Standardized tests are characterized by their lack of variability, but essential information (e.g. lethality, alterations of growth rate) can only be obtained with such single species test. The influence of the main environmental variables on the interaction of suspended particulates and organisms should also be determined under simulated field conditions (Ahlf 1987). A great deal of research was done to find suitable organisms for bioassays. Natural phytoplankton size assemblages have been used succesfully and extensively in the North American Great Lakes and have proven to be very sensitive indicators of contaminant stress (Munawar et al. 1987d; Munawar and Thomas 1987).

However, short- and long-term bioassays give only a first estimation for determining potential environmental impacts of the dredging activities. Use of natural algal populations, in situ bioassays, flow-through bioassays and biomonitoring are recommended over static procedures. Also, bioassays should include at least three different trophic levels. Laboratory ecosystems have good potential for experimental research with dredged material and should be used in conjunction with in situ bioassays.

In general, the bioassay results must be interpreted in relation to the field situations and site-specific data. Therefore, bioassays should be refined to the point where the main question can be answered: "Are laboratory tests an adequate surrogate for natural ecosystems?" (Cairns 1981). A preliminary proposal to overcome this problem was a comparison of field and laboratory studies (Tagatz and Deans 1983; Chapman 1986). Benthic bioassay procedures, due to these recent developments, are relatively important and should continue to be evaluated to estimate the relationship between laboratory and field impacts.

Finally, all work for screening potential impacts of contaminated sediments will not be useful if not preceded by an extensive investigation to establish sediment quality criteria for dredging and disposal activities. The proposed testing strategy is presented mainly as a guideline for standard procedures.

Acknowledgements. We would like to thank Lynda McCarthy, Warren Norwood and Janice Jones for their assistance in the preparation of this manuscript.

References

Ahlf W (1983) Biologische Verfügbarkeit von Schwermetallen in der Elbe. In: Gewässerschutz an der Elbe. Okologie-Forum 1983 12:55–58

Ahlf W (1985) Behaviour of sediment-bound heavy metals in a bioassay with algae: bioaccumulation and toxicity. Vom Wasser 65:183–188

Ahlf W (1987) Transfer of heavy metals from suspended sediment to the alga *Stichococcus bacillaris* in estuarine water medium. Proc. Conf. Coastal and Estuarine Pollution. Fukuoda, Japan

Ahlf W, Weber A (1981) A simple monitoring technique to determine the heavy metal load of algae in aquatic ecosystems Environ Technol Lett (2) 317–322

Ahlf W, Calmano W, Förstner U (1986) The effects of sediment-bound heavy metals on algae and importance of salinity. In: PG SLY (ed) Sediment and water interactions Springer-Verlag New York pp 319–324

Arimoto R, Feng SY (1983) Histological studies on mussels from dredge spoil dump sites. Estuar Coast Mar Sci 17:535–546

Berghahn R, Karbe L, Seidel U, Burchert S, Zeitner R (1986) The ecotoxicology of dredged, fluviatile sediment in seawater and in brackish water. Vom Wasser 66:211–224

Cairns JJr (1981) Biological monitoring, Part VI-Future needs Water Res 15:941–952

Cairns JJr, van der Schalie WH (1980) Biological monitoring. Part I. Early warning systems. Water Res 14:1179–1196

Calamari D, Marchetti R (1978) Relevance of studies on development and young stage of *Salmo gairdneri* in establishing water quality criteria. Commission of the European Communities. Bericht 10 UBA

Calamari D, Galassi S, DA Gasso R (1979) A system of tests for the assessment of toxic effects on aquatic life: an experimental preliminary approach. Ecotoxicol Environm Saf 3:75–89

Chapman PM (1986) Sediment quality criteria from sediment quality triad: an example. Environ Tox Chem 5:957–964

Chapman PM, Fink R (1984) Effects of Puget Sound sediments and their elutriates on the life cycle of *Capitella capitata.* Bull Environ Contam Toxicol 33:451–459

Crow ME, Taub FB (1979) Designing a microcosm bioassay to detect ecosystem level effects. Intern J Environmental Studies 13:141–147

Diks DM, Allen HE (1983) Correlation of copper distribution in a freshwater-sediment system to bioavailability. 30:37–43

Engler RM (1980) Prediction of polluted potential through geochemical and biological procedures: development of regulatory guidelines and criteria for the discharge of dredged fill material. In: Baker RA (ed) Contaminants and sediments. Ann Arbor Sci Publ. Ann Arbor, Mich. 1:143–169

Gannon JE, Beeton AM (1969) Studies on the effects of dredged materials from selected Great Lakes harbors on plankton and benthos. Center for Great Lakes Studies, Milwaukee, Wisconsin, Spec. Rept. No. 8

Gannon JE, Beeton AM (1971) Procedures for determining the effects of dredge sediments on biota-benthos viability and sediment selectivity tests. J Wat Pollut Control Fed 43:392–398

Hakanson L (1984) Aquatic contamination and ecological risk Water Res 18:1107–1118

Halter MT, Johnson HE (1977) A model system to study the desorption and biological availability of PCB in hydrosoils Proc Amer Soc Testing Materials 1st Ann Symp, Aquat Toxicol, Memphis, Tennessee. 178–195

Hannan PJ, Patouillet CE (1979) Results of an algal toxicity test applied to sediment elutriates. Naval Research Lab, Washington, DC. Rept No NRL-MR-3952. 28pp

Hegemann DA, Keenan JD, (1983) Bioassay techniques for soil-nutrient availability. Water Air Soil Poll 19:259–276

Hendricks CW (1971) Enteric bacterial metabolism of stream sediment elutriates. Can J Microbiol 17:551–556

Hendrix PF (1981) Microcosms as test systems for the ecological effects of toxic substances: an appraisal with cadmium. U.S Environmental Protection Agency, EPA-600/3-81-036

Lalancette PLM (1984) The effects of dredging on sediments, plankton and fish in the Vauvert area of Lake St. Jean, Quebec. Arch. Hydrobiol 99(4). 463–477

Langston WJ (1984) Availability of arsenic to estuarine and marine organisms: a field and laboratory evaluation. Mar Biology 80:143–154
Laube V, Ramamoorthy S, Kushner DJ (1979) Mobilization and accumulation of sediment-bound heavy metals by algae. Bull Environm Contam Toxicol 21(6):763–770
Liu D, Strachan WM (1981) A field method for determining the chemical and biological activity of sediments. Water Res 15:353–359
Luoma SN (1983) Bioavailability of trace metals to aquatic organisms. A review. Sci Total Environ 28:1–22
Luoma SN, Bryan GW, Langston WJ (1982) Scavenging of heavy metals from particulates by brown seaweed. Mar Pollut Bull 13:394–396
Malins DC, McCain BB, Brown DW, Chan SL, Myers MS, Landahl JT, Prohaska PG, Friedman AJ, Rhodes LD, Burrows DG, Gronlund WD, Hodgins HO (1984) Chemical pollutants in sediments and disease of bottom-dwelling fish in Puget Sound, Washington. Environ Sci Tech 18:705–713
Malueg KW, Schuytema GS, Gakstatter JH, Krawczyk DF (1984) Toxicity of sediments from three metal-contaminated areas. Environ Toxicol Chem 3:279–291
Marquenie JM, Simmers JM, Birnbaum E (1985) An evaluation of dredging in the western Scheldt (the Netherlands) through bioassays. TNO, division of technology for society, Delft, Rep No R85/075
Mayes RA, Mcintosh AW, Anderson VL (1977) Uptake of cadmium and lead by a rooted aquatic macrophyte (*Elodea canadensis*) Ecology 58(5):1176–1180
McCain BB, Hodgins HO, Gronlund WD, Hawkes JW, Brown DW, Myers MS (1978) Bioavailability of crude oil from experimentally oiled sediments to English sole and pathological consequences. J Fish Res Bd Can 35(5):657–664
Miller WE, Peterson SA, Greene JC, Callahan CA (1985) Comparative toxicology of laboratory organisms for assessing hazardous waste sites. J Environ Qual 14(4):569–574
Moore JW (1979) Diversity and indicator species as measures of water pollution in a subarctic lake. Hydrobiologia 66(1):840–847
Moore JW, Beaubien VA, Sutherland DJ (1979) Comparative effects of sediment and water contamination of benthic invertebrates in four lakes. Bull Environm Contam Toxicol 23(6):840–847
Munawar M, Munawar IF (1982) Phycological studies in Lakes Ontario, Erie, Huron and Superior. Can J Bot 60:1837–1858
Munawar M, Munawar IF (1987) Phytoplankton bioassays for evaluating toxicity of in situ sediment contaminants. In R L Thomas, R Evans, A Hamilton, M Munawar, T Reynoldsen, H Sadar (Eds.), Proc Internat Workshop on ecological effects of in situ sediment contaminants. Hydrobiologia
Munawar M, Norwood WP (1987) Bioavailability of in-place pollutants, Toronto Harbour, Ontario, Canada. Report, pp 47. Great Lakes Laboratory for Fisheries & Aquatic Sciences. Fisheries & Oceans Canada. Burlington, Ontario, Canada.
Munawar M, Thomas RL (1987) Phytoplankton bioassays of sediment from two areas of concern in the North American Great Lakes: Toronto (Ontario) and Toledo (Ohio) Harbours. In Proc 4th Internat Symp Sed Water Interactions. Melbourne, Australia
Munawar M, Munawar IF, Culp LR, Dupuis G (1978) Relative importance of nannoplankton in Lake Superior phytoplankton biomass and community metabolism. J Great Lakes Res 4(3–4):462–480
Munawar M, Mudroch IF, Munawar IF, Thomas RL (1983) The impact of sediment-associated contaminants from the Niagara River and mouth on various size assemblages of phytoplankton. J Great Lakes Res 9:303–313
Munawar M, Thomas RL, Shear H, McKee P, Mudroch A (1984) An overview of sediment-associated contaminants and their bioassessment. Can Tech Rep Fish Aqua Sci No 1253, 136pp
Munawar M, Thomas RL, Norwood WP, Mudroch A (1985) Toxicity of Detroit River sediment-bound contaminants to ultraplankton. J Great Lakes Res Internat Assoc Great Lakes Res 11(3):264–274
Munawar M, Thomas RL, Norwood WP, Daniels SA (1986) Sediment toxicity and production-biomass relationships of size-fractionated phytoplankton during on-site dredging experiments in a contaminated pond. In PG Sly (Ed.), Sediments and water interactions. Springer-Verlag, New York. pp 407–426
Munawar M, Gregor D, Daniels SA, Norwood WP (1987a) A sensitive screening bioassay technique for the assessment of small quantities of contaminated bottom or suspended sediments. In Proc 4th Internat Symp Sed Water Interactions. Melbourne, Australia
Munawar M, Wong PTS, Rhee GY (1987b) The effects of contaminants of algae: an overview. In N Schmidke (Ed.). Proc World Conference on Large Lakes, Mackinac, Michigan.

Munawar M, Munawar IF, Norwood W, Mayfield C (1987c) Significance of autotrophic picoplankton in the Great Lakes and their use as early indicators of contaminant stress. In M Munawar (Ed.) Proc Internat Symp on Phycology of Large Lakes of the World. Arch Hydrobiol Beih Ergebn Limnol/Advances in Limnology. No 25

Munawar M, Munawar IF, Ross PE, Mayfield C (1987d) Differential sensitivity of natural phytoplankton size assemblages to metal mixture toxicity. In M Munawar (Ed.) Proc Internat Symp on Phycology of Large Lakes of the World. Arch Hydrobiol Beih Ergebn Limnol/Advances in Limnology. 25:123–139

Oviatt CA, Pilson ME, Nixon SW, Frithsen JB, Rudnick DT, Kelly JR, Grassle JF, Grassle J (1984) Recovery of a polluted estuarine system: a mesocosm experiment. Mar Ecol Prog Ser 16:203–217

Phillips DJH (1976) The common mussel *Mytilus edulis* as an indicator of pollution by zinc, cadmium, lead and copper. Mar Biology 38:59–69

Porcella DB, Adams VD, Medine AJ, Cowan PA (1982) Using three-phase aquatic microcosms to assess fates and impacts of chemicals in microbial communities. Water Res 16:489–496

Prater BL, Anderson MA (1977) A 96-hour sediment bioassay of Duluth and Superior harbor basins (Minnesota) using *Hexagenia limbata, Asellus communis, Daphnia magna* and *Pimephales promelas* as test organisms. Bull Environm Cotam Toxicol 18(2):159–169

Prater BL, Hoke RA (1980) A method for the biological and chemical evaluation of sediment toxicity. In: Contaminants and sediments Vol. I, Baker (ed) Ann Arbor Science Pub, Ann Arbor

Ray S, McLeese D, Pezzack D (1980) Accumulation of cadmium by *Nereis virens.* Arch Environm Contam Toxicol 9(1).1–8

Ray S, McLeese DW, Peterson MR (1981) Accumulation of copper, zinc, cadmium and lead from two contaminated sediments by three marine invertebrates. Bull Environm Contam Toxicol 26:315–322

Rossi SS (1977) Bioavailability of petroleum hydrocarbons from water, sediments and detritus to the marine annelid, *Neanthes arenaceodentata.* Proc. Oil Spill Conf. New Orleans, Amer. Petrol Inst., Washington, DC: 621–625

Samolloff MR, Bell J, Birkholz DA, Webster GRB, Arnott EG, Pulak R, Madrid A (1983) Combined bioassay-chemical fractionation scheme for the determination and ranking of toxic chemicals in sediments. Environ Sci Technol 17(6):329–334

Severn SRT, Munawar M, Mayfield CI (1987) Effects of sediment-elutriate material on the growth and activity of picoplankton from Lake Ontario. In Proc 4th Internat Symp Sed Water Interactions. Melbourne, Australia

Shuba PJ, Tatem HE (1977) Guidance for performing liquid phase and suspended particulate phase animal bioassays. In: U.S. Environm Protect Agency/Corps Engin, Techn Comm Criteria Dredged and Fill Material. Implementation Manual for Section 103 of Public Law 92–532

Shuba PJ, Carroll JH, Wong KL (1977a) Biological assessment of the soluble fraction of the standard elutriate test. Technical Rep D-77-3, US Army Engineer Waterways Exp Stat, Vicksburg

Shuba PJ, Tatem HE, Gentile JH (1977b) Guidance for performing liquid phase and suspended particulate phase phytoplankton bioassays. In: U.S. Environm Protect Agency/Corps Engin, Techn Comm Criteria Dredged and Fill Material. Implementation Manual for Section 103 of Public Law 92–532

Skaar DR, Johnson BT, Jones JR, Huckins JN (1981) Fate of kepone and mirex in a model aquatic environment: sediment, fish, and silt. Can J Fish Aquat Sci 38:931–937

Sodergren A (1984) The effect of sediment dredging on the distribution of organochlorine residues in a lake ecosystem. Ambio 13:206–210

Stainken DM (1984) Organic pollution and the macrobenthos of Raritan Bay. Environ Toxicol Chem 3:95–101

Tagatz ME, Deans CH (1983) Comparison of field- and laboratory-developed estuarine benthic communities for toxicant-exposure studies. Water Air Soil Pollut 20:199–209

Tagatz ME, Plaia GR, Deans CH, Lores EM (1983) Toxicity of creosote-contaminated sediment to field- and laboratory-colonized estuarine benthic communities. Environ Toxicol Chem 2:441–450

Tessier A, Campbell PGC, Anclair JC (1983) Relationships between trace metal partitioning in sediments and their bioaccumulation in freshwater pelecypods. Proc Heavy Metals in the Environment. Heidelberg

Tietjen JH, Lee JJ (1984) The use of free-living nematodes as a bioassay for estuarine sediments. Mar Environm Res 11:233–251

Titus JA, Pfister RM (1982) Effects of pH, temperature, and E_h on the uptake of cadmium by bacteria and an artificial sediment. Bull Environm Contam Toxicol 28:697–704

United States Environmental Protection Agency (1973) Biological field and laboratory methods for measuring the quality of surface waters and effluents. Rept No EPA 670/4-73-001

VanDolah RF, Calder DR, Knott DM (1984) Effects of dredging and open-water disposal on benthic macroinvertebrates in a South Carolina Estuary. Estuaries 7(1):28–37

Wainwright M, Duddridge JE (1982) Effects of heavy metals on enzyme synthesis in substrate-amended river sediments. European J Appl Microbiol Biotechnol 15:241–245

Wentsel RA, McIntosh A, Atchinson G (1977) Sublethal effects of heavy metal contaminated sediment on midge larvae (*Chironomus tentans*). Hydrobiologia 56(2):153–156

Soil Pollution by Metals from Mining and Smelting Activities

TERUO ASAMI[1]

1 Introduction

Soil pollution by metals such as Cd, Zn, Pb, Cu, As, and Hg has been recognized in many countries. In recent years, it has become one of the most important environmental problems in the developed countries and perhaps in the developing countries.

Metal pollution of soil occurs from the following sources:

1. Emission from metal mines.
2. Emission from metal smelters and refineries.
3. The disposal of wastes from metal mines, smelters, and refineries.
4. Emissions from metal processing factories.
5. Emissions from coal or oil combustion in power stations and various factories, etc.
6. Emissions from municipal incinerators.
7. Emissions from moving sources, mainly automobiles.
8. The use of agricultural materials, such as agricultural chemicals, fertilizers, or soil amendments.
9. Biomass burning such as combustion of fuel wood, forest fires, and agricultural burning.
10. Natural sources such as mineralizations and volcanic activities.

Soil pollution by metals (1–3) is the most important of these ten pollution sources. Therefore, the author will review the soil pollution problems by metals from mining and smelting activities in Japan and some other countries.

2 Soil Pollution by Cadmium and Other Metals in Japan

Japan is famous for environmental metal pollution including soil pollution. The reasons why such severe pollution has occurred are derived from the following features (Yamane 1981).

1. High population density (geographical feature): Japan has a large population, though its plain area is very small and the population density is very high in the

[1]Faculty of Agriculture, Ibaraki University, Ami, Ibaraki, Japan

plains. As the industrial areas are located there, toxic substances emitted from the factories are immediately present in the residential areas.

2. Existence of a large number of metal mines (geological feature): Since the Japanese Islands have been very active in diastrophism and igneous activity, there are many metal mines. This is one of the most important causes of severe metal pollution in Japan.
3. Rapid economic development (economic feature): Since the Meiji Restoration (1868), the Japanese Government has developed the policy to increase metal production for economic prosperity and military aggrandizement. After World War II, especially in the high-development period in economics after 1960, the metal production increased abruptly. The rapid increase in metal production spread toxic substances including metals from mines, smelters, and metal-processing factories.
4. Rice farming as the main form of agriculture (agricultural feature): One of the reasons why metal pollution in agricultural land has been severe is that the main form of Japanese agriculture has been rice farming which uses river water for irrigation. If river water is polluted by metals emitted from metal mines and smelters, which exist usually in hilly sides, metals are introduced into paddy fields with irrigation water.

2.1 Metal Pollution of Agricultural Soil and Its Effect on Human Health

Maximum allowable limits of metals in soil have been set for Cd, Cu, and As. The requirements for the designation of agricultural land soil pollution policy area under the Agricultural Land Soil Pollution Prevention Law are based on the maximum allowable limit of Cd (1.0 μg/gFM unpolished rice) in rice, and on the concentration of Cu (125 μg/gDM soil by 0.1 N HCl extraction) and of As (15 μg/gDM soil by 1 N HCl extraction) in soils. The determination of the maximum allowable limit of Cd in soil is based on the concentration of Cd in rice , thus set so as not to increase the Cd concentration in rice to a level hazardous to human health. Areas, in which rice contaminated with over 1.0 μg/gFM of Cd is produced, or which lie in the vicinity of some such area and thus are highly probable of producing rice contaminated with the same amount, are designated as agricultural land soil pollution policy areas. Maximum allowable limits of Cu and As are set so as not to decrease the production of rice (for review see Asami 1981b).

By the end of December 1984, 126 areas (6,710 ha) were polluted beyond the maximum allowable limits by Cd, Cu, and/or As, and 56 areas (5,610 ha) were designated under the law. In addition, 42 other areas (about 330 ha) have been budgeted as independent prefectual projects. The details are shown in Table 1. Figure 1 shows the distribution of the 56 designated areas in Japan. Toyama Prefecture has the largest area, i.e., 1,630.1 ha in three areas including the Jinzu River Basin where Itai-itai disease and renal dysfunction have been prevailing. Akita Prefecture has the second largest area, i.e., 1,290.2 ha in 16 areas including the Kosaka area where renal dysfunction has been prevailing. A few areas are still designated every year.

Table 1. Agricultural land soil pollution policy areas (December 31, 1984)

Toxic metals	Areas beyond the maximum allowable limits[a]	Designated areas[b,c]					Undesignated areas	Countermeasures were completed under the independent prefectual projects
		Countermeasures were formulated	Countermeasures were not formulated	Designations were cancelled: All	Designations were cancelled: Partly	Total		
Cd	90 areas (6,150 ha)	25 areas (3,030 ha)	8 areas (1,410 ha)	16 areas (770 ha)	4 areas (310 ha)	49 areas (5,520 ha)	20 areas (320 ha)	31 areas (310 ha)
Cu	37 (1,440)	7 (960)	– (–)	4 (100)	1 (130)	11 (1,190)	12 (220)	14 (30)
As	13 (400)	5 (130)	1 (30)	– (–)	– (–)	6 (160)	7 (240)	– (–)
Total	126 (6,710)	30 (3,100)	8 (1,410)	18 (790)	4 (310)	56 (5,610)	37 (770)	42 (330)

[a] Areas beyond the maximum allowable limits are based on detailed surveys from 1971 to 1983.

[b] As the soils in the same areas are polluted by two elements, the sum of the figures in each vertical column is not equal to the total.

[c] As the same polluted areas were divided and designated as the agricultural land soil pollution policy areas and four designated areas were cancelled partly after rehabilitation, the number of the areas beyond the maximum allowable limits is not equal to the sum of the figures of each horizontal line (Environment Agency 1985a).

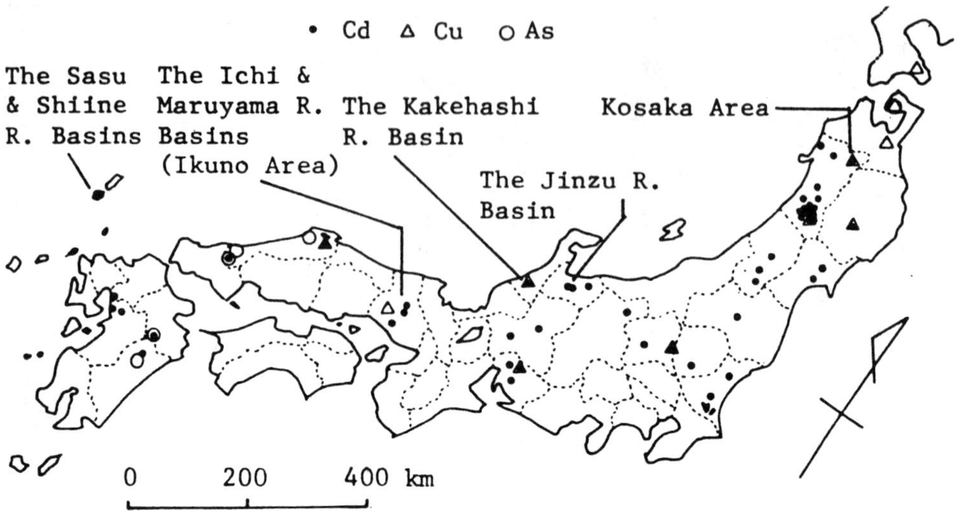

Fig. 1 Distribution of agricultural land soil pollution policy areas (December 31, 1984)

In addition to the detailed survey, a basic survey to monitor the soil pollution by Cd, Zn, and Cu has been conducted at the same points in agricultural lands all over Japan. The paddy fields cover about 3,000,000 ha and the upland fields, including orchard fields, cover about 2,500,000 ha in Japan. Sampling has been made on every 1,000 ha of paddy fields and on every 2,000 ha of upland fields. The results of the basic survey proved that 9.5% of paddy soil, 3.2% of upland soil, and 7.5% of orchard soil were polluted more or less by Cd. These soils were also polluted by Zn and Cu (Agricultural Production Bureau, Ministry of Agriculture and Forestry 1974).

In the designated areas shown in Fig. 1, the health effects derived from the pollution of the soil-plant system were reported in five Cd-polluted areas; patients suffering from Itai-itai disease have been found in one area, patients suffering from a similar disease in three areas, and patients with renal dysfunction in five areas (Table 2). The health effects were mainly caused by intake of Cd in the rice grains from Cd-polluted paddy fields and drinking water. The patients with Itai-itai

Table 2. The cadmium-polluted areas where health effects have been reported[a]

Areas	Itai-itai disease	Renal dysfunction
The Jinzu River Basin, Toyama Prefecture	118 cases	+
The Ichi River Basin, Hyōgo Prefecture	5	+
The Sasu and Shiine River Basins, Nagasaki Prefecture	3	+
The Kakehashi River Basin, Ishikawa Prefecture	2	+
Kosaka Area, Akita Prefecture	0	+

[a]References are cited in the text.

disease were certified officially by the Pollution-Related Health Damage Compensation Law only in the Jinzu River Basin (Environment Agency 1985b). The author will describe the soil pollution in these five areas, the pollution sources of which were mines and/or smelters with very long operating histories.

To evaluate soil pollution by metals, it is important to know the metal concentrations in unpolluted soil. Mean metal concentrations in unpolluted, surface paddy soils of Japan are 0.45 μgCd/gDM, 99 μgZn/gDM, 29 μgPb/gDM, and 32 μgCu/gDM. The mean Cd concentration of unpolished rice produced in unpolluted paddy fields in Japan is 0.09 μg/gFM (Iimura 1981). Cd concentrations in the so-called unpolluted rice in Japan are two or three times higher than those in other countries (Suzuki et al. 1980). This is derived from the fact that many Japanese paddy fields have been polluted by Cd, as seen from the results of the basic survey.

2.1.1 The Jinzu River Basin, Toyama Prefecture

The ore deposits rich in Zn, Pb, Au, Ag, Bi, and Cd were found about 1,200 years ago. Since the 1890s, mining activity in this area has been rapidly developed by the Kamioka Mine of Mitsui Mining and Smelting Co. Ltd. At the highest period of production about 4,000 t of ore were mined daily, and about 40,000 t of Zn, 10,000 t of Pb, and 100 t of Cd were smelted annually (Morishita 1981). The pollution source of the paddy soil in the Jinzu River Basin is the Kamioka Mine which exists about 40 km upstream of the Jinzu River from the polluted area.

Yanagisawa et al. (1984) surveyed the concentrations of Cd, Zn, and Cu in the paddy soil and the unpolished rice from 3,128 ha of the designated area of Itai-itai disease from 1971 to 1976. Heavy metal concentrations [mean(range)] in the surface paddy soils (N = 1,667) extracted by 0.1 N HCl were 1.35(0.18–6.88) μgCd/gDM, 59(8.1–865) μgZn/gDM, 6.3 (0.1–44.1) μgCu/gDM. Heavy metal concentrations in the unpolished rice (N = 2,570) were 0.37(0.00–5.20) μgCd/gFM, 25.2(9.8–78.3) μgZn/gFM, and 4.5(0.9–19.5) μgCu/gFM. Heavy metal concentrations in the surface paddy soils of each soil group are shown in Table 3. Total amounts of heavy metals were determined by the HNO_3-$HClO_4$ digestion method. As seen from Table 3, 67.0% of Cd, 17.0% of Zn, and 29.3% of Cu are extracted by 0.1 N HCl.

Based on the investigation by Yanagisawa et al. (1984) 1,500.6 ha of paddy field was designated as agricultural land soil pollution policy area under the Agricultural Land Soil Pollution Prevention Law. In the designated area, Cd concentrations in the surface paddy soils (N = 544) extracted by 0.1 N HCl were 1.12(0.46–4.85) μg/gDM and those in the unpolished rice (N = 544) were 0.99(0.25–4.23) μg/gFM.

Table 4 shows the vertical distribution of heavy metals at two sites (T. Asami, unpublished). Heavy metals were translocated to the lower layers as the pollution history is very long.

The Cd level in the paddy soils in the Jinzu River Basin is classified as a lower case of Cd pollution in Japan, in spite of the higher level of Cd in rice grain. One of the reasons is that most of the soils in the basin have a higher permeability of

Table 3. Heavy metal concentrations of the surface paddy soil in the Jinzu River Basin[a] μg/g DM

Soil group	Number of samples	0.1 HCl extraction Cd	Zn	Cu	HNO_3-$HClO_4$ digestion Cd	Zn	Cu
Strong-gley soils	252	0.86	38.1	7.4	1.17	238	30.0
Gley soils	24	0.82	36.9	7.7	1.17	258	28.7
Gray soils	183	1.03	64.7	6.8	1.57	364	29.6
Grayish-brown soils	180	1.31	54.2	6.8	2.08	335	15.1
Black soils	285	2.13	78.3	3.5	3.46	411	13.0
Yellowish-brown soils	39	2.34	77.0	6.2	3.83	472	33.0
Gravelly layer soils	141	1.13	57.1	8.1	1.78	395	23.2
Gravelly soils	53	1.30	58.7	7.3	2.06	363	26.4
Mean	1,157	1.37	59.3	6.3	2.13	349	21.5

[a]According to a detailed survey on 2,982.5 ha conducted from 1971 to 1973 (Yanagisawa et al. 1984).

Table 4. Vertical distribution of heavy metals in the polluted paddy soils of the Jinzu River Basin[a]

Locality	Depth (cm)	Soil texture	Metal concentrations in soil (μg/gDM) Cd	Zn	Pb	Cu
Oshiage,	0–11	SCL	3.32	780	280	41.1
Fuchū	11–14	SL	2.34	794	271	35.4
Town	14–25	S	2.08	540	154	25.8
Hagishima,	0–12	LiC	3.59	1,010	448	54.0
Fuchū	12–15	SiL	3.08	920	386	48.0
Town	15–35	SiL	2.26	358	45	23.6
	35–	SL	2.02	258	21	18.4

[a]Sampling and analysis were made in 1972 (T. Asami, unpublished). HNO_3-$HClO_4$ digestion method was used.

irrigation water, frequently over 100 mm/day of the sink depth of the water level in the field, because of the coarse texture of soil throughout the soil profile, frequent occurrence of sandy or gravelly sublayers, and a steeper gradient of agricultural lands, a characteristic of an alluvial fan. Therefore, Cd in the paddy soils is easily absorbable by the rice plant due to the oxidative condition of the paddy soil. More details on the Cd behavior in the soil-plant system were described by Asami (1984).

The prevalence of Itai-itai disease and renal dysfunction in relation to hamlet-average rice-Cd concentration has been reported (Nogawa and Ishizaki 1979; Nogawa et al. 1983a, b). Household unpolished rice was collected from 32 hamlets in the polluted area. As shown in Table 5, there were no patients with Itai-itai disease in the control area. The prevalences of Itai-itai disease and renal dysfunction (proteinuria with glucosuria was used as an index of renal dysfunction) increased with the increasing Cd concentrations in rice.

A rehabilitation plan of Cd-polluted paddy soils of 108 ha in the Jinzu River Basin was made in 1978 in which the polluted surface paddy soils were covered with a 25-cm-deep layer of unpolluted soil. The cost was estimated at \$82,546 ha^{-1} (here, 200 yen is converted to 1 dollar) (Toyama Prefecture 1978). After the rehabilitation,

Table 5. Prevalences of Itai-itai disease and proteinuria with glucosuria in relation to hamlet-average rice-cadmium concentration

Cd in rice (μg/g)	N[a]	Itai-itai disease	Females over 50 years of age N[b]	Females over 50 years of age Proteinuria with glucosuria
Control	446	0 (0[d])	408	20 (4.9[d])
0.05–0.32 (0.15[c])				
Polluted hamlet				
0.30–0.49 (0.40)	280	19 (6.8)	258	56 (21.7)
0.50–0.69 (0.61)	255	43 (16.9)	240	70 (29.2)
0.70–1.05 (0.82)	105	22 (21.0)	100	45 (45.0)

[a]Population of hamlet on July 1st, 1967.
[b]Number of women examined.
[c]Means in parentheses.
[d]Percentage of people with indicated condition (Nogawa et al. 1983b).

the Cd concentrations of the unpolished rice were below 0.1μg/gFM, which is the Cd concentration produced in unpolluted paddy fields under usual agricultural practice. The measures to decrease the emission of Cd and related metals have also been made at the pollution source.

2.1.2 The Ichi and Maruyama River Basins, Hyōgo Prefecture

Some paddy fields in Ikuno-, Ōkōchi-, and Asako-Town in the Ichi and Maruyama River Basins were designated as agricultural land soil pollution policy areas. Ikuno Mine, Akenobe Mine, and Nakase Mine are found in the basins; the Ikuno Mine is the largest of these three mines.

An outcrop of Ikuno Mine was discovered in 807 AD. The ores contained Cd, Cu, Zn, Pb, As, Ag, Sn, W, etc. Ag smelting began as early as 1542. Afterwards, Cu and Pb were also smelted. After 1868, modern mining techniques were introduced from abroad. Zn was smelted from 1916. Ikuno Mine ceased operation in 1973. Over 10,000,000 t of raw ore were mined from 1896 to 1973 (Asami 1981a).

Heavy metals such as Cd, Zn, Pb, and Cu were contained in the mine water and the waste water from the mines, the smelters, the dressing factories, and the waste heaps, etc. These wastes have flowed into the Ichi and Maruyama Rivers and their tributaries, and then into paddy fields with irrigation water.

There are three large waste heaps around Ikuno Mine, the slime of which contains many heavy metals. Kyūhō waste heap (469×10^3 m^3) was used from 1930 to 1937; the slime contained 48.8 μgCd/gDM, 6,670 μgZn/gDM, 2,510 μgPb/gDM, 1,710 μgCu/gDM, and 29,400 μgAs/gDM. Daisen waste heap (410×10^3 m^3) was used from 1937 to 1941; the slime contained 45.6 μgCd/gDM, 5,830 μgZn/gDM, 1,070 μgPb/gDM, 2,000 μgCu/gDM, and 5,660 μgAs/gDM. Miyanotani waste heap ($2,849 \times 10^3$ m^3) has been used since 1941; the slime contained 9.78 μgCd/gDM, 1,880 μgZn/gDM, 583 μgPb/gDM, 2,300 μgCu/gDM, and 585 μgAs/gDM (Asami 1981a).

Heavy metal concentrations of the river sediments, the paddy soils, and the rice grains in the area were determined (Asami 1981a; Asami et al. 1981, 1982, 1983). Waste water from Ikuno Mine was discharged mainly into the Ichi River which flowed south and drained into the Seto Inland Sea and some parts into the Maruyama River which flowed north and drained into the Japan Sea. Table 6 shows the concentrations of Cd, Zn, Pb, and Cu in the paddy soils and of Cd in the unpolished rice produced in the paddy fields along the main streams of the Ichi and Maruyama Rivers in relation to the distances from Ikuno Mine. In paddy soils along the Ichi River, the heavy metal concentrations were higher in the upstream area and decreased gradually downstream. In the paddy soils along the Maruyama River, the heavy metal concentrations were also higher in the upstream area and decreased abruptly downstream. The unpolished rice produced in Ikuno- and Ōkōchi-Town in the Ichi River Basin, and Ikuno- and Asako-Town in the Maruyama River Basin showed higher Cd concentrations than the downstream areas.

The vertical distributions of Cd, Zn, Pb, and Cu in the polluted paddy soils are shown in Fig. 2. In the polluted paddy soils, even the subsoils were seriously polluted by heavy metals. In some soils, the subsoils were more polluted than the surface soils. These tendencies, as in the paddy soils in the Jinzu River Basin, are derived from the very long period of the mining and smelting operations.

According to the Hyōgo Prefectural Government, Cd concentrations of the surface paddy soils extracted by 0.1 N HCl were 8.20(3.90–12.2) μg/gDM and those of the unpolished rice were 0.69(0.15–1.43) μg/gFM in the designated area (Hyōgo Prefecture 1972, 1973).

In Ichi River Basin, Nogawa et al. (1975a) found five patients whose clinical features were completely in accord with those of patients with Itai-itai disease of the Jinzu River Basin, and concluded that they also suffered from Itai-itai disease. Renal dysfunction patients were also found in this area (Nogawa et al. 1975b). Three Itai-itai patients had been living in towns which were not designated as an agricultural land soil policy area.

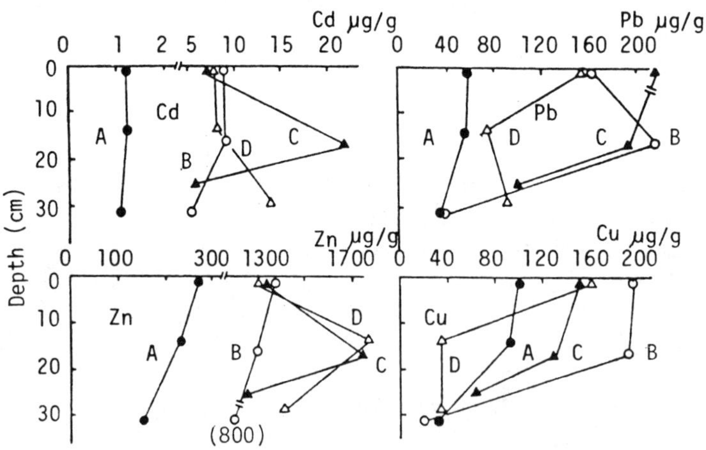

Fig. 2. Vertical distribution of heavy metals in paddy soils in the Ichi and Maruyama River Basins; *A* Habuchi, Asako-Town; *B* Minamimayumi, Ikuno-Town; *C* Kawajiri, Ikuno-Town; *D* Okawa, Ōkōchi-Town (Asami et al. 1982)

Table 6. Heavy metal concentrations of the surface paddy soils and the unpolished rice along the main streams of the Ichi and Maruyama Rivers

River	Distance from Ikuno Mine (km)	Locality	Heavy metal concentrations of the surface paddy soils (μg/gDM)					Cd concentrations in unpolished rice (μg/gFM)		
			No. of Samples	Cd	Zn	Pb	Cu	No. of samples	Cd	Maximum
Ichi	1	Ikuno	5	7.17±1.67[a]	1,205±279	231± 28	192± 69	–	–[b]	>1[c]
	7	Ōkōchi	5	7.65±2.74	1,096±462	238±193	206±146	7	0.61±0.64[d]	1.73
	12	Kanzaki	6	5.59±1.39	947±337	220± 47	169± 64	40	0.43±0.25	1.00
	16	Ichikawa	7	4.23±1.85	780±272	151± 57	118± 33	78	0.28±0.22	1.42
	24	Fukuzaki	4	2.97±1.43	515±277	168± 86	117± 45	22	0.24±0.11	0.51
	28	Kodera	8	2.42±0.53	534±214	199±100	147±106	50	0.24±0.17	0.72
	38	Himeji	44	1.93±2.03	393±215	132±156	66± 33	148	0.25±0.20	1.50
Maruyama	2	Ikuno	4	9.15±2.20	1,020±401	371±387	128± 73	7	0.69±0.48	1.64
	8	Asako	29	3.89±2.01	512±316	101± 53	46± 20	157	0.45±0.37	2.19
	20	Wadayama	8	1.88±0.49	132± 37	49± 7	37± 11	15	0.29±0.31	0.97
	26	Yabu	4	1.83±0.33	142± 45	58± 28	28± 13	5	0.13±0.08	0.24
	32	Yōka	8	2.13±0.72	172± 63	57± 14	44± 20	9	0.09±0.08	0.28
	38	Hidaka	2	1.75±0.78	126± 48	53± 24	34± 12	5	0.22±0.13	0.33
	50	Toyooka	11	2.20±1.18	186± 68	70± 24	50± 26	11	0.23±0.17	0.56

[a]Mean±SE.
[b]Unpolished rice was not taken since the production of rice was prohibited because of Cd pollution.
[c]Therefore, maximum concentration of unpolished rice would be higher than 1.0 μg/gFM.
[d]A part of the paddy field was prohibited to produce rice in Okōchi-Town (Asami et al. 1982, 1983).

Paddy fields of Ikuno-Town (6.05 ha) and Ōkōchi-Town (16.5 ha) were rehabilitated in 1974, where a 30-cm-deep layer of unpolluted soil was placed on the polluted surface soil. The cost was estimated at \$70,290 ha^{-1} (Hyōgo Prefecture 1972). After the rehabilitation, the Cd concentrations in unpolished rice were practically below 0.1 μg/gFM (Asami 1981a; Asami et al. 1983). Rehabilitation must be made on the polluted paddy fields outside the designated area, because Itai-itai disease and renal dysfunction patients are found even in towns in which paddy fields were not designated.

2.1.3 The Sasu and Shiine River Basins, Nagasaki Prefecture

The Sasu and Shiine River Basins are on the Tsushima Islands, Nagasaki Prefecture, which are located between Kyushu Island and the Korean Peninsula. In this area, mining started 1,300 years ago (674 AD). Pb and Zn were mined from the beginning of this century. Tōhō-Aen Co. Ltd. obtained Taishu Mine in 1940. Modern mining and dressing techniques were introduced in 1948. This mine ceased operation in 1973. From 1948 to 1972, 3,585,893 t of raw ore were mined. In this period, 208,899 t of Pb concentrate and 457, 729 t of Zn concentrate were produced. Heavy metals such as Cd, Zn, and Pb were contained in the mine water and the waste water from the mine, the dressing factory, and the waste heaps, etc. These wastes have flowed into the Sasu and Shiine Rivers and then into paddy fields with irrigation water. Moreover, the paddy fields in this area were produced on the soil contaminated by the slag in old age, and afterwards, they were polluted by the fine material from slag, refuse, and slime as well as the waste water (Nagasaki Prefecture 1973).

New refuse in the modern technique period contained 38(trace-187) μgCd/gDM, 5,000(650–19,790) μgZn/gDM, and 2,650(190–12,120) μgPb/gDM. Old refuse contained 158(6–259) μgCd/gDM, 19,440(800–38,930) μgZn/gDM, and 7,380(300–27,400) μgPb/gDM. Slime of the Okutomi waste heap contained 40 μgCd/gDM, 6,870 μgZn/gDM, and 3,000 μgPb/gDM (Nagasaki Prefecture 1973).

The agricultural land soil pollution policy area in the Sasu and Shiine River Basins is composed of 57.8 ha paddy field. About 20% of the surface paddy soil contained above 8 μgCd/gDM and about 85% of the surface paddy field contained above 2 μgCd/gDM in the basins. As mentioned above, part of the heavy metals in the paddy soil was derived from soil materials containing heavy metals, whereas the rest flowed into the paddy soil with irrigation water. This fact is obvious, because the mean concentrations of heavy metals in the surface paddy soil (Cd 8.6 μg/gDM, Zn 1,321 μg/gDM, Pb 2,165 μg/gDM) are higher than those of the neighboring upland field soil (Cd 4.9 μg/gDM, Zn 1,000 μg/gDM, Pb 1,870 μg/gDM) developed from the same soil materials (Nakashima and Ono 1979).

Cd concentrations in the surface paddy soil (0.1 N HCl extraction) and the unpolished rice in the designated area were 5.6(1.0–12.0) μg/gDM ($N=69$) and 0.90(0.20–3.64) μg/gFM ($N=400$), respectively (Agricultural Section, Tsushima Office, Nagasaki Prefecture 1979).

The vertical distribution of heavy metals in two paddy soil profiles is shown in Table 7. The paddy soils are polluted by heavy metals throughout the soil profiles.

Table 7. Vertical distribution of heavy metals in the paddy soils of the Sasu and Shiine River Basins (μg/gDM)[a]

Kashine				Shiine			
Depth (cm)	Cd	Zn	Pb	Depth (cm)	Cd	Zn	Pb
0–22	7.7	1,420	2,400	0–18	6.7	1,040	2,140
22–32	6.9	940	3,620	18–45	9.7	1,340	2,890
32–50	5.4	782	1,920	45–68	9.2	1,360	1,800
50–64	4.1	920	1,740	68–83	11.1	1,120	6,720
64–75	10.9	1,220	4,380	83–	11.8	2,080	6,240
75–	6.0	1,780	2,040				

[a]HNO_3-$HClO_4$ digestion method was used (Nagasaki Prefecture 1973).

In this area, three patients with a disease similar to Itai-itai disease (Takebayashi 1980, 1983, 1984), and 18 patients with renal dysfunction (Nagasaki Prefecture 1985) have been found.

A rehabilitation plan of the Cd-polluted paddy soils (55.75 ha) was made in which the polluted soils are covered with a 25-cm-deep layer of unpolluted soil. The cost was estimated at \$190,135 ha^{-1} (Agricultural Section, Tsushima Office, Nagasaki Prefecture 1979). The cost is higher than that of the other four areas, because this area is mountainous and perhaps, the countermeasure work will be difficult.

2.1.4 The Kakehashi River Basin, Ishikawa Prefecture

In the upper Kakehashi River area, there is the Ogoya Mine, the ore vein of which was discovered in 1682. After the Nippon Mining Co. Ltd. obtained the mine in 1931, the production of Cu increased rapidly, and the annual production of raw ore reached 220,000 t in 1961. In 1962, Hokuriku Mine Co. Ltd. obtained the mine. The mine ceased operation in 1971. The total amount of raw ore produced from 1931 to 1971 was 4,710,000 t. The main metals were Au, Ag, Cu, and Zn. Cd was not extracted and was dumped as mine refuse and slime (Ishikawa Prefecture 1976).

Heavy metals emitted from mining and smelting activities of the mine have flowed into the paddy fields in the Kakehashi River Basin with irrigation water.

The detailed survey based on the Agricultural Land Soil Pollution Prevention Law has been made on 705 ha of paddy field in this area, and 518.6 ha were designated as an agricultural land soil pollution policy area.

In the designated area, Cd concentrations in the surface paddy soil and the subsurface paddy soils extracted by 0.1 N HCl were 3.11(1.01–17.70) μg/gDM and 2.09(0.10–14.74) μg/gDM (N = 122), respectively, and those of the unpolished rice were 0.81(0.41–2.84) μg/gFM (N = 122) (Ishikawa Prefecture 1980).

In the patients with suspected Cd intoxication, Looser's zone was found in the bones of two patients (Ishikawa Prefecture 1982; Ishikawa Prefecture 1983). Looser's zone is a characteristic of osteomalacia. Therefore, these two patients were thought to be suffering from Itai-itai disease. A large number of renal dysfunction patients (7.8% in 2,614 residents) have also been found in this area (Ishikawa

Prefecture 1976; Nogawa et al. 1978). Retinal binding protein in urine was used as the index of renal dysfunction.

A rehabilitation plan (345.8 ha) has been made in which the polluted paddy soils are covered by a 20-cm deep layer of unpolluted soil with or without removing the polluted surface paddy soils. The cost was estimated at \$82,300 ha^{-1} (Ishikawa Prefecture 1980).

2.1.5 Kosaka Area, Akita Prefecture

Many mines exist in Akita Prefecture; 9 mines are now operating and 238 mines have ceased operation. Ores of these mines contained Au, Ag, Cu, Pb, and Zn, etc. (Akita Prefecture 1984). Therefore, soil pollution by heavy metals has been found in many areas in Akita Prefecture and 16 areas have been designated as agricultural land soil pollution policy areas.

Kosaka area (47.9 ha) is one of these designated areas. The majority of the area is in Hosogoe Hamlet, Kosaka-Town, the pollution source of which is Kosaka Mine with a long history of operation. It was discovered in 1861, and at present, it is regarded as one of the three largest Cu mines in Japan. The dust and the fume containing heavy metals have fallen around the smelter and have polluted the paddy soils and the forest soils in Kosaka-Town (Homma 1981).

The surface paddy soils (N = 41) of Hosogoe Hamlet were taken and analyzed by Homma et al. (1977). Heavy metal concentrations were determined as 6.6(2.9–14.5) μgCd/gDM, 437(252–762) μgZn/gDM, 509(208–912) μgPb/gDM, and 367(160–600) μgCu/gDM by the HNO_3-$HClO_4$ digestion method. Table 8 shows the vertical distribution of heavy metals in the paddy soils in Hosogoe Hamlet. Heavy metal concentrations were high even in the subsoils. The Cd concentration of unpolished rice produced in Hosogoe Hamlet was 0.78(0.16–4.81) μg/gFM (Ogawa et al. 1985).

Renal dysfunction patients who showed concurrent proteinuria and glucosuria were found in Hosogoe Hamlet (33 cases in 147 residents) (Saito et al. 1977a, b), but patients with the disease similar to Itai-itai disease have not been found in this area. Saito et al. (1981) determined the β_2-microglobulin in urine and found

Table 8. Vertical distribution of heavy metals in the paddy soils of the Kosaka area (μg/gDM)[a]

Paddy field A					Paddy field B				
Depth (cm)	Cd	Zn	Pb	Cu	Depth (cm)	Cd	Zn	Pb	Cu
0– 5	8.0	248	440	336	0– 5	5.4	232	488	302
5–10	6.5	240	408	312	5–10	7.0	240	538	332
10–15	6.5	234	404	296	10–15	7.0	236	536	332
15–20	6.6	230	360	272	15–20	6.0	234	520	322
20–25	6.6	202	304	232	20–25	4.0	230	370	268
25–30	4.0	184	200	150	25–30	1.1	176	52	70
30–40	0.9	134	40	22					
40 –	0.8	122	36	16					

[a] HNO_3-$HClO_4$ digestion method was used (S. Homma, 1972, personal communication).

that the concentration of β_2-microglobulin at the Kakehashi River Basin was nearly equal to that at the Sasu and Shiine River Basins and the concentration of β_2-microglobulin in the Kosaka area was comparatively lower. From these results, it is assumed that the health effect by Cd in the Kakehashi River Basin and the Sasu and Shiine River Basins was severer than in the Kosaka area.

A rehabilitation plan of the Cd-polluted paddy fields (42.9 ha) was made in which the polluted soils are covered by a 25-cm-deep layer of unpolluted soil. The cost was estimated at \$90,250 ha^{-1} (Akita Prefecture 1982).

In summarizing the available data in Japan, maximum daily intake may be under 200–290 μg of Cd, and the Cd concentration in rice, which would not cause renal dysfunction, may be set under 0.30–0.40 μgCd/gFM (Nogawa 1984).

In addition to these five areas, the soil pollution by Cd and related metals around Nisso Aizu Smelter (Bandai-Town, Fukushima Prefecture) and in the Usui River Basin (Annaka City and Takasaki City, Gunma Prefecture) was also reported (Asami 1972; Kobayashi et al. 1973; Hirata and Takenaga 1977; Takenaga and Hirata 1977; Hirata and Ōtsuka 1979; Tatekawa 1978; Hirata 1981).

2.2 Soil Pollution by Chromium from Chromium Slag

In Japan, 1,146,532 t of Cr slag have originated from Cr smelters; 54.4% from Nippon Chemical Industrial Co. Ltd., and 40.6% from Nippon Denko Co. Ltd. (Asami and Sampei 1977). Cr slag from Komatsugawa Plants of Nippon Chemical Industrial Co. Ltd. have been dumped in Tokyo and in Chiba Prefecture.

According to the Environment Protection Bureau, Tokyo (1977) the amounts of Cr slag dumped were estimated at over 330,000 t in 330,000 m^2 (172 sites) in Tokyo, especially in Kōtō and Edogawa Wards. The thickness of dumped slag reached 7 m in some places. A survey of 45 sites conducted in 1975 proved that 123,000 t slag and 1,192,000 m^3 CrVI-polluted soil existed in 185,000 m^2. This means that CrVI migrating from the slag polluted a large amount of soil around the dumped sites. Maximum concentration of CrVI was 15,600 μg/gDM. The river water and air in these areas were also polluted by Cr. Air samples contained 0.219–0.382 μgCr m^{-3}, which was much higher than the values of the usual sites in Tokyo (0.011–0.018 μgCr m^{-3}). When air sampling is made by a high-volume air sampler, CrVI is reduced to CrIII on the filter. Therefore, CrVI is not found in the usual cases. But in this area, 0.086 μgCrVI m^{-3} was found in an air sample. This means that much more CrVI existed in the air.

It is noteworthy that the relatively low-level impairments in the respiratory system, skin, and digestive organs have occurred in high frequency in the residents of the Cr-polluted areas (Environment Protection Bureau, Tokyo 1977).

Countermeasures were conducted in which the Cr slag and the polluted soils were mixed with $FeSO_4$ to reduce CrVI to CrIII. Contrary to CrVI, CrIII do not migrate to the soils around the slag and the polluted soils. However, some CrVI remained, and the unpolluted soils were protected not to be polluted by the migrated CrVI.

In Ichikawa City, Chiba Prefecture, about 60,000 t of Cr slag from the Komatsugawa Plants of Nippon Chemical Industrial Co. Ltd. were brought to fill the reclaimed land (about 5 ha) from 1970 to 1971. The slag contained 10,400–20,600 μgCrVI/g and 54,600 μgCr/gDM. Ichikawa City conducted a countermeasure in which the slag of the surficial 20-cm-deep layer was mixed with $FeSO_4$ to reduce CrVI to CrIII and then, the slag was covered with a 1-m-deep layer of unpolluted soil. But, CrVI existed as a chromate ion which is not adsorbed by the soil particles, and the remaining CrVI moves to the surface of the covered soil with capillary water (Asami 1973). Maximum concentration found in the surface soil was 4,560 μgCrVI/gDM (Asami and Sampei 1977). Therefore, Ichikawa City conducted another countermeasure in which 51,300 m^3 of Cr slag and 10,200 m^3 of polluted soil were removed to the land obtained by Nippon Chemical Industrial Co. Ltd. (Ichikawa City, Chiba Prefecture 1979).

2.3 Soil Pollution by Beryllium Discharged from Beryllium Factories

Be was smelted in four factories of three corporations in Japan. The largest one is NGK Insulators Corporation which had the ability to produce 1,000 t of Be-Cu mother alloy (Be 4%), 50 t of Be-Al mother alloy (Be 2–5%), 60 t of BeO powder, and 3.6 t or Be metal in 1977. Soil pollution by Be around two factories of the corporation was surveyed in 1972 and 1976 (Table 9). The Be concentration around Atsuta Factory was nearly the same level in either year, and that around Chita Factory in 1976 was twice as much as that in 1972, because the production of Be materials had mostly been conducted in Chita Factory in recent years. Sediment of drainage contained high levels of Be, the maximum value being 326 μg/gDM, and the drainage water contained 1.11 and 2.19 μg l^{-1} at Chita Factory (Asami and Fukazawa 1981a). Be concentrations in the plants around Atsuta and Chita Factories were 0.05 (trace-0.34) μg/gDM and 0.69 (0.04–3.42) μg/gDM, respectively (Asami and Fukazawa 1981b).

Be concentrations of soil, sediment, fresh water, and plant in unpolluted sites in Japan are shown in Table 10. Be concentrations in the soils, sediments, drainage water, and plants around the Be factories were much higher than those in unpolluted sites.

Total Be emission from Chita Factory was 153 g day^{-1} in 1980, which was much higher than the national emission standards for Be (10 g Be day^{-1}) in USA (Noma and Kawaguchi 1981).

Afterwards, Be emission was reduced, but countermeasures against Be in soils have not been made.

3 Soil Pollution by Heavy Metals in Countries Outside Japan

Soil pollution by heavy metals from mining and smelting activities has been studied in many countries, especially in Canada, USA, and UK. In most instances, the soil pollution by heavy metals emitted as dust and fume from the smelters has been reported. The author cannot find a report which deals with the soil pollution by irrigation water containing heavy metals from mines and smelters.

Table 9. Beryllium concentrations in the surface soils around two factories of NGK Inslators Corporation[a]

Factory	Locality	Distance from factory	1972 No. of samples	1972 Mean (range)	1976 No. of samples	1976 Mean (range)
			Be concentrations in surface soils (μg/gDM)			
Atsuta	Atsuta Ward, Nagoya City, Aichi Prefecture	<500 m	3	7.37(2.17–17.0)	14	8.56(1.54–55.0)
		>500 m	3	2.78(1.47–4.83)	7	2.22(1.65–3.21)
Chita	Handa City, Aichi Prefecture	<500 m	6	2.92(0.82–8.03)	9	6.53(2.70–11.5)
		>500 m	3	1.14(0.88–1.45)	4	2.07(1.48–3.01)

[a]Sodium carbonate fusion method was used (Asami and Fukazawa 1981a).

Table 10. Beryllium concentrations of unpolluted soils, sediments, fresh water, and plants

Environmental materials	No. of samples	Mean±SE (range)	References
Surface soils[a]	27	1.17±0.43(0.27–1.95) μg/gDM	
Subsoils	97	1.35±0.52(0.34–3.49) μg/gDM	
All soil horizons[a]	124	1.31±0.51(0.27–3.49) μg/gDM	
Surface sediments	8	1.36±0.60(0.65–2.40) μg/gDM	Asami and Fukazawa (1985)
Subsurface sediments	41	1.28±0.49(0.50–2.25) μg/gDM	
All sediment layers	49	1.29±0.50(0.50–2.40) μg/gDM	
Fresh water	19	0.017±0.012(0.005–0.040) μg l^{-1}	Asami and Fukazawa (1980b)
Plants	18	0.010±0.017(n.d.[b]–0.067) μg/gDM	Asami and Fukazawa (1980a)

[a]Except AO horizons.
[b]n.d.: Not detected.

3.1 Canada

Metal concentrations in the surface soils and some other data around mines and/or smelters are shown in Table 11.

According to Freedman and Hutchinson (1981) the fate and effects of the emission of pollutants from the Sudbury smelters into the surrounding environment are better documented than those of any other smelter in the world, with more than 125 published papers or reports dealing with various aspects of pollutant dispersal or environmental degradation in the Sulbury area. In about 104 km^2 surrounding the Sudbury smelters, the vegetation was almost totally removed and

Table 11. Metal pollution of surface soil by mining and smelting activities in Canada

Locality	Pollution source	Horizon or depth (cm)	Metal concentration in soil ($\mu g\ g^{-1}$) mean and/or range	Method	References
Sudbury, Ontario	Ni-Cu smelter	$O_1+O_2+A_1$	Ni 100–3,000, Cu 60–3,700 Zn 70–180 [3.0–76.5 km][a]	HNO_3-$HClO_4$ digestion	Freedman and Hutchinson (1980)
Sudbury, Ontario	Ni-Cu smelter	Surface soil	Maximum Ni 9,372, Cu 6,912	HNO_3-H_2O_2 digestion	Taylor and Crowder (1983)
Trail, British Columbia	Pb-Zn smelters	0–2.45	Pb 22–4,200 [0.4–97 km]		Schmitt et al. (1971)
Trail, British Columbia	Pb-Zn smelters	Mineral surface horizon (5–10)	Cd 5.6(0.1–21.5), Pb 194(12–1,002) Zn 148(38–391) [0.3–2.7 km]	1 N HNO_3 digestion	John et al. (1975)
Trail, British Columbia	Pb-Zn smelters	Mineral surface horizon	Cd 1.31[b](0.59–7.69), Pb 53.4(12.0–999.4) Zn 74.4(22.4–346.9) [6.4–78.8 km]	HNO_3-HF-$HClO_4$ digestion	John et al. (1976)
Toronto, Ontario	Secondary Pb smelter	Surface soil	Pb 2,626(355–8,750)		Roberts et al. (1974)
Toronto, Ontario	Secondary Pb smelter	0–5	Pb 133–21,200, As 8.7–533 [100–600 m]		Linzon et al. (1976)
Vancouver, British Columbia	Battery smelter	Surface soil	Pb 8,019(106–59,581) [15–300 m]	1 N HNO_3 extraction	John (1971)
Vancouver, British Columbia	Battery smelter	Surface soil	Cd 15.8(0.45–95.4) [15–300 m]	1 N HNO_3 extraction	John et al. (1972)

[a]Distance from pollution source.
[b]Geometric mean.
Unless otherwise stated, atomic absorption spectrophotometry was used after acid digestion or extraction in the reports cited in Tables 11–14.

soil was eroded from slopes, leaving blackened, naked bedrock. In Table 11, only the data of Freedman and Hutchinson (1980) and of Taylor and Crowder (1983) are shown. Data of heavy metal concentrations of soil surrounding the Sudbury smelters are also reported by Whitby and Hutchinson (1974), Hutchinson and Whitby (1974, 1977), Rutherford and Bray (1979), Lozano and Morrison (1981), and Hazlett et al. (1984).

Pb poisoning in horses (Schmitt et al. 1971) and accumulation of Cd, Pb, and Zn in soils (John et al. 1975, 1976) were studied around the Pb-Zn smelter complex at Trail, British Columbia.

The relationship between environmental Pb pollution and human health around the secondary Pb smelters (Roberts et al. 1974; Linzon et al. 1976), and Pb and Cd contamination of agricultural soils near battery smelters (John 1971; John et al. 1972) were also investigated.

3.2 USA

Metal concentrations in the surface soils near metal smelters in USA are shown in Table 12.

Pb in air, dust, food, tap water, and soil around Pb-Cu-Zn smelters and Pb in the blood of the residents were analyzed. The obtained data suggested that particulate Pb in dust and air accounted for most of the Pb absorption in El Paso children (Landrigan et al. 1975).

Soil cores, grasses, and ambient air aerosols were sampled in the vicinity of a Pb-smelting complex in Kellogg, Idaho. Of 34 elements analyzed, Cd, Sb, Ag, Pb, Au, Zn, Se, As, In, Ni, Cu, and Hg were the most highly enriched in surface soils (Ragaini et al. 1977).

Zn, Cd, Cu, and Pb which escaped from two Zn smelters have highly contaminated the soil and vegetation near the smelters. The total numbers of bacteria, actinomycetes, and fungi were greatly reduced in the most severely Zn-contaminated soils, and the litter decomposition rate was decreased (Buchauer 1973; Jordan and Lechevalier 1975).

Cu, Zn, and Cd contamination of soil with respect to distance from the Cu smelter, the chemical fractionation of these metals, and the mobility of these metals in soils have been studied (Kuo et al. 1983).

The vegetation of an upland community near a recently inactivated Cu smelter was studied. The most probable cause of the observed vegetational changes was effluents, particularly SO_2 and Cu, emitted by the Cu smelter (Wood and Nash III 1976).

3.3 UK

Metal concentrations in the surface soils near mines and/or smelters in UK are shown in Table 13.

Pb, Zn, and Cd concentrations in grass, moss, lichen, and soil near the Zn-Pb smelter were determined (Burkitt et al. 1972). Little and Martin (1972) also surveyed these metals in soil and natural vegetation in the same area.

Table 12. Metal pollution of surface soil by mining and smelting activities in USA

Locality	Pollution source	Horizon or depth (cm)	Metal concentration in soil mean and/or range ($\mu g\ g^{-1}$)	Method	References
El Paso, Texas	Pb, Cu, Zn smelter	Surface soil(?)	Pb 3,457(560–11,450) [<200 m][a]		Landrigan et al. (1975)
Kellogg, Idaho	Pb smelter	0–2	Cd 18–140, Sb 5–260, Ag 2.7–31 Pb 170–7,900, Zn 200–13,000 As 24–110 [1.1–19.0 km]	NAA and XFA[b]	Ragaini et al. (1977)
Palmerton, Pennsylvania	Zn smelter	A_1	Zn 50,000–80,000, Cd 900–1,500 Cu 600–1,200, Pb 200–1,100 [<1 km]	$HClO_4$ digestion	Buchauer (1973)
Palmerton, Pennsylvania	Zn smelter	A_1	Zn 110–60,700, Cd 3–750, Cu 18–400, Pb 113–1,350 [2–46 km]	HNO_3-$HClO_4$ digestion	Jordan[c] and Lechevalier (1975)
Ruston, Washington	Cu smelter	0–15	Cu[d] 1,550–2,480, Zn 430–1,370 Cd 5.0–14.3 [<1 km]	HNO_3-H_2O_2 digestion	Kuo et al. (1983)
Superior, Arizona	Cu smelter	Surface soil	Cu 24–9,618, Zn 10–299, Pb 4–323, Cd 0.6–5 [0.4–8.1 km]	0.05 N HCl + 0.025 N HNO_3 extraction	Wood and Nash III (1976)

[a]Distance from pollution source.
[b]Neutron activation analysis and X-ray flourescence analysis.
[c]Formerly Buchauer.
[d]Data were obtained from a figure.

Table 13. Metal pollution of surface soil by mining and smelting activities in UK

Locality	Pollution source	Depth (cm)	Metal concentration in soil ($\mu g\ g^{-1}$) mean and/or range	Method	References
Avonmouth, England	Zn-Pb smelter	0–5	Pb 60–600, Zn 90–5,000 Cd 1.1–32 [0.3–12.7 km][a]	Conc.HNO_3 digestion	Burkitt et al. (1972)
Swansea Valley, Wales	Zn, Pb, Cd, Ni Cu smelter	<5	Zn 20–543, Pb 6–263, Cd 0.3–26.0 Ni-2–234, Cu 35–214 [1.5–16 km]	[b]	Gordon et al. (1971)
Ystwyth Valley, Wales	Old mining area	0–15	Pb 1,419(90–2,900), Zn 455(95–810) Cu 30(17–42), Cd 2.5(1.2–4.0)	Conc.HNO_3 and aqua regia digestion	Alloway and Davies (1971)
Conway Valley, Wales	Old mining area	0–15	Pb 57(30–88), Zn 226(114–400) Cu 19(14–25), Cd 1.8(0.8–2.9)		
Parys Mountain, Wales	Old mining area	0–15	Pb 379(171–1,002), Zn 335(190–650) Cu 642(116–1,405), Cd 1.8(1.0–3.9)		
Derbyshire, England	Old mining area	0–15	Pb 190–30,000 [<2 km]	HNO_3-$HClO_4$ digestion	Colbourn and Thornton (1978)
Halkyn, Wales	Old mining area	0–15	Pb 886(35–47,995), Zn 728(10–49,393) Cu 18(2.3–252), Cd 6.1(0.4–540)	Conc.HNO_3 digestion	Davies and Roberts (1978)
Derbyshire, England	Old mining area	0–12	Pb 1,927(137–3,779), Cd 16.3 (1.9–62)	Conc.HNO_3 digestion	Davies et al. (1981)
Halkyn, Wales	Old mining area	0–12	Pb 1,597(285–3,055), Cd 6.4(0.3–21)		
Ceredigion, Wales	Old mining area	0–12	Pb 1,678(61–8,355), Cd 0.48(0.05–2.4)		
Mendip, England	Old mining area	0–12	Pb 1,738(103–7,940), Cd 18.0(1.2–94)		
Aberystwyth, Wales	Old mining area	0–10	Pb 1,759(38–14,910), Zn 96(11–641) Cu 32(2.1–42.4), Cd 1.35(0.60–2.90)	Conc.HNO_3 digestion	Davies and White (1981)
Shipham, England	Old mining area	0–15	Cd 24–440, Pb 475–7,800, Zn 2,040–50,000, Cu 22–49	Conc.HNO_3 digestion	Matthews and Thornton (1982)
London, England	Secondary metal smelter	0–2	Pb 635(304–1,169) [45–150 m]	HNO_3-$HClO_4$ digestion	Muskett et al. (1979)

[a]Distance from pollution source.
[b]0.5 N CH_3COOH extraction; Cu was extracted by EDTA.

Gordon et al. (1971) used plants and soils as indicators of metals in the air near a nonferrous metal smelter.

According to Thornton and Abrahams (1984), in Britain, the oldest and most extensive sources of metal pollution are related to the history of metalliferous mining which commenced in Roman and earlier times, thrived until the end of the 19th century, and then declined rapidly. It has been estimated that over 4,000 km^2 of agricultural land in England and Wales is contaminated with one or more metals due to this legacy.

Therefore, many case studies on soil pollution around old mine areas have been conducted in UK.

Alloway and Davis (1971) studied heavy metal concentrations in soils affected by base metal mining in three areas in Wales, and found enhanced levels within the mining areas of total Cd, Pb, Ag, Zn, and sometimes, Cu in alluvial soils and high Pb in slope soils.

Colbourn and Thornton (1978) studied Pb pollution in agricultural soils near old mining and smelting sites in an area of approximately 600 km^2.

Davies and Roberts (1978) analyzed surface soil samples from an area 260 km^2 in and around the Hylkyn Mountain Pb-mining district, and found extensive areas of Cd, Pb, and Zn contamination and a smaller area of Cu contamination.

Davies et al. (1981) studied Cd and Pb levels of soils and Cd and Pb uptake by radish in field garden plots in four mining areas in England and Wales, and found that the Halkyn area has the severest Pb problem and the Mendips (including the village of Shipham) the severest Cd problem.

Davies and White (1981) investigated the environmental Pb pollution by windblown material from a spoil heap at a derelict Pb mine in West Wales, and found a marked movement of dust from the heap to nearly 1,800 m down the valley. Pb in the dust polluted the soils and plants.

Matthews and Thornton (1982) found that very high Cd and Zn levels existed in the soils reclaimed from old mine workings at Shipham, England. Nevertheless, little of the metals is transferred into the pasture herbage.

Cd and Pb in house dust and air samples and Pb in soil samples around a secondary metal refinery in London were analyzed (Muskett et al. 1979).

3.4 Western Europe, Australia, and Zambia

Metal concentrations in the surface soils around mines and/or smelters are shown in Table 14.

The distribution and availability of Cd, Zn, Cu, and Pb in soils at two highly polluted areas located near Zn smelters in Belgium were investigated (Scokart et al. 1983).

Zn, Pb, and Cu concentrations in arable top soils, lichens, and bryophytes related to distance from a secondary smelter in Copenhagen were determined. The heavy metal gradients from the pollution source were steeper in soil compared with lichens and bryophytes (Andersen et al. 1978).

Chemical analysis of samples of needles of Allepo pine, humus, and mineral soil collected in the vicinity of a Pb smelter in Lavrion, Greece, showed that an area

Table 14. Metal pollution of surface soil by mining and smelting activities in Western Europe, Australia, and Zambia

Locality	Pollution source	Depth (cm)	Metal concentration in soil ($\mu g\ g^{-1}$) mean and/or range	Method	References
Balen-Neet, Belgium	Zn smelter	[a]	Cd 3.2–85, Zn 66–4,550 Pb 137–4,351, Cu 16–356 [<2 km][b]	$HCl-HNO_3$ digestion	Scokart et al. (1983)
Engis, Belgium	Zn smelter	[a]	Cd 11–1,781, Zn 1,340–180,000 Pb 178–14,000, Cu 40–1,089 [<2 km]	$HCl-HNO_3$ digestion	Scokart et al. (1983)
Copenhagen, Denmark	Secondary smelter	0–10	Zn[c] 100–600, Pb 100–1,500 Cu 20–260 [<0.8 km]	1 N HNO_3 digestion	Andersen et al. (1978)
Lavrion, Greece	Pb smelter	0–10	Pb 1,250–18,500, Zn 550–16,000 [50–1,500 m]	HNO_3-HClO_4 digestion	Nakos (1979)
Arnhem, Netherlands	Secondary Pb smelter	0–5	Pb 370 (21–2,309), Cd 1.60(0.28–10.47) [<1 km]	3M HCl digestion	Keizer et al. (1982)
West Germany	Zn smelter	Surface soil	Zn 640–4,500, Cd 17.0–40.8 [0.25–1.2 km]	$HCl-HNO_3$ digestion	Ernst (1972)
Port Pirie, Australia	Pb smelter	0–5	Pb[c] 23–3,700, Zn 10–3,700 Cd 0.5–11 [2–26 km]	0.1 M EDTA extraction	Cartwright et al. (1976)
Kabwe, Zambia	Zn–Pb smelter	0–10	Cd 9.0(0.55–46.5), Pb 862(92–2,580) Zn 1,050(180–3,500), [1–5 km]	$HNO_3-H_2O_2$ digestion	Nwankwo and Elinder (1979)

[a]Maximum amounts of heavy metals in the soil horizons including AO horizon.
[b]Distance from pollution source.
[c]Data were obtained from a figure.

of about 2 km around a Pb smelter was heavily polluted with Pb and also with Zn, S, and F^-. Most of the Pb was accumulated in the humus and the top layer of the mineral soil (Nakos 1979).

Pb and Cd pollution of soil and street dust in the vicinity of a secondary Pb smelter in Arnhem, Netherlands was studied (Keizer et al. 1982).

Airborne Zn and Cd pollution of soils and plants by a Zn smelter in West Germany was investigated. The heavy metal pollution of soils and plants inhibited the vitality of higher plants and the activity of soil organisms (Ernst 1972).

The concentrations of EDTA-extractable Pb, Zn, and Cd in the surface soils surrounding an isolated Pb smelter in Port Pirie, South Australia were determined (Tiller et al. 1975; Cartwright et al. 1976).

Cd and Pb pollution of soil and food grown close to a Zambia Pb and Zn smelter was studied (Nwankwo and Elinder 1979).

3.5 Eastern Europe, USSR, and the People's Republic of China

Metal concentrations in the surface soils around mines and/or smelters are shown in Table 15.

Data of Eastern Europe and USSR are cited from Kabata-Pendias (1984) and/or Kabata-Pendias and Pendias (1984) except for two areas.

Roszyk and Strojek (1978) found that the accumulation of Pb and Cu in the soils around a Cu smelter in Poland did not affect the concentrations of these metals in the cereal grain.

Table 15. Metal pollution of surface soil by mining and smelting activities in Eastern Europe, USSR, and the Peoples Republic of China

Locality	Pollution source	Metal concentration in soil ($\mu g\ g^{-1}$) mean or range	References
Bulgaria	Nonferrous metal smelter	Cd 2–5, Cu 24–2,015	Kabata-Pendias (1984)
Hungary	Pb smelter	As 2,000	Kabata-Pendias (1984)
Poland	Zn-Pb smelter	Cd 6–40, Zn 1,665–5,576	Kabata-Pendias (1984)
Poland	?	Pb 72–1,350	Kabata-Pendias (1984)
Poland	Cu smelter	Cu 72–125	Kabata-Pendias and Pendias (1984)
Poland	Cu smelter	Pb 25–215, Cu 65–305, Zn 110–230	Roszyk and Strojek (1978)
USSR	Cu-Zn smelter	Zn 400–4,245, Cu 720–2,900	Kabata-Pendias (1984)
USSR	?	Pb 3,000	Kabata-Pendias (1984)
USSR	Pb-Zn smelter	Pb 21–3,044	Kabata-Pendias and Pendias (1984)
Zerjav, Yugoslavia	Pb smelter	Pb 191–24,880	Djurić et al. (1971)
Guizhou Province, China	Abandoned Hg mine	Tl 20–80	Dai-xing and Shan-sheng (1982)

Djurić et al. (1971) determined Pb concentrations in air, water, soil, and local vegetation and the urinary excretion of δ-aminolevulinic acid around a Pb mine and smelter in Yugoslavia, and suggested that the absorption of Pb by human beings in this area might be hazardous.

Tl from an abandoned Hg mine in the People's Republic of China polluted the agricultural soil and then, Tl was absorbed by crops. About 200 cases of chronic Tl intoxication via Tl-polluted foods were found from 1960 to 1977 (Dai-xing and Shan-sheng 1982).

References

Agricultural Production Bureau, Ministry of Agriculture and Forestry (1974) Results of survey for agricultural land soil pollution policy, 1973. Agric Prod Bur Minist Agric For, Tokyo (J)[2]

Agricultural section, Tsushima Office, Nagasaki Prefecture (1979) Outline of the method of the paddy soil improvement against Cd pollution in the Sasu Area (J)

Akita Prefecture (1982) Countermeasure plans against the agricultural soil pollution in Kosaka Area (J)

Akita Prefecture (1984) White paper on environment in Akita Prefecture, 1983, pp 119–130 (J)

Alloway BJ, Davies BE (1971) Trace element content of soils affected by base metal mining in Wales. Geoderma 5:197-208

Andersen A, Hovmand MF, Johnsen I (1978) Atmospheric heavy metal deposition in the Copenhagen area. Environ Pollut 17:133–151

Asami T (1972) The pollution of paddy soils by cadmium, zinc, lead, and copper in the dust, fume, and waste water from Nisso Aizu Smelter. J Sci Soil Manure Jpn 43:339–343 (J)

Asami T (1973) Environmental pollution by chromium slag brought to the reclamated land by filling. J Sci Soil Manure Jpn 44:223–225 (J)

Asami T (1981a) The Ichi and Maruyama River Basins: Soil pollution by cadmium, zinc, lead, and copper discharged from Ikuno Mine. In: Kitagishi K, Yamane I (eds) Heavy metal pollution in soils of Japan. Japan Scientific Societies Press, Tokyo, pp 125–136

Asami T (1981b) Maximum allowable limits of heavy metals in rice and soil. In: Kitagishi K, Yamane I (eds) Heavy metal pollution in soils of Japan. Japan Scientific Societies Press, Tokyo, pp 257–274

Asami T (1984) Pollution of soils by cadmium. In: Nriagu JO (ed) Changing metal cycles and human health. Springer, Berlin Heidelberg New York Tokyo, pp 95–111

Asami T, Fukazawa F (1980a) Determination of beryllium in plants by atomic absorption spectrophotometry. J Sci Soil Manure Jpn 51:323–327 (J)

Asami T, Fukazawa F (1980b) Determination of beryllium in water by atomic absorption spectrophotometry. J. Sci Soil Manure Jpn 51:345–347 (J)

Asami T, Fukazawa F (1981a) Beryllium pollution of soils surrounding beryllium factories. J Sci Soil Manure Jpn 52:1–7 (J)

Asami T, Fukazawa F (1981b) Beryllium pollution of plants growing near beryllium factories. J Sci Soil Manure Jpn 52:8–10 (J)

Asami T, Fukazawa F (1985) Beryllium contents of uncontaminated soils and sediments in Japan. Soil Sci Plant Nutrit 31:43–53

Asami T, Sampei H (1977) Environmental pollution by chromium slag brought to the reclamated land by filling (Part 2). J Sci Soil Manure Jpn 48:323–328 (J)

Asami T, Homma S, Tanabe T, Hata A (1981) Pollution of the sediments of the Ichi and Maruyama Rivers by heavy metals discharged from Ikuno Mine, Hyōgo Prefecture. Jpn J Soil Sci Plant Nutrit 52:433–438 (J)

Asami T, Homma S, Tanabe T, Hata A (1982) Pollution of the paddy soils in the Ichi and Maruyama River Basins by cadmium, zinc, lead, and copper discharged from Ikuno Mine, Hyōgo Prefecture. Jpn J Soil Sci Plant Nutrit 53:507–512 (J)

[2]J: In Japanese.

Asami T, Homma S, Wada T, Nakashima K, Kubota M (1983) Pollution of unpolished rice produced in the Ichi and Maruyama River Basins by cadmium discharged from Ikuno Mine, Hyōgo Prefecture. Jpn J Soil Sci Plant Nutrit 54:30–36 (J)

Buchauer MJ (1973) Contamination of soil and vegetation near a zinc smelter by zinc, cadmium, copper, and lead. Environ Sci Technol 7: 131–135

Burkitt A, Lester P, Nickless G (1972) Distribution of heavy metals in the vicinity of an industrial complex. Nature (London) 238:327–328

Cartwright B, Merry RH, Tiller KG (1976) Heavy metal contamination of soils around a lead smelter at Port Pirie, South Australia. Aust J Soil Res 15:69–81

Colbourn P, Thornton I (1978) Lead pollution in agricultural soils. J Soil Sci 29:513–526

Dai-xing Z, Shan-sheng L (1982) Preliminary investigation on control and prevention of soil pollution by thallium. Acta Pedol Sin 19:409–411 (in Chinese)

Davies BE, Roberts LJ (1978) The distribution of heavy metal contaminated soils in northeast Clwyd, Wales. Water Air Soil Pollut 9:507–518

Davies BE, White HM (1981) Environmental pollution by wind blown lead mine waste: A case study in Wales, UK. Sci Total Environ 20:57–74

Davies BE, Cinnever RC, Lear JM (1981) Cadmium and lead contaminated soils in some British metal mining areas. In: Hemphill DD (ed) Trace substances in environmental health-XV. Univ Missouri, Columbia, pp 323–332

Djurić D, Kerin Z, Graovac-Leposavić L, Novak L, Kop M (1971) Environmental contamination by lead from a mine and smelter. Arch Environ Health 23:275–279

Environment Agency (1985a) White paper on environment, 1985. Printing Bureau, Ministry of Finance, Japan, Tokyo, p 322 (J)

Environment Agency (1985b) White paper on environment, 1985. Printing Bureau, Ministry of Finance, Japan, Tokyo, p 338 (J)

Environment Protection Bureau, Tokyo (1977) Report on the countermeasures against the soil pollution by hexavalent chromium slag (J)

Environment Protection Bureau, Tokyo (1979) The countermeasure against the pollution by chromium slag around Komatsugawa South and North Plants and Komatsugawa Second Plant (J)

Environment Protection Bureau, Tokyo (1980a) The countermeasure against the pollution by chromium slag in Horie Area (J)

Environment Protection Bureau, Tokyo (1980b) The countermeasure against the pollution by chromium slag in Horie Area (Part 2) (J)

Environment Protection Bureau, Tokyo (1981) The countermeasure against the pollution by chromium slag in Kōto Ward (a primary school and a fill-up land) (J)

Environment Protection Bureau, Tokyo (1982a) The countermeasure against the pollution by chromium slag in Horie Area (Part 3) (J)

Environment Protection Bureau, Tokyo (1982b) The countermeasure against the pollution by chromium slag in Komatsugawa Area (Part 2) (J)

Ernst W (1972) Zink- und Cadmium-Immissionen auf Böden und Pflanzen in der Umgebung einer Zinkhütte. Ber Dtsch Bot Ges 85:295–300

Freedman B, Hutchinson TC (1980) Pollutant inputs from the atmosphere and accumulations in soils and vegetation near a nickel-copper smelter at Sudbury, Ontario, Canada. Can J Bot 58:108–132

Freedman B, Hutchinson TC (1981) Sources of metal and elemental contamination of terrestrial environments. In: Lepp NW (ed) Effect of heavy metal pollution of plants, vol 2. Metals in the environment. Applied Science Publisher, London New Jersey, pp 35–94

Gordon T, Goodman GT, Roberts TM (1971) Plants and soils as indicators of metals in the air. Nature (London) 231:287–292

Hazlett PW, Rutherford GK, Loon GW van (1984) Characteristics of soil profiles affected by smelting of nickel and copper at Coniston, Ontario, Canada. Geoderma 32:273–285

Hirata H (1981) Annaka: Land polluted mainly by fumes and dust from zinc smelter. In: Kitagishi K, Yamane I (eds) Heavy metal pollution in soils of Japan. Japan Scientific Societies Press, Tokyo, pp 149–163

Hirata H, Ōtsuka T (1979) Cadmium and zinc pollution of arable lands due to smoke from zinc smelter – Actual condition of pollution around the zinc smelter after the improvement of discharge facilities. Man Environ 5(2):25–33(J)

Hirata H, Takenaga H (1977) Cd and Zn contamination in arable soil and upland rice due to fumes from Zn-refinery – An example at Annaka City, Gunma Prefecture (Part 1). J Sci Soil Manure Jpn 48:406–411 (J)

Homma S (1981) The Yoneshiro River Basin: Soil pollution by heavy metals discharged from various mines. In: Kitagishi K, Yamane I (eds) Heavy metal pollution in soils of Japan. Japan Scientific Societies Press, Tokyo, pp 137–148

Homma S, Yahata H, Nakatani T (1977) Heavy metal pollution of soil in the vicinity of copper smelter in Kosaka Town, Akita Prefecture. Man Environ 3(1)22–27 (J)

Hutchinson TC, Whitby LM (1974) Heavy-metal pollution in the Sudbury mining and smelting region of Canada, I. Soil and vegetation contamination by nickel, copper, and other metals. Environ Conserv 1:123–132

Hutchinson TC, Whitby LM (1977) The effects of acid rainfall and heavy metal particulates on a boreal forest ecosystem near the Sudbury smelting region of Canada. Water Air Soil Pollut 7:421–438

Hyōgo Prefecture (1972) Countermeasure plan against the agricultural land soil pollution in the vicinity of Ikuno Mine (J)

Hyōgo Prefecture (1973) Results of studies on the health effects of cadmium on the population in the vicinity of Ikuno Mine (2) (J)

Ichikawa City, Chiba Prefecture (1979) Explanation of the removal of chromium slag in Saiwai-2-chome (Giheisinden), Ichikawa City (J)

Iimura K (1981) Background contents of heavy metals in Japanese soils. In: Kitagishi K, Yamane I (eds) Heavy metal pollution in soils of Japan. Japan Scientific Societies Press, Tokyo, pp 19–26

Ishikawa Prefecture (1976) Report on health examination of the inhabitants in the Kakehashi River Basin (J)

Ishikawa Prefecture (1980) White paper on environment in Ishikawa Prefecture, 1979, pp 285–293 (J)

Ishikawa Prefecture (1982) White paper on environment in Ishikawa Prefecture, 1981, pp 180–182 (J)

Ishikawa Prefecture (1983) White paper on environment in Ishikawa Prefecture, 1982, pp 189–192 (J)

John MK (1971) Lead contamination of some agricultural soils in Western Canada. Environ Sci Technol 5:1199–1203

John MK, Chuah HH, VanLaerhoven CJ (1972) Cadmium contamination of soil and its uptake by oats. Environ Sci Technol 6:555–557

John MK, VanLaerhoven CJ, Cross CH (1975) Cadmium, lead and zinc accumulation in soils near a smelter complex. Environ Lett 10:25–35

John MK, VanLaerhoven CJ, Bjerring JH (1976) Effect of a smelter complex on the regional distribution of cadmium, lead, and zinc in litters and soil horizons. Arch Environ Contamin Toxicol 4:456–468

Jordan MJ, Lechevalier MP (1975) Effects of zinc-smelter emissions on forest soil microflora. Can J Microbiol 21:1855–1865

Kabata-Pendias A (1984) Progress in trace element research in Poland and neighbouring countries. In: Hemphill DD (ed) Trace substances in environmental health-XVIII. Univ Missouri, Columbia, pp 576-591

Kabata-Pendias A, Pendias H (1984) Trace elements in soils and plants. CRC, Boca Raton, Fl

Keizer MG, Hooghiemstra-Tielbeek M, de Haan FAM (1982) Contamination of soil and street dust with lead and cadmium near a lead smelter at Arnhem, Netherlands. Neth J Agric Sci 29:227–235

Kobayashi J, Morii F, Muramoto S, Nakashima S, Urakami Y, Nishizaki H (1973) Distribution of Cd, Pb, and Zn contained in soils polluted by a zinc refinery in Annaka City, Gunma Prefecture. J Sci Soil Manure Jpn 44:471–485 (J)

Kuo S, Heilman PE, Baker AS (1983) Distribution and forms of copper, zinc, cadmium, iron, and manganese in soils near a copper smelter. Soil Sci 135:101–109

Landrigan PJ, Gehlbach SH, Rosenblum BF, Shoults JM, Candelaria RM, Barthel WF, Liddle JA, Smrek AL, Staehling NW, Sanders JF (1975) Epidemic lead absorption near an ore smelter. The role of particulate lead. New Engl J Med 292:123–129

Linzon SN, Chai BL, Temple PJ, Pearson RG, Smith ML (1976) Lead contamination of urban soils and vegetation by emissions from secondary lead industries. J Air Pollut Control Assoc 26:650–654

Little P, Martin MH (1972) A survey of zinc, lead, and cadmium in soil and natural vegetation around a smelting complex. Environ Pollut 3:241–254

Lozano FC, Morrison IK (1981) Disruption of hardwood nutrition by sulfur dioxide, nickel, and copper air pollution near Sudbury, Canada. J Environ Qual 10:198–204

Matthews H, Thornton I (1982) Seasonal and species variation in the content of cadmium and associated metals in pasture plants at Shipham. Plant Soil 66:181–193

Morishita T (1981) The Jinzu River Basin: Contamination of soil and paddy rice with cadmium discharged from Kamioka Mine. In: Kitagishi K, Yamane I (eds) Heavy metal pollution in soils of Japan. Japan Scientific Societies Press, Tokyo, pp 107–124

Muskett CJ, Roberts LH, Page BJ (1979) Cadmium and lead pollution from secondary metal refinery operations. Sci Total Environ 11:73–87

Nagasaki Prefecture (1973) Report on the cause of environment pollution by cadmium and other heavy metals in the Sasu and Shiine River Basins, Izuhara Town, Nagasaki Prefecture (J)

Nagasaki Prefecture (1985) White paper on environment in Nagasaki Prefecture, 1984, p 97 (J)

Nakashima S, Ono S (1979) Counterplans of paddy soils contaminated by cadmium and other heavy metals in Tsushima Islands. Bull Nagasaki Agric For Exp Stn (Sec Agric) 7:339–385 (J)

Nakos G (1979) Lead pollution – Fate of lead in the soil and its effects on *Pinus Halepensis.* Plant Soil 53:427–443

Nogawa K (1984) Cadmium. In: Nriagu JO (ed) Changing metal cycles and human health. Springer, Berlin Heidelberg New York Tokyo, pp 275–284

Nogawa K, Ishizaki A (1979) A comparison between cadmium in rice and renal effects among inhabitants of the Jinzu River Basin. Environ Res 18:410–420

Nogawa K, Ishizaki A, Fukushima M, Shibata I, Hagino N (1975a) Studies on the women with acquired Fanconi syndrome observed in the Ichi River Basin polluted by cadmium – Is this Itai-itai disease? Environ Res 10:280–307

Nogawa K, Ishizaki A, Kobayashi E, Inaoka H, Shibata I (1975b) Studies on renal dysfunction of inhabitants in Cd-polluted district. Jpn J Hyg 30:549–555 (J)

Nogawa K, Ishizaki A, Kawano S (1978) Statistical observations of the dose-response relationships of cadmium based on epidemiological studies in the Kakehashi River Basin. Environ Res 15:185–198

Nogawa K, Kawano S, Kato T, Sakamoto M (1983a) The prevalence of Itai-itai disease and the mean cadmium concentration in rice produced by individual villages. Jpn J Hyg 37:843–847 (J)

Nogawa K, Yamada Y, Honda R, Ishizaki M, Tsuritani I, Kawano S, Kato T (1983b) The relationship between Itai-itai disease among inhabitants of the Jinzu River Basin and cadmium in rice. Toxicol Lett 17:263–266

Noma M, Kawaguchi Y (1981) An investigation of beryllium emission from beryllium alloy factory. Rep Aichi Environ Prot Cent 9:14–21 (J)

Nwankwo JN, Elinder C-G (1979) Cadmium, lead, and zinc concentrations in soils and in food grown near a zinc and lead smelter in Zambia. Bull Environ Contamin Toxicol 22:625–631

Ogawa B, Taguchi K, Uemura R (1985) Studies on the migration of damage for crop by heavy metal in soils. Bull Akita Agric Exp Stn 27:1–28 (J)

Ragaini RC, Ralston HR, Roberts N (1977) Environmental trace metal contamination in Kellogg, Idaho, near a lead smelting complex. Environ Sci Technol 11:773–781

Roberts TM, Hutchinson TC, Paciga J, Chattopadhyay A, Jervis RE, Loon J van (1974) Lead contamination around secondary smelters: Estimation of dispersal and accumulation by humans. Science 186:1120–1123

Roszyk E, Strojek Z (1978) Effect of dustiness on the content of Pb, Cu, and Zn in soils and grains in the vicinity of a copper smelter. Pol J Soil Sci 11:95–101

Rutherford GK, Bray CR (1979) Extent and distribution of soil heavy metal contamination near a nickel smelter at Coniston, Ontario. J Environ Qual 8:219–222

Saito H, Shioji R, Hurukawa Y, Arikawa T, Saito T, Nagai K, Furuyama T, Yoshinaga K (1977a) Chronic cadmium poisoning induced by environmental cadmium pollution. Renal lesions (multiple proximal tubular dysfunctions) identified in residents of cadmium-polluted Hosogoe, Kosaka Town, Akita Prefecture, Japan. Jpn J Med 16:2–13

Saito H, Shioji R, Hurukawa Y, Nagai K, Arikawa T, Saito T, Sasaki Y, Furuyama T, Yoshinaga K (1977b) Cadmium-induced proximal tubular dysfunction in a cadmium polluted area. Contrib Nephrol 6:1–12

Saito H, Nakano A, Sugihara N, Mitane Y, Wakisaka I (1981) Health examination of the residents in the soil pollution areas by cadmium. Kankyo Hoken Rep 47:173–174 (J)

Schmitt N, Brown G, Devlin EL, Larsen AA, McCausland ED, Saville JM (1971) Lead poisoning in horses. An environmental health hazard. Arch Environ Health 23:185–195

Scokart PO, Meeus-Verdinne K, De Borger R (1983) Mobility of heavy metals in polluted soils near zinc smelters. Water Air Soil Pollut 20:451–463

Suzuki S, Djuangshi N, Hyodo K, Soemarwoto O (1980) Cadmium, copper, and zinc in rice produced in Java. Arch Environ Contamin Toxicol 9:437–449

Takebayashi S (1980) Report on the autopsy of a patient, suspicious of cadmium intoxication in the cadmium polluted area in Tsushima. Kankyo Hoken Rep 46:259–262 (J)

Takebayashi S (1983) Report on the clinical and pathological study of the second case of osteomalacia found in the cadmium polluted area in Tsushima (The second case of autopsy). Kankyo Hoken Rep 49:207–211 (J)

Takebayashi S (1984) Report on the clinical and pathological study of the third case of osteomalacia found in the cadmium polluted area in Tsushima. Kankyo Hoken Rep 50:238–241 (J)

Takenaga H, Hirata H (1977) Cd, Zn, and Pb contamination in arable soil and barley due to fumes from Zn-refinery – An example at Annaka City, Gunma Prefecture, pt 2. J Sci Soil Manure Jpn 48:412–418 (J)

Tatekawa H (1978) Studies on the analysis and countermeasures for improvement of paddy fields polluted by heavy metals, particularly cadmium, in Fukushima Prefecture. Spec Bull Fukushima Agric Exp Stn 1:1–64 (J)

Taylor GJ, Crowder AA (1983) Accumulation of atmospherically deposited metals in wetland soils of Sudbury, Ontario. Water Air Soil Pollut 19:29–42

Thornton I, Abraham P (1984) Historical records of metal pollution in the environment. In: Nriagu JO (ed) Changing metal cycles and human health. Springer, Berlin Heidelberg New York Tokyo, pp 7–25

Tiller KG, Merry RH, Cartwright B, Bartlett NR (1975) Dispersal of lead emissions from an isolated lead smelter within an agricultural region of South Australia. Search 6:437–439

Toyama Prefecture (1978) Countermeasure plan against the agricultural land soil pollution in the Jinzu River Basin (J)

Whitby LM, Hutchinson TC (1974) Heavy metal pollution in the Sudbury mining and smelting region of Canada. II. Soil toxicity tests. Environ Conserv 1:191–200

Wood CW, Nash III TN (1976) Copper smelter effluent effects on Sonoran desert vegetation. Ecology 57:1311–1316

Yamane I (1981) Features of heavy metal pollution in Japan. In: Kitagishi K, Yamane I (eds) Heavy metal pollution in soils in Japan. Japan Scientific Societies Press, Tokyo, pp 3–15

Yanagisawa M, Shinmura Y, Yamada N, Segawa A, Kida K (1984) Heavy metal pollution and methods of restoration of polluted soils in the Jinzu River Basin. Bull Toyama Agric Exp Stn 15:1–110 (J)

Geochemistry of Priority Pollutants in Anoxic Sludges: Cadmium, Arsenic, Methyl Mercury, and Chlorinated Organics

M. KERSTEN[1]

1 Introduction

It has been stressed in the preceding chapter on the "Assessment of metal mobility in dredged material and mine waste by pore water chemistry and solid speciation" that redox change of the sludges is one of the most important factors controlling pollutant mobility. Thus, particular attention will be given to the question whether anoxic conditions alone are suitable for long-term retention of pollutants. However, such an evaluation even on the limited number of priority pollutants will be beyond the scope of this book. Thus, it has been decided to present case studies on some of the most important pollutants covering a broad range of characteristics, pathways and fates within natural environments: the trace metal cadmium, the trace metalloids arsenic and mercury with the latter in its methylated form, and the group of xenobiotic organic chemicals such as PCBs have been selected to be treated within this review. This is due to (1) their highest pollution potentials in terms of accumulation, toxicity and ecological risk index (Håkanson 1980) among all pollutants discovered to date; (2) their quite different, but well-investigated properties and behaviour as to sediment-water systems; and last not least (3) their citation in every priority pollutant list known to the author. Thus, as to the four different groups of pollutants and the wide variety of other pollutants for which they stand, this brief review will be well suited for an evaluation of the secondary pollution impact of typical anoxic dredged sludges and mine tailings introduced into aquatic systems.

2 Evolution and Classification of Biogeochemical Reactions Relevant to Pollutant Mobility in Sediments

The most intensive biogeochemical reactions, the so-called early diagenetic reactions, are usually found in sediments where variations of the redox conditions take place. Characteristic, well-investigated examples have been described both from the still "natural" sediments of the deep sea and from highly polluted environments, e.g. deposits of dredged materials. Examples of redox processes, involving the major elements C, N, O, S, H, Fe and Mn, are arranged in Fig. 1

[1]Technological University of Hamburg-Harburg, Department for Environmental Science and Technology, P.O. Box 901403, D-2100 Hamburg 90, FRG

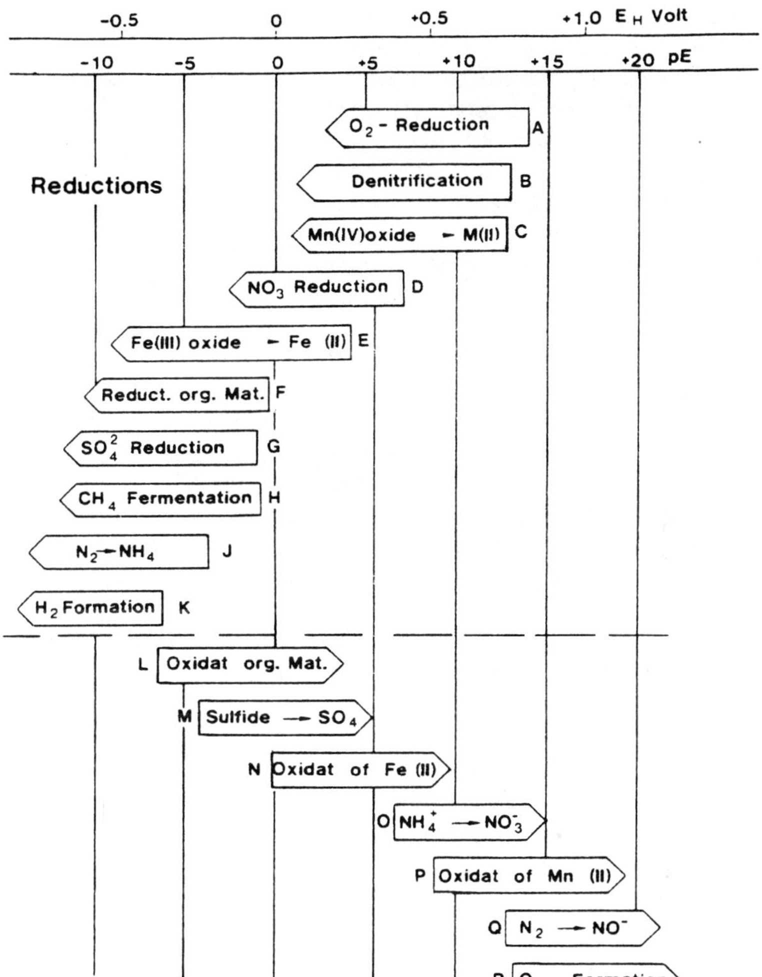

Fig. 1. Sequence of early diagenetic redox reaction in sediments (Stumm and Baccini 1978)

(Stumm and Baccini 1978) in the sequence of reactions observed in principle in an aqueous system at various redox potential (Eh-) values. Because the reactions considered are biologically mediated (with a more or less intense catalytical abiotic component), the chemical reaction sequence is paralleled by an ecological succession of sedimentary microorganisms: aerobic heterotrophs, denitrifiers, sulphate reducers, fermentators and methane bacteria (Stumm and Morgan 1981). Driving force of reduction processes is the decomposition of organic matter by non-photosynthetic microbes, thereby obtaining a source of energy for their metabolic needs. The energy yields can be evaluated from the combination of processes given in Fig. 1, indicating that this succession follows the order of decreasing energy yield of the respective reactions. The sedimentary environments corresponding to these processes are respectively: "oxic", "post-oxic", "sulphidic" and "methanic" (Berner 1980).

The sequence of events affecting these transformations can be evaluated both from pore water analyses and from mineral assemblage characterization typical of the different biogeochemical zones within short sediment cores. The succession of typical pore water constituents has been depicted by Froelich et al. (1979) from the study of such *undisturbed* (as to, e.g. bioturbation by macrobenthic animals) deep-sea sediment cores from the eastern equatorial Atlantic (Fig. 2). They found that oxygen is consumed in the uppermost sediment layer during decomposition of organic matter. This process, which involves dissolution of carbonate, releases ammonia, which was found to be subsequently oxidized to nitrate; NO_3 diffuses down to be denitrified in the lower part of the core. Not far below the nitrate maximum organic carbon starts to become oxidized by simultaneous reduction of MnO_2, upward diffusing Mn^{2+} ions will be precipitated in the stronger oxidizing part of the sediment sequence, an observation made first by Lynn and Bonatti (1965) on pelagic sediments. Because MnO_2 reduction involves a small but characteristic effect on carbonate precipitation, the concentration of Mn^{2+} in pore waters of both lacustrine and marine sediments may be controlled by the solubility of Mn carbonate rodochrosite (Callender et al. 1974, Holdren et al. 1975). At greater sediment depth, the process of iron reduction to Fe^{2+} begins, thus indicating the persistence of iron oxides, but not manganese oxides, under post-oxic conditions (Berner 1981). It has been inferred, however, by Froelich et al. (1979) that only a relatively small portion of Fe is available for the reduction process, namely Fe in oxidic coatings, not in the silicate compounds. As for the MnO_2 reduction, the change from Fe^{3+} to Fe^{2+} usually involves a small effect on the carbonate solubility, but also phosphate solubility. Under certain conditions,

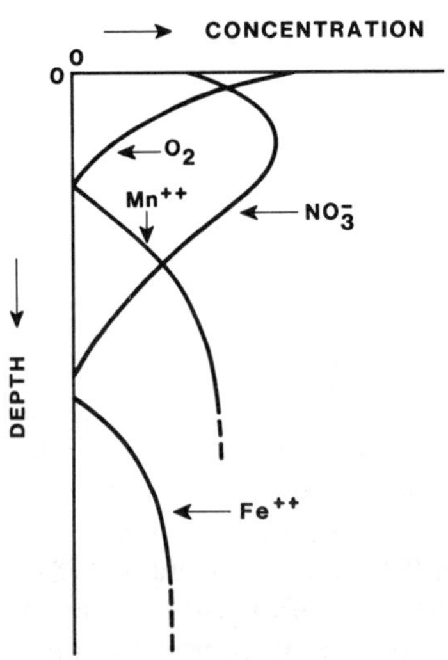

Fig. 2. Schematic succession of concentration profiles of oxygen, nitrate, manganese and iron in pore waters from short, deep-sea sediment cores (Froelich et al. 1979)

FeOOH and $FeCO_3$ or $Fe_2(PO_4)_3$ might be expected to co-exist (Jones and Bowser 1978, Hearn et al. 1983).

In stronger reducing environments the metal distribution should be controlled primarily by the formation of (highly insoluble) sulphides. At lower free-S^2 concentrations such as found in freshwater systems, it should be controlled by the stability of carbonate and phosphate phases. As a consequence, a scheme for the recognition and distinction of early diagenetic environments has been proposed by Berner (1981) on the basis of typical mineral associations as illustrated in Table 1. The mathematical modelling of early diagenetic processes in sediments also involving bioturbation and other disturbing effects has been given by Berner (1980).

Table 1. Geochemical classifications of sedimentary (diagenetic) environments (Berner 1981)[a]

Environment	Characteristic phases
I. Oxic ($C_{O_2} = 10^{-6}$)	Hematite (Fe_2O_3); goethite (FeOOH); MnO_2-type minerals, no organic matter
II. Anoxic ($C_{O_2} - 10^{-6}$)	
A. Sulphidic ($C_{H_2S} \leqslant 10^{-6}$)	Pyrite, marcasite (FeS_2); rhodochrosite ($MnCO_3$); alabandite (MnS); organic matter
B. Non-sulphidic ($C_{H_2}S \leqslant 10^{-6}$)	
1. Post-oxic	Glauconite and other Fe^{+2}-Fe^{+3}-silicates (also siderite, vivianite, rhodochrosite); no sulphide minerals; minor organic matter
2. Methanic	Siderite ($FeCO_3$); vivianite ($Fe_3(PO_4)_2 \cdot 8H_2O$); rhodochrosite; earlier formed sulphide minerals; organic matter

[a] "C" refers to concentrations in moles per litre. Total sulphide is represented by H_2S.

The early diagenetic reactions involving stronger reduction processes have been studied experimentally in the field by Salamons and Förstner (1984). Three large pits (80 × 30 × 6 m) were dug in the Rhine estuary (The Netherlands) below the water table, and were filled with fresh water (Waalhaven), brackish water (Botlek) and marine (Maasmond) sediment. The consolidation rate and the change in pore water composition were studied over a 1-year period (Fig. 3). The changes in the element concentrations with time can be divided into three categories:

1. The concentrations show a decrease. This is the case for sulphate in all cases, nickel in the sediments from the Maasmond and arsenic from the Botlek and Waalhaven sediments.
2. The concentrations show a continuous increase. This is the case for o-PO_4, NH_4, silicon, chromium, BOB_5 and DOC.
3. The concentrations show a peak decreasing subsequently. This is the case for iron and manganese in all three examples, for copper from the Botlek and Waalhaven sediments and cadmium in the Maasmond sediments.

The evolution of metal mobility within these pits follow more or less the evolution of the biogeochemistry succession and early diagenesis of the sediments

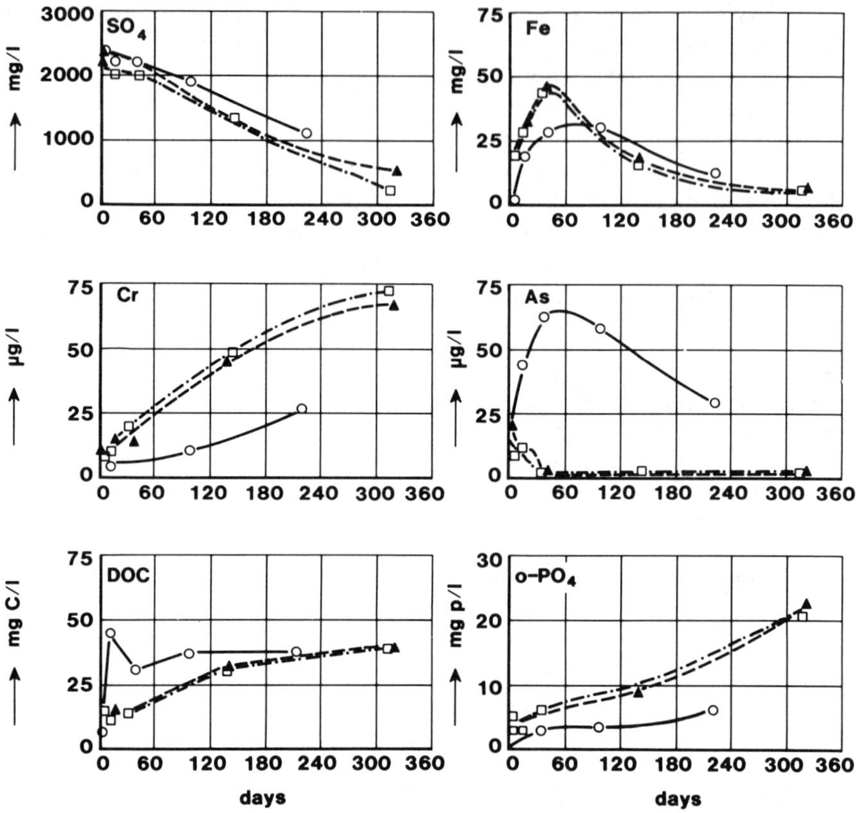

Fig. 3. Changes in composition of pore waters in original fresh water (□), brackish water (▲) and marine (○) sediments dumped into a marine environment over a 360-day period (Salomons and Förstner 1984)

with time. A rapid increase in manganese and iron concentrations is observed in the first few weeks (Fig. 2). During this period no sulphate reduction takes place as is shown by the constant levels of sulphate. The electron acceptors for the degradation of the organic matter are apparently oxygen and nitrate. The low oxygen levels cause a reduction of iron and nitrate and a reduction of iron and manganese and the subsequent appearance of Fe^{2+} and Mn^{2+} in the pore waters. Sulphate concentrations start to decrease after about 2 months. The decrease is most rapid in the sediments with the highest organic matter contents (Botlek and Waalhaven), but is slower in the Maasmond sediment. With the decrease in sulphate concentrations, and the production of sulphide, the iron and manganese levels start to decrease, possibly as the result of the formation of sulphidic minerals. Silicon and ortho-phosphorus concentrations show a continuous increase. The phosphorus release probably takes place as a consequence of the breakdown of the organic matter. About 10–20% of the total P in the sediments is organically bound (Salomons and Gerritse 1981).

The early diagenetic environment is characterized by the highly complex interactions of metal ions, metal-organic complexes, sulphur compounds, organic and inorganic solids. Further generalizations of trace metal behaviour on the basis of simple geochemical terms are complicated by the overwhelming importance of specific chelation with inorganic and organic ligands arising under anoxic conditions. Metals with low ionic potentials tend to form primarily simple ions in solution, while those with intermediate and high ionic potentials tend to form soluble complexes (e.g. Cd). Common inorganic complex-forming ions in anoxic pore waters include PO_4^{3-}, CO_3^{2-}, HCO_3^-, HS^-, S^2 and Cl^-. Recent evidence indicates that anaerobic microorganisms may play an important role in metabolic complexation of non-nutrient elements, thereby generating metastable metal and metalloid species. Information on organic ligands in anoxic pore waters is especially limited. Organic ligand complexation may result in orders of magnitude metal solubility increases, thereby increasing their apparent solubility beyond the limitations imposed by the solubility product of the most solids stable under anoxic conditions. In addition, solubility may be particularly pronounced if the species stable in pore water is of oxy-anionic form (e.g. As), which is not fixed tightly by the predominantly negatively charged sediment particles. Moreover, disproportion, microbially-mediated methylation and precipitation/dissolution reactions may occur concurrently in an anoxic environment (e.g. Hg).

Another important interaction of metals in the early diagenetic environment is with the various sulphur species. Figure 4 (after Andrews 1977) shows the interactions of sulphidic species with the iron species in the systems affected by organic matter, mainly designed to demonstrate the possible pathway of acid mine drainage. The characteristic process is the reduction of the Me^{3+} by oxidation of additional MeS_2 (s), in the presence of H_2O, but without need of O_2 as an initiation agent (reaction "d"). Completion of this step allows initiation of a cycling between Fe^{2+}, Fe^{3+} and FeS_2 (s). It has been demonstrated that with FeS_2 (pyrite or markasite) the rate-determining step in this system is the oxidation of the ferrous to ferric iron form. In an abiotic system, the oxidation of ferrous iron is roughly by a factor of 10^{-6} slower than in a system mediated by microbes (Singer and Stumm 1970).

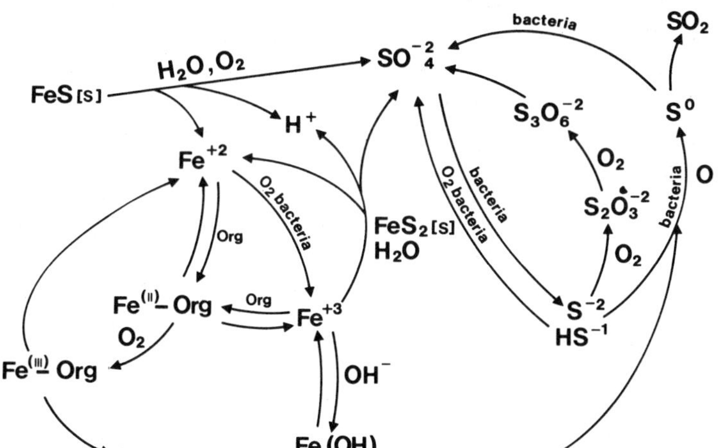

Fig. 4. Model of geochemical leaching of FeS_2 as it may take place in a tailings pile (After Andrews 1977)

3 Cadmium

The migration of trace metals such as cadmium in sediment can usually be inferred from the knowledge of the sediment composition and environmental physico-chemical condition due to the appreciate amount of research attention that has been devoted to these contaminants in the last decade (Nriagu 1980, 1987). Studies conducted on particulate trace metal associations involved in different early diagenetic sediment environments (Kersten and Förstner 1986), as well as literature reviews providing the basis from which an assessment on the specific behaviour and fate of cadmium in anoxic sludges can be made will be presented in the following section.

The redox potential of a sediment-water system is a controlling factor in regulating the chemical association of particulate cadmium, whereas pH and salinity influence the stability of its various forms. This fact has been confirmed by a number of earlier publications based on sequential extraction studies (reviews given by Jenne and Luoma 1977, and Khalid 1980). In the sequential extraction pattern of cadmium in the anoxic harbour sludge given in the preceding chapter (Förstner and Kersten, this Vol.) the H_2O_2-oxidizable sulphidic/organic fraction dominates with proportions up to 98%. The remainder of particulate cadmium from this contaminated freshwater sediment sample was found to be associated with oxyhydrates or residual sediment components (Kersten and Förstner 1986). Similar data have been reported from the marine San Francisco Bay sediments, which indicate that 90–94% of total particulate Cd was complexed by insoluble organic matter or associated with sulphides (Serne and Mercer 1975). The distribution of Cd in the Ashtabula and Bridgeport Harbor sediments as analyzed similarly by sequential extraction experiments also showed dominating proportions in the sulphidic/organic fraction (Brannon et al. 1976).

Lee and Kittrick (1984) succeeded recently in directly recording Cd-rich particles in anoxid harbour mud by electron beam microprobe. In ca. 90% of these particles Cd was found to be associated with sulphur. The authors concluded from their data that cadmium was precipitated as the single phase CdS (greennokite) rather than being adsorbed on the surface of other sulphide minerals. With a $K = 27.0$ (Lindsay 1979), CdS is one of the least soluble metal sulphides (following HgS and CuS). Figure 5 shows the redox at which CdS (greennokite) can form in sediments relative to the many other sulphides. Clearly, reduced conditions ($pe + pH < 4.73$; Lindsay 1979) are required, yet greennokite forms before sphalerite (ZnS). In fact, Lee and Kittrick (1984) observed their Cd + S-rich particles to be associated significantly with Zn, but not with Fe.

In the preceding chapter on solid speciation assessment by sequential chemical extraction, it was demonstrated that following the application of the short elutriate test using oxidized water, the oxidizable sulphidic/organic portion of Cd decreases drastically. This drastic changing of Cd forms in the short aeration experiment is most likely caused by a rapid oxidation of the sulphidic association leading to the formation of Cd^{2+} and SO_4^{2-} ions. Co-precipitation and adsorption of Cd with the precipitated oxyhydrates, however, may have removed the liberated metal from solution as evidenced by the high Cd proportions in the easily reducible fraction

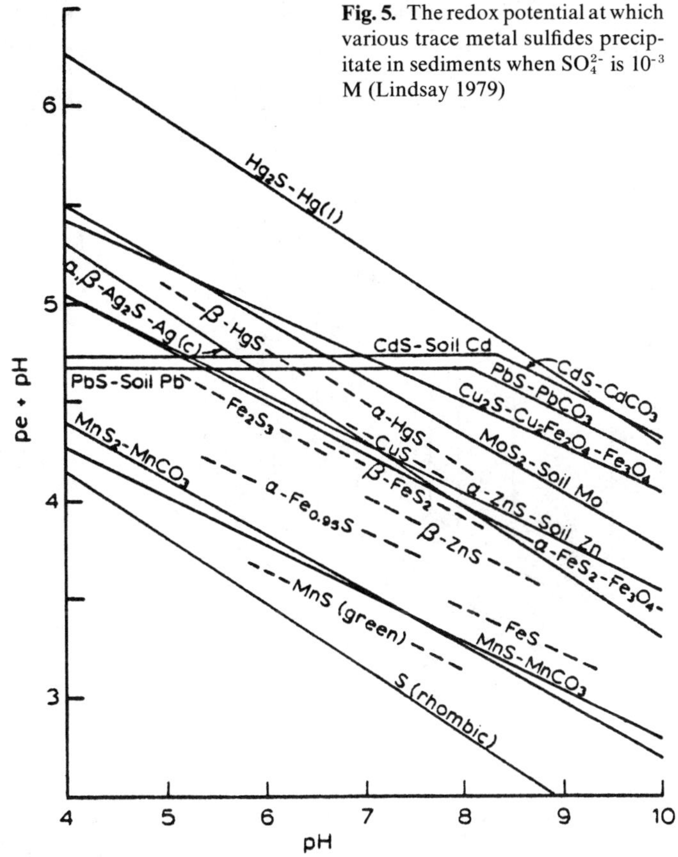

Fig. 5. The redox potential at which various trace metal sulfides precipitate in sediments when SO_4^{2-} is 10^{-3} M (Lindsay 1979)

upon oxidation in the elutriate test. Freshly precipitated oxyhydrates are much more effective in scavenging high concentrations of trace metals because of greater reactive surface area than aged crystalline materials (Jenne 1977). After freeze- and oven-drying of the initially anoxic samples, cadmium proportions were found even in the most mobile, operationally defined carbonate and exchangeable fractions. The formation of a carbonate association of cadmium is probably promoted by the decomposition of $MnCO_3$ (rhodochrosite), which is also indicated from the extraction results (Kersten and Förstner 1986).

This simple laboratory experiment supports the fact that oxidation of reduced sediments results in the transformation of insoluble sulphide-bound Cd into more mobile and potentially available hydroxide, carbonate and exchangeable forms as shown in a couple of earlier studies (Holmes et al., 1974; Lu and Chen 1977). Isotope studies performed by Gambrell et al. (1977) with Mississippi River freshwater sediment samples indicated that exchangeable Cd-109 levels are strongly pH-redox potential-dependent. A much higher proportion of the incubated cadmium isotope was recovered in the readily bioavailable forms than for any other potentially toxic heavy metal studied. Figure 6 represents these results

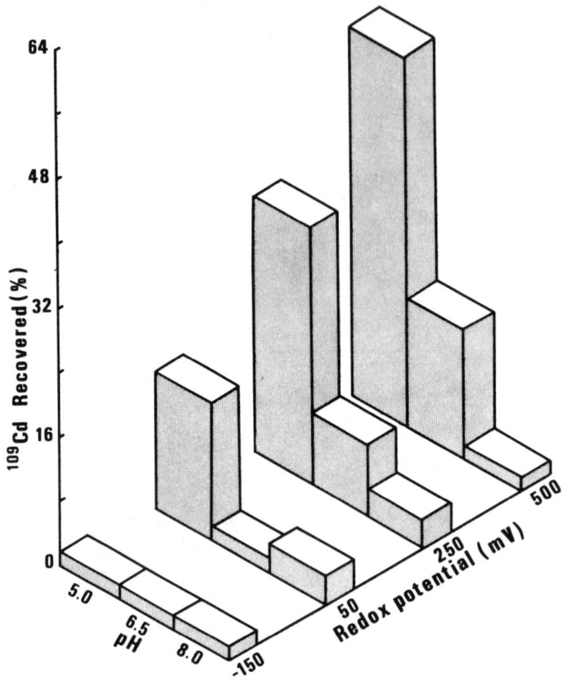

Fig. 6. Effects of pH and redox potential on exchangeable Cd-109 in Mississippi River sediment suspension (After Gambrell et al. 1977)

and suggests that considerable cadmium release to relatively mobile forms may occur as the Cd-contaminated sediment is transported from reducible, near-neutral pH conditions to a moderately acid, oxidized environment. Under the latter conditions, cadmium levels of subsurface drainage water from landfill disposal of dredged material may be increased, and cadmium availability to plants growing on the material enhanced (Gambrell et al. 1977). The results of these authors were different to ours in that free and total sulphide was not detectable under reduced conditions at any pH level in the Mississippi River sediment suspensions, and thus sulphide precipitation was probably not a controlling factor for Cd in this study. Instead, Gambrell et al. (1977) stressed that Cd-organic insoluble complexes are more stable under their reduced conditions and, as a result of oxidation, this organic Cd fraction was mobilized to bioavailable water-soluble, exchangeable or chelated fractions. The influence of pH-redox potential conditions on dissolved Cd levels was investigated extensively by batch experiments with marine sediment suspensions conducted by Herms and Brümmer (1979), and is shown in Fig. 7. Here, the solubilities of CdS (greennokite) and $CdCO_3$ (octavite) are inserted. It appears that $CdCO_3$ provides a limit on the solubility of Cd^{2+} in oxidized sediments of near-neutral to slightly alkaline pH. This result agrees well with the studies conducted with soils (Street et al. 1978) and with our sequential extraction results of oxidized sediment samples as discussed in other sections.

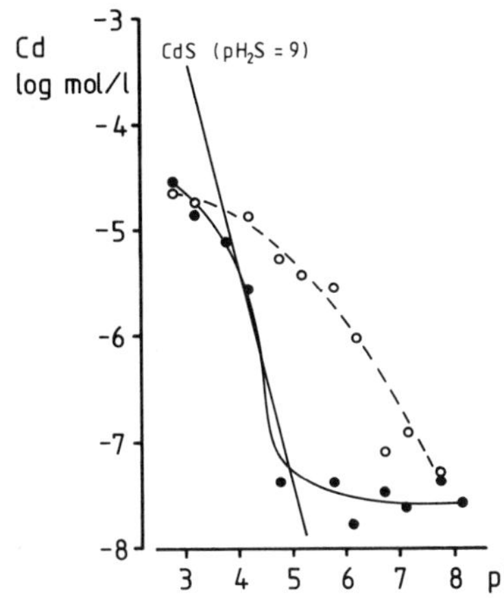

Fig. 7. Influence of pH on Cd solubility in salt marsh sediments, (•) reducing and (○) oxidizing conditions (Herms and Brümmer 1979)

Comparisons between Cd levels in anoxic pore waters and appropriate sediment concentrations are shown in Fig. 8. Solid-phase precipitation is again evidenced, since no direct relation is found between Cd concentrations in the sediment and in pore waters. In addition, if one would calculate distribution coefficients from these data, the values would be much higher than the ones normally characteristic for oxic sediment/water systems (Salomons 1985).

Very little direct information is available on the speciation of trace metals in anoxic pore waters. Preliminary results on pore waters from the Rhine estuary show that between 25–75% of cadmium may be present as non-labile complexes. Modelling studies on the possible importance of organic complexation in sulphidic pore waters gave stability constants several orders of magnitude higher than those reported for surface waters (cf. Chap. 11, this Vol.). Although it was found that the overall K-values for dissolved complexes in pore waters (complexation measurements) are higher than those reported for surface waters,they are still too low to make organic complexation a significant factor in the speciation of sulphidic pore water at natural pH values. Jacobs and Emerson (1982) classified two kinds of metals with regard to speciation in a sulphidic environment: (1) transition metals (Fe, Mn, Co, Ni) which form sulphide precipitates but do not form strong soluble sulphide complexes with increasing free-sulphide concentrations; (2) "class B metals" (Hg, Cu, Cd, Pb, Zn) which form sulphide precipitates but also form strong complexes with sulphide, causing increasing dissolved metal levels in pore water with increasing free-sulphide concentrations. From these theoretic findings it follows that Cd concentrations in pore water should increase with decreasing pH at relatively low values. At pH levels characteristic for seawater, however, cadmium is almost invariant to changes in sulphide concentrations. This probably also explains the nearly constant low levels for cadmium in Fig. 8. In applying these modelling results to sediments one should be aware of additional environmental conditions to be taken into account. The fraction of the produced free sulphide depends on the sediment type and environmental conditions. If there is a high rate of sulphate reduction provided by high degradable organic matter and sulphate

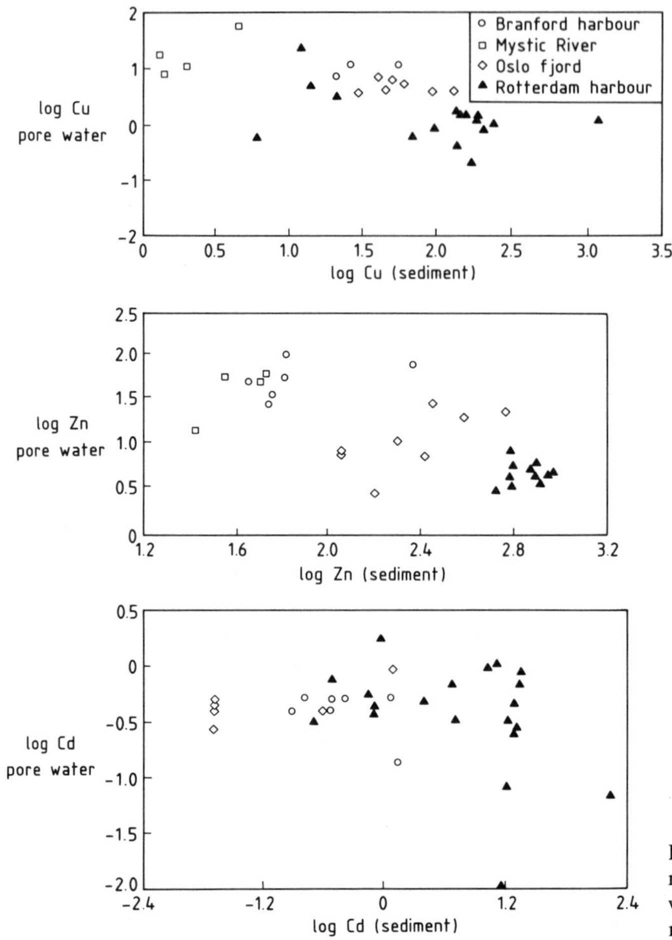

Fig. 8. Relationship between metal concentrations in pore waters and in sediment (Salomons et al. 1985)

impact and thus a high H_2S level in the pore water, then most of the H_2S may be lost by diffusion to the surface sediment where it is oxidized (Jörgensen 1977). Bioturbation may contribute significantly to this loss since the macrobenthos ventilate zones with a high sulphate reduction rate. This is the situation in unconfined dredged material deposits in coastal shallow water environments where only a minor fraction of the produced H_2S goes into the iron sulphide pool. In situations where dredged material is deposited in capped under-water deposits, sulphate will be significantly less available to form free sulphide. Generally, there will be sufficient reducible iron pool buffering H_2S activity at low levels. Williamson and Bella (1980) developed a procedure to measure the capacity of a sediment to form insoluble sulphides while avoiding the complexities of the iron cycle, which they termed the reduced sulphide capacity test. This test, originally developed in view of the biological importance of hydrogen sulphide toxicity in estuarine sediments, may be usefully adopted to situations where secondary pollution potential of class B metals to marine environments is to be predicted.

An important finding from the fractionation studies described in the preceding chapter relates to the dominant role of the non-residual sulphidic fraction of cadmium in anaerobic sediments. Due to these associations of low solubility even strong accumulations of anthropogenic Cd are rendered less bioavailable. This is particularly effective since bottom sediments, especially those which are contaminated by Cd and organic wastes, always tend to become more reduced because of the continuous release of electrons to the environment through the respiration processes of microorganisms as discussed in the second section. From all this experimental evidence we conclude that equilibrium concentrations of cadmium at the sediment-water interface and its bioavailability are regulated via adsorption-desorption and precipitation-dissolution mechanisms according to the suggestions made by Jenne and Luoma (1977) and Salomons (1985). The continuous transition of electrons accounts for the fact that most polluted bottom sediments remain reduced once these processes are initiated. A reversal of these processes and hence the oxidation of the stable sulphidic Cd forms needs an energy input such as provided by both drastic changing of environmental conditions and dredging operations.

In the marine environment several observations have been made suggesting similar, but significantly slower, oxidation processes on the release of cadmium from polluted sediments. Field evidence for changing cadmium mobilities was reported by Holmes et al. (1974) from Corpus Christi Bay Harbor. During summer when the harbour water is stagnant Cd is precipitated as CdS at the sediment-water interface. In the winter months, on the other hand, the intrusion of oxygen-rich water into the bay results in the release of some of the precipitated metal. Lu and Chen (1977) studied the migration between the interface of seawater and polluted surficial sediments under different redox conditions. Their experimental results showed that over a period of 4–5 months of sediment/seawater contact time, the released Cd amount increased as the environment became more oxidizing. The geochemical partitioning of Cd in the surface sediment (top 1 cm) showed that cadmium associated initially with organic or sulphide solids (H_2O_2-extractable fraction) was gradually changed to carbonate forms under aerobic conditions. From this the authors stressed that the concentrations of cadmium in the interfacial waters were limited by the prevailing solids under the different redox conditions due to the change of the controlling mechanism from solubilities of sulphides to carbonate or adsorption sites (Lu and Chen 1977). Gendron et al. (1986) also found evidence for different release mechanisms of cobalt and cadmium near the sediment/water interface in the St. Lawrence Estuary. The hydroxylamine extractable fraction of the sediment surface layer in the Lawrentian Trough was enriched in cobalt, but depleted in Cd relative to the subsurface sediment. While in this layer the highest dissolved Cd concentrations were detected, the concentrations of dissolved Co were negligible. The profiles for Co resemble those for Mn and Fe with increased levels downwards, suggesting a release in the reduced zone and reprecipitation in the surface layer. On the other hand, cadmium appears to be released in the surface, probably as a result of the aerobic remobilization as described above. The authors found in addition that the upward flux of Cd across the sediment/water interface is one order of magnitude higher than the downward flux into the sediment, and stressed that the Cd released near the interface is efficiently recycled back into the water column. Pore water analyses in shallow water sediments of the Puget Sound estuary indicate higher concentrations of Cd in the upper oxidized centimetres of sediment covering reduced H_2S-containing deposits (Emerson et al. 1984). This metal remobilization has been explained by the removal of sulphide from the pore waters via biologically mediated ventilation of the upper sediment layer with oxic overlying water, allowing the enrichment of dissolved cadmium which might otherwise be very low in concentration due to insoluble sulphide formation. The authors suggested a greatly enhanced flux of the metal to the bottom waters by this mechanism. These processes have been shown to be more effective during spring and summer than during the winter months (Hines et al. 1984). Because the oxidation process of metal sulphides takes place at sea-water pH levels, this reaction is probably a chemical oxidation and not the fast microbially mediated oxidation of sulphides which occurs at much lower pH values (Calmano et al. 1983). From enclosure experiments in Narragansett Bay (Rhode Island, USA) it has been estimated that by mechanisms such as oxidation of organic and sulphidic material the anthropogenic proportion of Cd in marine sediments is released to the water within approx. 3 years: for remobilization of Cu and Pb approx. 40 and 440 years are needed respectively, according to these extrapolations made by Hunt and Smith (1983).

There are still conflicting results with regard to the mechanisms and kinetics of cadmium release from solids to the interfacial water. While Hoff et al. (1982) found no significant Cd release from silicate-binding forms of mine tailings dumped into sea-water, both Salomons (1980) and recently Prause et al. (1985) observed in their batch experiments with dredged sediments suspended in estuarine water that a considerable release of Cd occurs in long-term experiments. No enrichment of Cd in solution could be observed during resuspension experiments, lasting between 12 and 24 h. In most cases even a decrease of the metal concentration was found, probably resulting from adsorption to freshly precipitated Fe/Mn oxyhydrates. Both authors have extended their experiments to several weeks, and found a maximum remobilization rate during the second and fourth week, the maximum of release occurring after more than 1 month. Prause et al. (1985) stressed that simple cation-exchange mechanisms obviously have no importance for the remobilization of Cd, because a new equilibrium at the surface of sorbents will normally be established in less than this time. In fact, our chemical fractionation results have also shown that the oxidation of sulphidic Cd associations and subsequent adsorption to oxyhydroxides lasts less than 1 h. In order to test the influence of bacteria on the rate of Cd remobilization, the authors added antibiotics to the system on the 21st day. This resulted in an interruption of the Cd release for a period similar to the initial lag phase as shown in Fig. 9. From these results Prause et al. (1985) suggested that bacterial activity can stimulate the Cd remobilization by a complex web of substrate decomposition and potential organic ligand liberation or formation. In addition, the results of both authors emphasize strongly the fact that the time spans of these release processes are much longer than the typical short-term dredging operations. Thus, a Cd-contaminated dredged spoil should be speedily entombed in a stable anoxic sulphide-bearing environment rather than released to weathering and long-term oxidation.

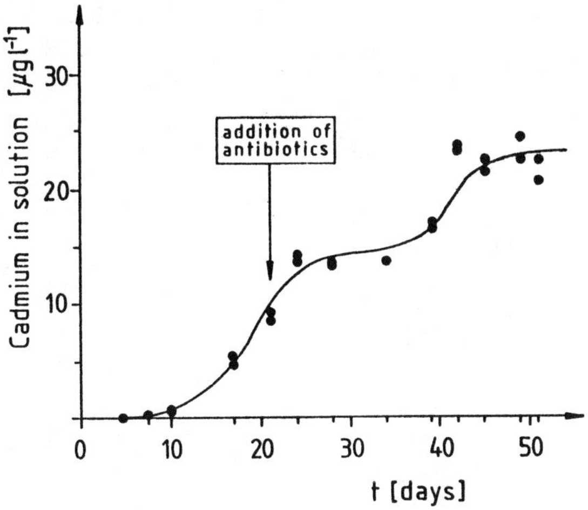

Fig. 9. Remobilization of Cd in a batch experiment as a function of time from Bremen Europahafen sludge. The bacterial activity in the suspension was suppressed by addition of antibiotics (Streptomycine and Chloramphenicole, 0.2 g l^{-1} each) on day 21 of the equilibration period. Salinity of the suspension (15 g solids l^{-1}): 32‰, pH = 7.9, T = 7°C, O_2 = 9 mg l^{-1} (Prause et al. 1985)

4 Mercury

4.1 Sources and Pathways of Mercury in Sediment-Water Systems

Mercury exists in natural sediment-water systems in three valence states: in elemental Hg(0), the mercurous Hg(I) and the mercuric Hg(II) state. Well-aerated water (Eh > 0.5 V) containing traces of stable organic ligands speciates mercury between different chloro- and hydroxy-mercuric complexes as a function of pH and pCl (Lockwood and Chen 1973). Mercuric chlorocomplexes tend to be more stable than for any other trace metals. Even higher is the capacity of the polarizable Hg(II) to form strong complexes with donor atoms such as sulphur, and sulphur may therefore be expected to play an especially important role in the environmental chemistry of mercury (Craig and Moreton 1985). Under reducing conditions and in the presence of free-sulphide ligands, mercury is stabilized in the mercuric state as extremely insoluble sulphide precipitates or bound as surface complexes with organic material containing sulphur donor atoms. An important transformation mechanism of mercuric species in natural environments is methylation, leading to volatile Hg forms, thus closing the biogeochemical mercury cycle as summarized schematically in Fig. 10.

In a comprehensive study Reimers et al. (1975) have analyzed the retention capacity of various natural and synthetic materials for mercury. It has been shown that all sediments or synthetic scavengers containing sulphur possess high affinity to the mercury compounds. The greatest uptake of mercury has been observed in organic mercaptans. Sulphide-containing materials were next in retention capacity, but when high, free-sulphide ligand concentrations occurred, inorganic mercury formed soluble polysulphide complexes. This behaviour is characteristic of "class B metals" discussed by Jacobs and Emerson (1982). As mentioned in Section 3, however, the occurrence of reducible iron in sediments usually

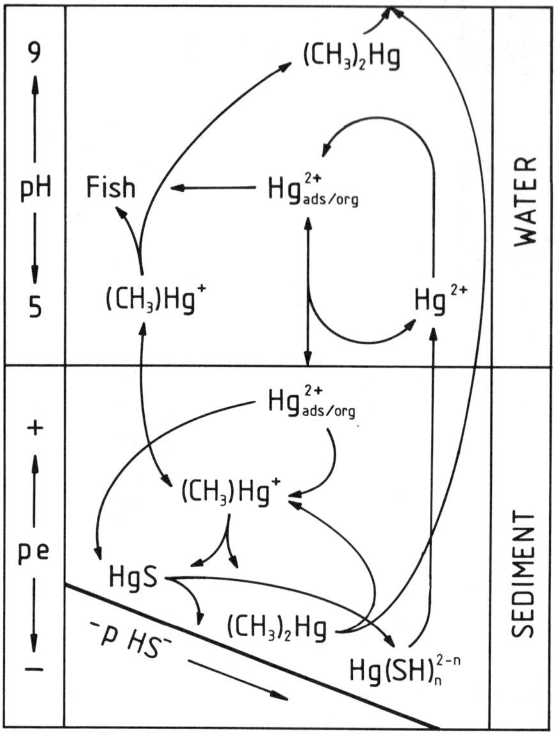

Fig. 10. Geochemical cyle of mercury in sediment-water systems

prevents the formation of enhanced sulphide levels. Lindberg and Harriss (1974) have suggested that the formation of polysulphide complexes may account for greater levels of dissolved Hg (up to 3.6 μg l^{-1}) found in their study of Florida Everglades swamp sediments rich in organic matter (up to 70%), poor in iron oxides and containing dissolved sulphide levels up to 140 mg l^{-1}.

Environmental pollution by mercury arises mainly by discharging into the aquatic environment, thereby releasing inorganic mercuric species, Hg(0) and phenyl mercuric acetate (Moore and Ramamoorthy 1984). Strong accumulation of mercury in sediments indicates that sediments can serve as a sink for industrially discharged mercury. Partition coefficients for mercury between suspended matter and water have been reported from both field and laboratory data to be in the order of 10^5. Adsorption is considered to be the main process for the enrichment of mercury onto suspended sediments (Reimers and Krenkel 1974). The chemical form of its dissolved, inorganic species determines the degree of association to suspended material and residence time of mercury in natural waters. The chloride ligand concentration, e.g. at the mineral-water interface, influences the adsorptive capacities for Hg because of the formation of chloromercurial complexes. The adsorptive capacity of sediments was shown to decrease in the order $Hg(OH)Cl > Hg(OH)_2 > HgCl_3^- > HgCl_4^{2-} > HgCl_2$ (Liu Ching-I and Tang Hong-xiao 1985). In addition, it was reported that at chloride concentrations less than 10^{-4} mol l^{-1}, humic matter, MnO_2 and illite have a considerable adsorption capacity for mercury. At a chlorinity of 0.56 mol l^{-1} characteristic of seawater (in which mercuric chlorocomplexes predominate), only the organic matter retained its adsorption capacity for Hg (Liu Ching-I and Tang Hongxiao 1985).

These results are corroborated by the data obtained from sediment extraction experiments (Gambrell et al. 1977, Eganhouse et al. 1978, Peng An and Wang Zijian 1985). The sequential chemical extraction experiments carried out on contaminated and natural sediments indicate that most of the particulate mercury found in aquatic environments is associated with humic materials soluble in KOH, and with organics and sulphides oxidizable with hot 30% H_2O_2. Crecelius et al. (1975) showed that more than 82% of the mercury in oxic sediments of Bellingham Bay (Washington, USA) was leachable with hydrogen peroxide and assumed that this Hg was bound to organic matter rather than to authigenic sulphides. Insignificant levels of mercury were found to be water soluble and exchangeable or associated with carbonates and oxides even in highly polluted sediments (Peng An and Wang Zijian 1985).

Significant proportions of total mercury forms extractable only with aqua regia, i.e. in residual minerals, are found when mine tailings are the pollutant source (Liu Ching-I and Tang Hongxiao 1985). It is reasonable to assume that in such cases a large fraction of the mercury will be present in inorganic forms, perhaps bound with lithogenic cinnabar. With increasing distance downstream from the source of this pollution in the Ji Yun River (China), however, the percentage of mercury associated with more or less degradable organics was found to increase significantly. The authors suspected humic acids in the interstitial water of the sediments to mobilize mercury slowly from the sediments by a complexation-dissolution mechanism. Similarly, Bothner and Carpenter (1973) showed that the mercury concentration in the oxidizing sediments of Bellingham Bay (Washington, USA) decreased with time since the high discharge of Hg was stopped in 1970. The rate of mercury loss appeared to follow first-order kinetics with a "half-life" of about 1.3 years in these oxic sediment layers that had initial mercury concentrations of 2–10 mg kg^{-1}, predominantly in organic forms (Creceluis et al. 1975).

The importance of suspended matter as a mercury transporting medium has been emphasized by a couple of studies, e.g. in the Thames estuary, UK, where about 98% of water-borne mercury was found associated with suspended matter following 0.45-μm filtration to define dissolved and particulate forms (Nelson 1981). In a study of influent and effluent levels of a number of pollutants at nine dredged sediment containment facilities representing upland, lowland and island sites, and a wide salinity range, Hoeppel et al. (1978) reported that the average removal efficiency of total mercury was as low as 46%. The effluent mercury in particular was suspected to be associated with very fine (< 0.45 μm), low-density particulate matter. Ramamoorthy and Rust (1976) showed that the sorption bonding constant maximum is in decreasing significance to organic content > grain size > cation exchange capacity > surface area. From these results it follows that much of the total Hg in oxic sediments is thus bound by non-specific chemical forms and is easily subject to transformation affecting its remobilization behaviour and bioavailability. Feick et al. (1972) reported accordingly that oxidation of freshwater sediments decreased their ability to adsorb mercury. In similar laboratory experiments conducted to investigate pH and oxidation effects on selected chemical forms of mercury using ^{203}Hg-labelled sediment suspensions, Gambrell et al. (1980) reported that at high-level additions there was a substantial increase in soluble and exchangeable mercury fractions with decreasing pH and increasing redox potentials. Greatest mobilization occurred under moderately acid, oxidized conditions. Organic matter oxidation in the course of oxic suspension incubation may explain the lower retention capacity for mercury. Dissolved and exchangeable mercury concentrations, which were significantly interrelated, tended to be inversely related to sediment organic matter. A more realistic reason for the increase in mercury solubility with increase in pH reported by Gambrell et al. (1980), however, may lie in the use of HCl to maintain pH at 6.5 and 5.0. Reasons for the increase in recovery of labelled soluble mercury under oxidized, weakly alkaline conditions are not provided. Under reduced and moderately oxidized conditions, essentially all the mercury was found to be sorbed by the sediment. The authors found evidence that upon oxidation of initially anoxic sediments, the mercury-binding strength to insoluble, large molecular weight humic materials was reduced.

4.2 Mercury Methylation in Anoxic Sludges

A number of large-scale collaborative efforts have gone into elucidating the behaviour of mercury in sediment-water systems during the past two decades outlining the height of concern about mercury pollution. More recent and detailed reviews of the environmental behaviour of especially organic mercury species in aquatic systems have been provided by two books edited by Nriagu (Nriagu 1979; Clarkson et al. 1984). The main conclusions of the long-lasting work of Jernelöv and his coworkers has also been presented recently (Lindquist et al. 1984). From these sources it is obvious that methylation of mercury in the environment is not only an academic question, but may give rise to serious problems, at least in localized areas.

There is substantial evidence for natural biological production of the methylated forms of mercury, emerged from the intensive research on mercury speciation since the early work published by Jenssen and Jernelöv (1969). The methylated species of metals such as mercury low in the periodic table (i.e. with high Z), in which the metal-carbon bond is relatively labile, represent one of the most toxic group of pollutants in the biosphere. Thus methylation of mercury usually discharged as inorganic species increases its biohazard due to its water and especially lipid solubility (Jeffries 1982). Estimates for methylation rates in sediments are highly variable. Jenssen and Jernelöv (1969) recorded net methylation rates of 15–40 ng g^{-1} sediment day^{-1}. Windom et al. (1976) found higher rates up to 137 ng g^{-1} day^{-1} in organic-rich salt marsh sediments. Liu Ching-I and Tang Hongxiao (1985) recently reported mercury emissions from highly contaminated Chinese estuary sediments in the range from 0.4 to as much as 5.0 $\mu g\ m^{-2}$ day^{-1}. However, following methylation, methyl mercury does not usually build up to more than about 1.5% of the total mercury present even in contaminated sediments. This in an approximate equilibrium level between its formation and removal (Craig and Bartlett 1978). Various factors govern the rate of methyl mercury release to the overlying water column (Wright and Hamilton 1982). Decrease in water pH does increase release of methyl mercury and may account for elevated methyl mercury levels in fish from low pH ponds and lakes (Miller and Akagi 1979).

Conflicting results arose from the question whether methyl mercurials are prone to biomagnification through the aquatic food chain, similarly to chlorinated hydrocarbons. Accumulating factors for mercury reported for the ecosystem of the Ji Yun river (China) are 400–700 for plankton and water weeds, approximately 1000 for the macrobenthos, 2000–3000 for fish, and 6000–7000 for water birds (Liu Ching-I and Tang Hongxiao 1985). Fujiki et al. (1977) found mercury accumulating in organisms to be derived by sorption from water. Exposed organisms were found to accumulate little methyl mercury from spiked food sources. Moreover, the very high methyl mercury levels found in fish are probably due to absorption from outside more than to internal methylation. Hence the low yields occurring naturally may still have significant environmental consequences owing to bioaccumulation. Despite the complex and unresolved questions in regard to the mechanisms of bioaccumulation, there is enough significant data available to date to assess the capacity of especially anoxic sediments to resupply methyl mercury removed from the overlying water by biotic and abiotic processes.

Natural production of methyl mercury ($CH_3\,Hg^+$), first reported by Jenssen and Jernelöv (1969), occurs primarily under anaerobic conditions. Mercury methylation is to date supposed to be mediated by microbes involving a carbanion transfer from methyl cobalamin ($CH_3\,CoB_{12}$) to mercurial species in either aquated, ionic or weakly complexed forms. Such "vitamin B_{12}" derivatives are present at concentration levels of up to 1 mg kg^{-1} in anoxic sediment material, particularly in areas subject to sewage effluent disposal and other sources of pollution (Dolphin 1982). Essentially, the reaction proceeds as follows (Craig and Moreton 1985):

$$CH_3\,CoB_{12} + Hg^{2+} H_2O \rightarrow CH_3\,Hg^+ + H_2OCoB_{12}{}^+ . \qquad (1)$$

Although some abiotic methylation by humic compounds and alkylated metallic species has been reported, these processes account for less than one-tenth of the methyl mercury formed in sediments (Berman and Bartha 1986). Wollast et al. (1975) found that maximum rates of microbial methyl mercury formation seem to occur under slightly reducing conditions in the redox potential range just above that favouring generation of free sulphide ligands (−100 to +150 mV). Compeau and Bartha (1984) conducted a series of inhibition-stimulation tests directly on low-salinity anoxic estuarine sediment samples, which have unequivocally identified sulphate-reducing bacteria as the principal methylators of Hg(II) in such sediments. These microbial processes, however, interfere at higher sulphide levels with precipitation of HgS.

The interfering role of sulphide in the methylation of mercury has been considered ever since the early experiments of Fagerström and Jernelöv (1971). Bartha and co-workers (Blum and Bartha 1980, Compeau and Bartha 1983, 1984) demonstrated HgS to be marginally or not at all available for direct biomethylation. Berman and Bartha (1986) found in sediments of Berry's Creek estuary (New Jersey, USA), contaminated with up to 1000 mg kg^{-1} mercury from past refining activity, methyl mercury levels, which remained at levels below 10 $\mu g\ kg^{-1}$. The authors established a causal connection between elevated sulphide levels and inhibition of methylation efficiency. They found it inlikely that an unusually high demethylation rate contributed to the low levels of monomethylmercury traced in the sediments. These low levels were found to be in good agreement with the low levels of Hg found in Berry's Creek biota.

Uptake of mercury by benthic organisms has been reported to be reduced as salinity levels increase (Neff et al. 1978). The interference of salinity with Hg methylation may be at first explained by the fact that chlorine bonding to mercury has an inhibiting effect on CH_3Hg^+ production in sediments (Craig and Moreton 1985). In an earlier study, Craig and Moreton (1984) proposed additionally that depletion of monomethylmercury results from a disproportion reaction with H_2S producing HgS and dimethylmercury $(CH_3)_2Hg$, which proceeds in the following way:

$$2CH_3Hg^+ + S^{2-} > (CH_3Hg)_2S > HgS + (CH_3)_2Hg. \quad (2)$$

This second form of the organic mercurials is non-polar due to the complete covalently bound mercuric ion, and is almost insoluble in water. Because dimethylmercury is volatile, it tends to evaporate into the atmosphere. Thus, high sulphide levels obviously not only inhibit the synthesis of methyl mercury, but may also cause transformation and thus removal of pre-existing CH_3Hg^+ in the sediment-water system. Moreover, Craig and Moreton (1984) verified the effectiveness of this removal mechanism under realistic estuarine conditions with natural CH_3Hg^+ levels. They found that the disproportion reaction in sediments containing appreciable free-sulphide concentrations (> 1.8 mg kg^{-1}) may account for a loss of about 12% annually of monomethylmercury from the sediments.

Compeau and Bartha (1986) stressed that their findings concerning the role of sulphate reducers in mercury methylation do not contradict the observed inhibition of this activity in estuarine sediments by high sulphide concentrations. They showed in fact that formation of monomethylmercury can proceed in the presence of high sulphate concentrations, but not to the same extent as when sulphate is at lower concentrations. Typically, conditions of high sulphide are found in marine, fine-grained sediments rich in organic matter. The methylating potential of the sulphate reducers, however, are fully expressed only when sulphate is limiting and fermentable organic substrates are still available that can also be utilized in the absence of sulphate. Compeau and Bartha (1986) added that sulphate-reducing microbes metabolize organic substrate by a cobalamin-dependent pathway, which accords with previous results, suggesting that methyl cobalamin is the most likely methyl donor. Sulphate-reducing potential is considerable even in low-sulphate freshwater sediments. In fact, Lovley and Klug (1986) indicated recently that at sulphate concentrations higher than 30 μmol a sulphate-reducing zone develops, because sulphate reducers maintain acetate concentrations too low for methanogens to grow. At these low levels, however, free-sulphide ligand generation is not effective in inhibiting mercury methylation, whereas even moderate numbers of sulphate reducers may produce considerable amounts of CH_3Hg^+. This microbial population is adapted to in situ pH values in the range of 5–7, which is the reason of peak methylation rates found within this narrow pH range (Baker et al. 1983).

In addition, complexation by sulphur leads to weakening of other bonds from the same mercury atom, e.g. a methylation-inhibiting Hg-Cl bond (Craig and Moreton 1985). It might then follow that the rate and extent of methylation of Hg(II) complexes in the anoxic environment (where they may be complexed by sulphur) would be greater than for the same complexes in the water column (where sulphur ligands are absent). If sulphate is abundant, however, organic substrates are utilized more effectively than in its absence, but the free sulphide ligands generated depress methylation of Hg.

Of particular relevance are disposal options that may prevent or minimize any increase in the Hg methylation capacity. With respect to Hg-contaminated anoxic sulphidic sediments, which obviously show the best Hg retention capacity, the question arises whether the short aeration period concurring with any dredging operations is critical in Hg methylation following chemical oxidation of HgS. Elutriate tests on 32 sediments representing a wide range of environmental properties and pollution conditions have in fact shown an insignificant mercury release for most sediments (Gambrell et al. 1978). Berman and Bartha (1986) demonstrated that aeration of sulphidic, highly mercury-polluted sediments for a 1-week period, which was suspected to be roughly comparable to the situation occurring with dredging and disposal operations, resulted in only very moderate increases in methylation activities. From their findings, the authors stressed that "a speedy entombment of the Hg-contaminated dredge spoils in an anoxic sulphide-rich state, with provisions for permanent exclusion of oxygen and oxygenated ground water appears to be safe long-range disposal option".

5 Arsenic

5.1 Mobility of Arsenic in Reduced Environments

Although arsenic is not a heavy metal, it is usually grouped with these elements due to its similar toxic properties and biogeochemical cycling. Elevated arsenic levels in aquatic sediments were found to be associated both with adjacent man-made sources of environmental pollution such as mine tailings and ore smelters (e.g. Creceluis et al. 1975; Brookds et al. 1982; Andreae et al. 1983), and with natural geothermal and early diagenetic fractionation processes (Aggett and Roberts 1986; Farmer and Lovell 1986). Additionally, elevated arsenic concentrations in reducing groundwaters has been used as a good pathfinder element for a wide variety of solid waste and dredged material deposits. This suggests a high mobility of arsenic in anoxic sediment-water systems.

Unlike the aforementioned potentially hazardous trace metals, arsenic is characterized by an essentially aqueous non-cationic chemistry. As group VI elements, arsenic and phosphorus are chemically similar. Thus, arsenate interferes harmfully with phosphate metabolism, which is common in the biosphere. In fact, the oxy-anionic arsenic forms act as weak acids. While the absence of free-cationic arsenic rules out direct complexation with humic or fulvic acids (Wagemann 1978), intracellular retention and microbial reduction of arsenate forms has been found with some microorganisms. The ecological logic of microbial arsenate reduction has been explained in terms of detoxification strategy aimed at avoiding the consequences of incorporating arsenate instead of phosphate, and has been encountered in both mixed heterotrophic communities (e.g. Freeman et al. 1986) and pure cultures (e.g. Vidal and Vidal 1980). As(III) (arsenite) oxidation to As(V) (arsenate), which has also been encountered (e.g. Scudlark and Johnson 1982), is attributed to a co-oxidation mechanism incidental to normal heterotrophic organic matter oxidation.

Though the alkylation of As compounds has been known for almost 100 years, this mechanism is in fact thermodynamically not favourable in the aqueous environment. It is therefore most likely of minor importance for the natural arsenic cycle in sediment-water systems (McBride et al. 1978; Chau and Wong 1978). In the field, methylarsenics were found neither in oxic nor in anoxic marine pore waters within arsenic-rich sediments (Andreae 1979; Froelich et al. 1985). An explanation of this controversy may evolve from an inhibition of arsenic methylation by sea salts similar to that observed for mercury. A more reasonable explanation may be found by comparing the methods by which the natural methylation of arsenic has been evidenced. Manipulating the carbon source in culture flasks and laboratory chemostat enrichment experiments may provide a strong selective pressure which challenges the microorganisms to evolve the ability to methylate arsenic. Although simple studies using pure cultures of microorganisms and single substrates are valuable, if not essential, for determining biochemical pathways, they cannot always be used in predicting species transformation in more natural situations. Because the methylation of inorganic As species is essentially a reduction reaction in oxic sediment-water systems, the abiotic and especially biotic oxidation of arsenite would tend to counteract any methylation process. On the other hand, Holm et al. (1980) found a strong tendency of anaerobic sediment microorganisms to demethylate rather than to methylate arsenic. Thus, methylarsenic concentrations frequently encountered in aquatic environments are not likely to be introduced by transformation of inorganic As species in underlying sediments, but are rather introduced by agricultural use of man-made organoarsenic pesticides (Froelich et al. 1985). Anaerobic as well as aerobic sediment microbes are capable of transforming these organoarsenics and their metabolites ultimately to volatile arsines (McBride et al. 1978). Cheng and Focht (1979) demonstrated accordingly that arsine was the major product released from soils contaminated by arsenate, arsenite, methylarsenate or methylarsenite. They suggested that demethylation proceeds through decarboxylation to arsenite, because no monomethylarsines were observed in their experiments. Neither the ubiquitously present microbe taxa *Alcaligenes* nor *Pseudomonas* were found to be able to produce methylated arsenic compounds (Cheng and Focht 1979).

The valence state of arsenic is critical both for its bio-availability and toxicity. Arsenite forms, As(III), are much more toxic for biological species and more mobile than those of arsenate, As(V) (Ferguson and Gavis 1972). Thus, any arsenic accumulation in natural environments is critical especially where anoxic conditions may potentially occur. The arsenic speciation in the aquatic environment, however, reflects a complex interplay of physico-chemical as well as abiotic and biotic processes. The arsenate-arsenite interconversion in response to Eh drop across oxic-anoxic interfaces results in drastically different As speciation and behaviour in oxidizing and reducing environments. Assuming thermodynamic equilibrium, the As (III) As (V) ratio can be calculated from the reaction describing interconversion of the two species:

$$HAsO_4^{2-} + 2H^+ + 2e^- = HAsO_3^{2-} + H_2O. \quad (3)$$

The pK of this reaction is −8.14 (Peterson and Carpenter 1983). Assuming further a pH of 7.3, which is typical of anoxic sediments, and no significant complexation of the arsenic oxy-anions, the As(III)/As(V) ratio can be expressed as a function of Eh:

$$\log (As^{3+}/As^{5+}) = [-Eh(mV)/29.6] - 6.46. \quad (4)$$

As(III)/As(V) ratios of 10 and 0.1, which are typical of natural sediment-water environments, occur at Eh values of −220 mV and −160 mV respectively. Theoretically, a change in Eh of only 60 mV may thus completely reverse the dominance of reduced versus oxidized arsenic species. From this it can be expected additionally that As(III) species dominance should be stable only within strongly reduced (methanogenic) environments.

Oxidation of As^{3+} to As^{5+} in natural environments does not concur with precipitation like Fe^{2+} and Mn^{2+} oxidation, suggesting that the interconversion itself has no discernible influence on As partitioning and mobility. In fact, near-constant dissolved As concentrations have been shown to exist across the water column O_2/H_2S interface of Saanich Inlet (Vancouver Island, Canada; Peterson and Carpenter 1983). Recorded arsenate/arsenite concentration ratios within the oxic water column at this site was shown to be 15:1, typical of those found in other oxic regimes, but incomplete in thermodynamic terms. This ratio reveals obviously a non-equilibrium state as the reduction of As^{5+} is likely the result of biological detoxification. Scudlark and Johnson (1982) predicted accordingly an As(V)/As(III) ratio of 20 in marine oxic environments from determining the ratio of the overall oxidation rate (chemical + biological yielding 2 nmol l^{-1} day^{-1}) to the rate of reduction by phytoplankton species (0.1 nmol l^{-1} day^{-1}).

Below the O_2/H_2S interface rapid As speciation changes were found in the Saanich Inlet water column, inverting the arsenate/arsenite ratio to 1:12 in the following anoxic zone (Peterson and Carpenter 1983). This valence state interconversion occurred at a depth shallower than that of the ferric-ferrous form, but in the range of Mn(IV)/Mn(II) interconversion. The relative order in dissolved species maxima with increasing depth in the water column was found to be Mn^{2+}—As^{3+}—Fe^{2+}—NH_4^+—H_2S. Though they showed suppression of the rate of arsenite oxidation by addition of antibiotics to seawater derived from the anoxic zone, Peterson and Carpenter (1983) suggested that arsenite oxidation may not be directly mediated by microbes. Because the particulate suspended matter was found to consist of bacteria up to 80% by weight, also Mn^{2+} oxidation and binding onto the solid phase is at least in part bacterially catalyzed. Thus, inorganically catalyzed oxidation of arsenite on Mn(IV) oxides (Oscarson et al. 1981, details discussed below), as well as biotic Mn^{2+} and As^{3+} oxidation, may have occurred with poison affecting directly only the latter processes. Accordingly, the authors suggested the presence of metastable Mn^{2+} as an explanation of why arsenite/arsenate interconversion was found to be inhibited above the O_2/H_2S interface in the oxic water layer.

The first comprehensive investigation of vertical distributions of both solid and dissolved As phases in recent sediments has been presented by Lovell and Farmer (1983). Such distributions have been found within sediment cores of Loch Lomond freshwater lake (Scotland, UK). Because these sediments were not polluted by anthropogenic sources, considerable enhancements of manganese, iron and arsenic observed in subsurface sections of sediment cores were attributed to post-depositional early diagenetic processes. Arsenic concentrations reached up to 675 mg kg^{-1} by dry weight, relative to typical background levels in the range of 15–50 mg kg^{-1} found below 30 cm in sediment depth. Peak iron levels were reported to coincide with peak As concentrations in sediments throughout the Loch Lomond area. Confirmation of concurring early diagenetic mobility of the three elements was provided by pore water profiles. In the anoxic deeper sediment strata pore water arsenic levels at least 200–400 times those of lake water were recorded. Sudden decreases in concentration found above the redox potential discontinuity were explained by removal to the solid phase by co-precipitation and/or adsorption processes in the oxic surface sediment zone.

More recently, Peterson and Carpenter (1986) presented arsenic species studies in sediment-water systems of four diverse regimes ranging from freshwater to seawater environments. The authors found pore water As profiles also exhibiting sub-surface peaks ranging up to $>$ 10 times the concentrations found in overlying waters. Pore water As peaks were found to be restricted to narrow depth ranges in sediments occurring consistently within the bottom of the bioturbated surface layer, which most likely represented post-oxic conditions. Near-surface sediment As peaks occurring above the pore water As peaks were explained by upward diffusion of dissolved As as the source of the surficial enrichments in a similar way as reported by Lovell and Farmer (1983). Marked differences in arsenic pore water profiles, however, were found between freshwater and marine environments by Peterson and Carpenter (1986). For marine pore waters, the As concentrations were found to coincide with solid-phase As concentrations at the depth of the pore water peak, indicative of As solid-solution partitioning controlled by adsorption/desorption mechanisms as illustrated in Fig. 11. Lacustrine pore water As peaks, on the other hand, were not found to show the same relationship, which possibly indicates control by dissolution-precipitation mechanisms. An explanation of the difference in the As pore water distributions between marine and freshwater as shown by Peterson and Carpenter (1986) may be found in the fact that at the depth of pore water As peaks, consistently higher As(III)/As(V) ratios were recorded in freshwater sediment pore waters than in those of the marine environment ($>$ 1.5 versus $>$ 0.3 respectively). Another important observation was that dissolved As concentrations in marine pore waters drop rapidly below the redox potential discontinuity, indicative of solid-phase precipitation or sorption processes. From the fact that a similar decrease in pore water As concentrations was not found in lacustrine As profiles, the authors suggested the formation of sulphidic

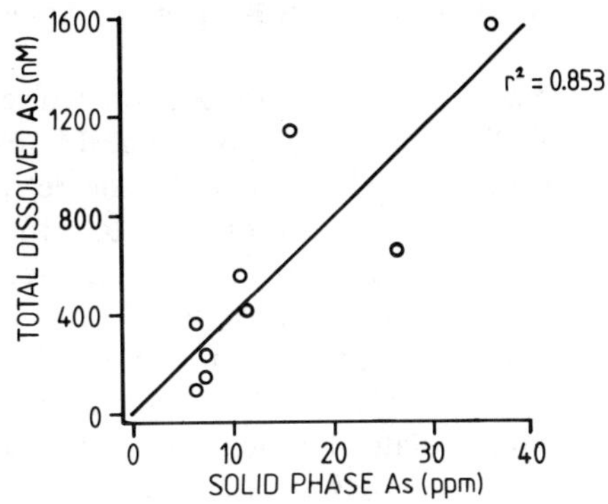

Fig. 11. Relationship between solid-phase and dissolved As concentrations at the depth of peak in a total As pore water profile of marine sediments of Puget Sound (Peterson and Carpenter 1986)

As precipitation or caption of As by Fe sulphides in the marine anoxic environment. Fe pore water data showed similar depletions at depth in the marine, but not in the lacustrine sediment cores.

The early diagenetic reactions in dredged sludge involving arsenic have been studied experimentally in the field by Kerdijk and Salomons (1983, see detailed discussion in the second section and Fig. 3). With the transformation of the dredged mud from oxic to anoxic conditions, the behaviour of arsenic was found to differ strongly between the three deposits studied. In the freshwater and brackish sediment bodies, the pore water As concentrations were found to decrease continuously from the value recorded subsequent to sediment disposal. On the other hand, in the marine sediment deposit, an As behaviour similar to that in Washington coast sediment cores reported by Peterson and Carpenter (1986) was found: As porewater concentrations first increased with development of post-oxic conditions probably due to dissolution of the iron oxide substrates, and subsequently decreased again with transition of the sediment body to sulphidic conditions.

In a laboratory study on the potential arsenic remobilization from dredged sludges, Brannon (1984) gained more indepth insight into the underlying mechanisms by determining the As speciation. He conducted aerobic leaching experiments of 6 months' duration to determine what As species would be released, the duration of the releases, and the sediment factors affecting releases. Arsenic usually persisted throughout the leaching period; releases were higher from amended compared with unamended sediments. Generally, As(III) release predominated initially, followed by AS(V) and organic As releases in the first 3 months. The final 3 months of leaching were characterized by predominant release of As(V). Conversely, leaching of the sediments under anaerobic conditions for 3 months resulted in almost exclusive release of As(III). Releases of As(III) were greatest when sediments were low in iron and high in interstitial water and exchangeable phase As(III). In this study, releases of As(V), As(III), and organic As from sediments were found to be relatively insensitive to the salinity of the mixing water. From these results, Brannon (1984) concluded that soluble As releases from Black Rock sediments are undesirable during the first months of aerobic leaching or at any time during anaerobic conditions because of the high toxicity as As(III).

5.2 Geochemical Barriers to Arsenic Remobilization

Besides the arsenite/arsenate interconversion concededly the main factor controlling arsenic behaviour and fate in soils and sediments is adsorption-desorption on sesquioxides of high isoelectric point such as am-$Fe(OH)_3$ or am-$Al(OH)_3$ (Questel and Scholefield 1953, Huang et al. 1982, Wauchope and McDowell 1984). The majority of arsenic in many sediment samples was found to be extractable with reducible Fe compounds (Crecelius et al. 1975, Brannon et al. 1976b, Kersten and Förstner 1985). Organic matter, carbonates and clays are relatively less important in the sorption of arsenic as evidenced by adsorption experiments with the appropriate chemically enriched soil components (Huang and Liaw 1979). These experiments indicated additionally that a considerable amount of arsenic from lithogenic origin resides occluded in apatite minerals not available to the aquatic environment. Brannon et al. (1976b) showed by their sequential extraction experiments of sulphidic marine sediments that these anoxic sediments are apparently ineffective in immobilizing arsenic as oxidizable sulphidic precipitates or organic matter. More recently, Belzile and Lebel (1986) showed, however, a strong correlation between arsenic and iron in the pyrite fraction of anoxic near-shore sediments of the Laurentian Trough, Canada, suggesting the capture of this element in the lattice sites of pyrite with a Fe/As ratio of 10^3. The relationship between As and sulphur minerals including pyrite is in fact an important pollution problem for sulphide ore roasting plants (Andreae et al. 1983). The pyrite fraction is somewhat problematic in sequential chemical extraction experiments, because it is not easily oxidizable, but rather extractable with the last residual step by strong acid digestion.

Once remobilized in the aquatic environment, As removals by re-adsorption onto oxides in the order of $> 90\%$ can be achieved by careful selection of the solid oxide to solution ratio and pH (Pierce and Moore 1982). The nature of the adsorbent species affects the adsorption-pH profiles in that peak As(V) adsorption was observed at pH 5.5, while for As(III) it was at pH 9.2 (Yoshida et al. 1978), which is the basis for analytical differentiation between both species. Under alkaline conditions both arsenite and arsenate show sharply decreased adsorption. Investigations in Lake Ijsselmeer (The Netherlands) confirmed the importance of environment pH to As adsorption (Salomons and Kerdijk 1983). A net efflux of dissolved arsenic and manganese was recorded from the sediment to the overlying water. While manganese was found to precipitate within the water column again, arsenic remained in solution. This was attributed to the high pH values of the overlying water column reaching up to pH 9. As a consequence little adsorption onto iron oxides was expected to limit the As mobility.

In fact, residence time and the solid-solution partition coefficients of arsenic in oxic sediment-water systems are governed primarily by the adsorption-pH profile of the predominating arsenate form. This is due to a rapid abiotic redox reaction between arsenite and excess manganese oxides to produce arsenate as evidenced by Oscarson et al. (1981):

$$MnO_2 + H_3AsO_3 \rightarrow HAsO_4^{2-} + Mn^{2+} + H_2O. \tag{5}$$

This reaction has both a negative change in standard free energy (–87.4 kJ mol^{-1}, recalculated from data given by Peterson and Carpenter 1983), indicating the thermodynamic feasibility of this inorganic oxidation process, and relatively fast kinetics at relevant pH values. The transformation between a readily mobilized form reduced to the oxidized arsenic form is completed within hours when contacted with MnO_2. Mn^{2+} initially present most likely replaces large monovalent cations in the manganese oxide lattice due to a marked preference for Mn^{2+} as charge-balancing cation (cf. Dymond et al. 1984). This was evidenced by the high concentrations of K in solution contacting the As(III)-Mn(IV) system (Oscarson et al. 1981). Accordingly, freshwater sediments from southern Saskatchewan (Canada), which were leached with hydroxylamin hydrochloride or acidified sodium acetate (both effective extractants for the selective removal of manganese oxides from sediments), showed greatly decreased arsenite oxidation capacities (Oscarson et al. 1981).

Though similar redox reactions between Fe(III) or NO_3^-, and As(III) are also thermodynamically favoured, they have not been observed, which was attributed to the relatively slow kinetics of these reactions. Organic matter and SO_4^{2-}, on the other hand, are not suitable electron acceptors in the oxidation of arsenite as evidenced by the negative equilibrium constants of the respective redox reactions (Huang et al. 1982). While carbonates as well as Fe and Al oxides do not oxidize arsenite, the oxides show a much greater sorption capacity for As than does MnO_2. This is obviously due to the fact that the point of zero charge e.g., of birnesite, is close to pH 2, resulting in a high negative oxide surface charge density at natural pH values. At pH values > 2.3 arsenic acid (H_3AsO_4) is considerably dissociated and negatively charged, but arsenous acid (H_3AsO_3) is predominantly a neutral complex below pH 9.2 (Oscarson et al. 1981). The effective sorption of arsenic in natural sediment-water systems at neutral to moderately acidic conditions is thus enhanced in the presence of manganese oxides, which maintain the predominance of As(V) forms. Development of more acidic conditions tends to release as large amounts of arsenic in soluble forms as is commonly found for Cd and other typical trace metals due to the decreasing affinity of both As forms for adsorption sites (cf. Calmano et al. 1986).

6 Xenobiotic Organic Contaminants

6.1 Compounds Relevant to Sediment-Water Systems

No contribution on the biogeochemical behaviour and fate of priority pollutants in anoxic sludges would be complete without considering organic pollutants. Many such organic compounds may be even much more toxic to mammals and the aquatic ecosystem than any trace metal. This is, among others, due to the fact that biological organisms in the ecosphere have not been exposed to an evolutionary history. Such "xenobiotic" compounds with structural characteristics not common in natural organic compounds are halogenated, particularly polychlorinated organic chemicals. They include such common, man-made, harmful chemicals as DDT (eggshell thinning in certain populations of birds), polychlorinated biphenyls (trace contaminants and combustion products like furans and dioxins), tetrachloroethylene in drinking water, etc. To date, myriads and an evergrowing number of xenobiotic chemicals are introduced into the environment by various pathways via accidental, unintentional or intentional, and transformation processes. In 1986, e.g. more than 100 accidents were reported by the German chemical industry along the Rhine river, most of them having released organic chemicals. Fortunately, no more than 100 compounds are of real environmental concern relevant to significant dredged material contamination, but these are not a trivial problem. For example, the US EPA priority pollutant list contains 114 such compounds, whereas only 60 organic compounds have been designated recently by the FRG BUA as priority pollutants.

Xenobiotic compounds may strongly biomagnified due to their lipophilic properties and persistence. Thus, it is important that we are able to assess and predict their fate once introduced into the aquatic environment by dredged spoils. Persistence, however, is not a property characteristic of xenobiotics. A number of naturally occurring organic compounds are persistent in geological time scales either because of their structure or external geological factors. Without such natural organic matter recalcitrant in the sedimentary environment, no oil would have been formed on earth.

The behaviour and fate of xenobiotics upon entering aquatic environments can be basically described by two mechanisms: solid-solution partitioning and transformation processes. The solid-solution partitioning is characterized by a wide range of solubilities to be considered for these more or less hydrophobic compounds. Simple relationships for the assessment of the environmental partitioning properties of a compound from its solubility, chemical structure and vapour pressure have been proposed by Briggs (1981). The very low aqueous solubilities for several of the relevant xenobiotics indicate that these compounds have a strong affinity for particulate surfaces in aquatic environments. The extremely hydrophobic nature of many of the xenobiotic compounds allow them to concentrate onto suspended matter which may accumulate to form contaminated bottom muds in streams and lakes. The sorption of organic chemicals on solid surfaces is dependent on their functional groups, size and shape of the molecule and, if there is any, their charge. Based on the properties characteristic for the organic chemicals, the following categories can be set up:

1. Cationic or basic compounds. They interact with negatively charged, natural, organically coated particles and are strongly to irreversibly bound.
2. Acidic compounds. They are repelled by the negative surficial charge of organic coatings. Thus, sorption is hardly to be expected. The phenoxy-acetic acids 2,4 D and 2,4,5 T are well-known examples of this group of compounds.
3. Non-polar, volatile substances, mainly used as dissolving agents. This group of compounds interacts with particulate matter through hydrophobic bonding. Only weak adsorption on sediments and solids has been observed, but strong biomagnification in marine organisms has been found since the early studies conducted by Dickson and Riley (1976).

4. Non-polar, non-volatile substances. The hydrophobic bonding forces of these compounds are much stronger than in the latter case, and increase with decreasing water solubility. Organic matter in sediments is primarily responsible for sorption of these essentially hydrophobic compounds.

Among the latter group, such toxic chlorinated pesticides and other xenobiotica like parathion, DDT, HCH, HCB and the PCBs are known. For their persistence and strong accumulation behaviour in dredged materials, this group is of particular concern. Thus, emphasis will be placed on this group within this limited review on the biogeochemical behaviour of xenobiotica in anoxic sediment-water systems.

6.2 Remobilization Mechanisms of Hydrophobic Chemicals from Anoxic Sediments

The partitioning of organic chemicals between sediments and pore water is important in evaluating the remobilization of these compounds in the course of diagenetic processes as well as during dredging and disposal operations. Removal of organic chemicals from solution and their partitioning onto particulates and colloids may not be so much an adsorption process as one of exclusion and solubilization of the organic chemical into an organic (lipidlike) surface layer on the particulates (Horzema and Ditoro 1983). Organic carbon content (or the quantity of organic coating) of the suspended and bottom particulates is in fact critical for the sorption of hydrophobic organics. Thus, the partition coefficient normalized to the organic carbon content of the sediment ($K_{oc} = K_D$/fraction of organic carbon) should be highly invariant over a wide range of sediment types (Karickhoff et al. 1979). In aquatic ecosystems the range from high (i.e. $> 10^5$) to low K_D follows a shift from non-polar (e.g. aromatic and aliphatic organochlorines) to semi-polar (e.g. heteropolycyclic organophosphates) to polar (e.g. methylcarbamates) compound characteristics which primarily reflect changes in solubility (Pavlou and Dexter 1980). A non- to semi-polar organic pollutant molecule will be sorbed by orientating itself so that the hydrophobic functional groups are oriented towards an organic colloid or coating surface, and the polarized structure towards the water phase. A universal sorption model for hydrophobic chemicals has yet to be found, mainly due to the problematic nature of the partition coefficients, and data must be treated on a case-by-case basis (Förstner 1987).

An example for non-polar, essentially hydrophobic chemicals are the polychlorinated biphenyls (PCBs). Larsson (1983) conducted laboratory microcosm experiments on the mobility of sediment-bound PCBs to water and air. He showed that these xenobiotic compounds were transported from sediment to water both with and without bioturbation processes and in systems fixed with $HgCl_2$. Adsorption of hydrophobic compounds on suspended colloidal organic matter in sediments, e.g. humic and fulvic acids, may exert significant influence on the environmental redistribution of xenobiotics (e.g. Means and Wijayaratne 1982, Carter and Suffet 1983). On the other hand, the presence of particulate organic matter as well as that of oil and grease in sediments was reported to effectively reduce remobilization of halogenated compounds into surface waters (Gambrell et al. 1978). Studies on the biogeochemistry of PCBs in pore waters of a coastal marine sediment conducted by Brownawell and Farrington (1986) showed an increase in pore water concentrations of total PCBs, expressed as Arochlor 1254, in the anoxic zone. Apparent distribution coefficients of most individual PCB congeners decreased with depth in sediment to levels far below those reported from surface waters. They were related to dissolved organic carbon levels, which are known to increase during anoxic diagenesis within sediments (Krom and Sholkovitz 1977). The Kd's measured are PCB congene-specific and tend to decrease with the increasing degree of chlorination and decreasing solubility. Brownawell and Farrington (1986) suggested that most PCBs in the pore waters of their core samples were sorbed to organic colloids, which were not separated by the conventional sediment-water fractionation methods. Means and Wijayaratne (1982) reported accordingly that it is the operationally defined high molecular weight fraction of DOC which is important in the partitioning of these components.

The migrational properties of natural colloidal organic matter in anoxic sediments is obviously critical to remobilization of hydrophobic chemicals. While diffusion of sediment-bound PCBs is effectively impeded by sorption and high K_p values, pore water advection or resuspension of anoxic sediments could be a more effective way of mobilizing PCBs from sediments. Consideration of these processes by a three-phase partitioning model has been shown to be successful in the prediction of the flux of hydrophobic chemicals from anoxic dredged material to the overlying water column (Farrington and Westall 1986). These authors stressed, however, that a better than first-order prediction of the fate and behaviour of those chemicals during dredged material dumping into the estuarine environment requires an understanding of the influence of salinity on solubility, on colloid-compound interaction and on degradation mechanisms and rates.

Brannon et al. (1984) conducted laboratory studies on the efficiency of PCB retention by capped underwater disposal of contaminated dredged material. Tissue analysis of *Rangia* suspended in the water column overlying contaminated, anoxic, fine-grained sediments capped with 50 cm of sand did not show elevated tissue concentrations in PCBs during the 40-day course of the experiment. On the other hand, the experiments showed that a 5-cm cap was ineffective in preventing transfer of the organic contaminants to biota due to the presence of severe bioturbation. This emphasizes the intimate link between benthic organisms and pollutant transport and the fate in uncapped underwater disposal sites. Karickhoff and Morris (1986) have recently shown that tubificid oligochaetes transported more than 90% of the chemicals contained in the biologically reworked zone to the sediment surface in laboratory microcosm experiments during a 30–50 day period. These experiments were initiated with a uniform depth distribution of the test chemicals: hexachlorabenzene, pentachlorobenzene and trifluralin. Diffusive transport of these chemicals to the surface from depth was shown to be much less than sediment-mediated transport via sand-sized fecal pellets egested from the macrobenthic species at the sediment-water interface. Fortunately, pollutant release into the water column was not found to be particularly enhanced by this process due to the high retardation by sorption to the surficial biodeposits. The authors stressed that their experiments were much simplified versions of natural systems with regard to factors such as advection and turbulence and more complete macro- and meiofauna community structures were absent. The importance of the latter biogeological factor to pollutant remobilization has been more recently reviewed by Kersten (1988).

6.3 Degradation of Chlorinated Compounds in Sediment-Water Systems

It is not easy to differentiate between degradation and the strong and often irreversible sorption of chemicals within sediment-water systems, because in both cases availability of those chemicals to measurement will be decreased. Both processes, however, may lead to opposite consequences. Considering an ecosystem as a whole, the strong binding to particulates does not contribute to the cause for reduction of the potential hazard of halogenated hydrocarbons. Xenobiotics eliminated from the aquatic ecosystem by sedimentation are not necessarily fixed permanently by the sediment, but may be recycled via chemical and biological agents, both within the sedimentary compartment and also back to the water column. Examples have been discussed in the previous section. Thus, degradation processes seem to be much more important, which ultimately lead to complete mineralization of xenobiotic chemical structures within sediment-water systems.

Concededly, the term "degradation" is used in a broad sense to refer to any measurable structural change in a pesticide molecule under natural environmental conditions. Clearly, any such structure transformations effectively remove a chemical compound from the system without implying necessarily a decrease in toxicity of its transformation products. Degradation can be found as beneficial or deleterious. In an optimal way, biodegradation can eliminate hazardous compounds by converting them to inorganic innocuous forms, degrading them by complete mineralization to carbon dioxide and water, or anaerobically decomposing them to carbon dioxide and methane. On the other hand, biotransformation of xenobiotica was reported to lead in numerous cases to other chemical products ("metabolites") with an even higher toxicity. Such a transformation of halogenated or polycyclic aromatic hydrocarbons into carcinogenic hydroxy and epoxy derivatives, e.g. heptachlor to heptachlor epoxide, aldrin to dieldrin, etc., has presently been an area of particular concern. A well-known example is the degradation pathway of DDT which shows a striking qualitative difference between anaerobic and aerobic biodegradation. Under anoxic conditions the reductive dehalogenation (an important step in biodegradation which will be discussed later) of DDT [1,1,1-trichloro-2,2-bis (p-chlorophenyl)ethane] into DDD [1,1-dichloro-2,2,-bis (p-chlorophenyl)ethane] leads to a compound with much higher toxicity and persistence in the environment (Alexander 1979), which has enforced restriction of DDT marketing.

6.3.1 Abiotic Degradation

Complete elimination of a chemical compound from the environment does not necessarily need microbial biodegradation. Abiotic catalytical transformations onto soil particles like humic matter as well as reactions coupled to inorganic species have been reported. Such an abiotic transformation can occur via reactive chemical forms such as nucleophiles, radicals, reduced iron and sulphur compounds, etc., or by photo- and hydrolysis. Photolysis is more or less coupled to the presence of free-radical oxidants in the euphotic zone, and are obviously of no relevance to anoxic sludges (Mill 1980). As an example of its overall importance to aerobic polycyclic aromatics degradation, however, we may cast a glance at the biodegradation pathway of anthracene. Rontani et al. (1985) found a natural marine microbial community that was unable to grow on anthracene as a carbon and energy source, but was able to metabolize anthraquinone, a common photoproduct of anthracene. Since anthracene is generally resistant to direct biodegradation and the anthraquinone is quite stable to further photo-oxidation, the combination of photolysis and biodegradation to completely decompose the anthracene is an important example for combined ecological processes relevant to the chemical fate assessment in dredged materials dispersed in oxidized waters.

In anoxic sediments similarly combined degradative processes have been encountered. There are indications, e.g. that degradation of such important xenobiotica as DDT and toxaphene are coupled to the reduction of Fe(III) to Fe(II) (Glass 1972, Williams and Bidleman 1978). In a more recent paper concerning groundwater pollution by volatile compounds, Schwarzenbach et al. (1986) showed that brominated and chlorinated alkanes undergo nucleophilic substitution reactions, which are especially important to abiotic degradation of halogenated compounds under anaerobic conditions, by hydrogen sulphide to produce very persistent and toxic sulphur-containing products under typical groundwater reducing conditions.

Hydrolytic processes have also been shown to be important steps in the degradation of many organic chemical compounds in sediment-water systems. Organophosphate pesticides such as malathion and parathion are well known for their susceptibility to hydrolytic degradation (e.g. cf. Chapman and Cole 1982, Macalady and Wolfe 1985). In the aquatic milieu, hydrolytic degradation is catalyzed mainly by hydroxyl radicals. While relationships between structure and hydrolysis rates have been shown to be reliable in the aqueous phase (Mabey and Mill 1978), hydrolysis in soils and sediments appear to be complicated by unusual pH relationships and by the possible incursion of metal -ion and especially humic and fulvic acid catalyzed processes (Mill 1980). Khan (1980) reported that the half-life of atrazine was shortened with increases in the amount of fulvic acid in solution. The humic matter catalyzed hydrolysis is mostly important in anoxic pore waters with their relatively high DOC levels.

6.3.2 Biological Degradation by Microorganisms

While abiotic degradation may be important for a number of organic chemicals, in sediment-water systems most of them are more effectively degraded by microbial attack. Among the reasons for this is that (1) most of the microbial activity and biogeochemical cycles in soil and sediment are concentrated on the same organic particles to which most of the chemicals are sorbed; (2) sediments provide accumulated concentrations of inorganic nutrients which promote microbial life; and (3) the sediment environment provides for an extremely broad range and dense accumulation of microbial communities and resulting acclimation.

In contrast to naturally occurring organic compounds, man-made chemicals are relatively refractory to biodegradation. One reason is that organisms that are present under prevailing environmental conditions have not adapted in producing the enzymes necessary to bring about transformation of the original chemical compound to a point at which the resultant intermediates can enter into common metabolic pathways and be completely mineralized. In addition, with the words of Matsumura (1974),

"it is hard to believe that microorganisms develop specific metabolic processes for chemicals which are present at only ppt to ppm (parts per trillion and million, respectively) levels in their surroundings. The answer appears to lie in the unique characteristics of pesticidal chemicals that make them to show high biological affinity. In other words, these chemicals come into contact with biological systems including microbial ones whether they are ready to metabolize and utilize or not... The microorganisms receiving low levels of pesticidal exposure are likely to utilize either already existing enzymatic systems to degrade these chemicals or else transport them to non-vital extra- or intra-cellular sites by changing the nature of these chemicals or, if possible, as they are. Thus such reactions are never meant to derive energy from the chemicals... 'Cometabolism' is defined as the microbial metabolic activities on chemicals from which microorganisms do not primarily receive energy".

Thus, the specific process of co-metabolism of xenobiotics introduced by Matsumura and co-workers is not concentration-, but primarily structure-dependent, i.e. microbial transformation of traces of toxic xenobiotics is accomplished together with natural substances with similar molecular structures by the appropriate microbes. According to Hutzinger and Veerkamp (1981), three factors are responsible for biodegradation or lack of biodegradation of a compound in a given contaminated site: (1) the structure of the chemical compound; (2) the physico-chemical condition of the environment; and (3) availability of appropriate microorganisms.

The chemical structure influences the biodegradability of an xenobiotic compound in two ways. On the one hand, the molecule structure may contain groups or substituents which cannot react with available or inducible microbially-derived enzymes (i.e. these chemicals cannot be co-metabolized). Generally, biodegradation is less likely for chemicals having larger molecules and structural characteristics never encountered in natural products. Attempts to relate chemical structure and biodegradability have led to lists of chemical substituents, that, when attached to the organic parent compounds, make these compounds more or less persistent. Such a list, as established by Hutzinger and Veerkamp (1981), is presented in Table 2. However, so many exceptions to these generalizations exist due to a variety of other important factors that such rules should be considered in practice only as broad guidelines. On the other hand, the structure may determine the compound to be in a physical state rendering it less susceptible to biodegradation. The compound's physical or chemical characteristics, such as solubility, volatility, hydrophobicity and octanol-water partition coefficient, contribute to the compound's availability to biodegrading microorganisms. A review of the literature concerning relationships of the K_D's of organic compounds and biodegradation rates shows that with increasing sorption strength, the biodegradation

Table 2. Influence of structure on biodegradability of xenobiotics (After Hutzinger and Veerkamp 1981)

Type of compounds or substituents	More degradable	Less degradable
Hydrocarbons	Higher alcanes (–12)	Lower alcanes High molecular weight Alcanes
	Straight-chain paraffinic	
	Paraffinic	Aromatic
	Mono- and bicyclic aromatic	Polycyclic aromatic
Aliphatic chlorine	-Cl more than six carbon atoms from terminal C	-Cl six or less carbon atoms from terminal C
Aromatic substituents	-OH, -COOH, $-OCH_3$, $-NH_2$	-F, $-CF_3$, NO_2, $-SO_3H$
	-nCl with $n<3$	-nCl with $n>3$
	Meta-halogens	Ortho- or para-halogens

rate generally decreases (Gordon and Millero 1985). Higgins and Gilbert (1978) observed, however, microorganisms capable of degradation of highly sorptive chemicals of high molecular weight simply through releasing surfactantlike reagents. From these results it can be concluded that any reaction which will change the sorption characteristics may determine the bioavailability and biodegradation rates of organic compounds.

The rate of biodegradation is largely dependent on the physico-chemical conditions of the environment, especially on the sediment redox conditions. Thus, it should be valuable to define the physico-chemical conditions for more effectively managing wastes and dredged sludges as to contamination with certain toxic compounds. This attempt seems timely since recently substantial progress in this field indicates that nature offers remarkable biodegradative capabilities with microorganisms acclimated under certain environmental conditons. Kobayashi and Rittmann (1982) conducted an in-depth evaluation of the potential for microorganisms to remove anthropogenic organic compounds, mainly priority pollutants and related compounds. Their results concerning the most favourable redox conditions in sludges for pollutant biodegradation are illustrated in Table 3, which shows that members of almost every class of man-made organic compound can be biodegraded under appropriate environmental conditions. It is well known that for most of the aliphatic and aromatic non-halogenated compounds biodegradation was shown to be most effective under aerobic conditions. More complicated are, however, the degradation pathways for halogenated compounds, which have been often regarded as persistent. Simple generalizations do not appear to be applicable as evidenced by Kobayashi and Rittmann (1982). The first and possibly limiting step to initiate biodegradation of these compounds is dehalogenation, for which in principle there are several biological mechanisms (Goldman 1972). In earlier studies, oxidative pathways were generally believed to be the typical means by which halogenated compounds are dehalogenated. Kobayashi and Rittmann (1982) stressed in their review that at present anaerobic, reductive dehalogenation, either biological or non-biological, is now recognized as

Table 3. Redox conditions under which examples of anthropogenic organic compounds can be biodegraded most efficiently in sludges (Kobayashi and Rittmann 1982, cf. references therein)

Compound	Aerobic	Anaerobic
Polycyclic aromatics [non-halogenated: benzo (a) pyrene, naphtalene, pyrene, fluoranthene, anthracene, phenanthracene, benzo (a) anthracene, dibenzanthracene, biphenyl]	X	
Nitrosamines (dimethylnitrosamine)		X
Technical compounds [phtalate esters, tetra- and trichloroethylene, mono- and dichlorobiphenyls (PCBs)]	X	X
Pesticides (toxaphene, lindane, dieldrin, parathion, phorate sulphoxide, aldrin, endrin, kepone, diuron)		X
Pesticides (DDT, heptachlorobornane, pentachloronitrobenzene, methoxychlor, acrolein, endosulfan, chlordimeform)	X	

the critical factor in the transformation or biodegradation of certain classes of more persistent compounds.

Sahm et al. (1986) presented two arguments which may represent the reasons for the biochemical predominance of reductive dehalogenation: (1) haloaromatic compounds tend to polymerize when degraded by ordinary aerobic bacteria, resulting in rather recalcitrant products as opposed to anaerobic conditions, where oxidation, and hence polymerization, is not possible; and (2) theoretical considerations show thata polyhalogenated aromatics should be more easily degraded by anaerobes than by aerobic microbes, because an increasing degree of halogenation of an aromatic ring decreases the electron density of the aromatic nucleus. Thus, electrophilic attack of oxygen on haloaromatic structures is less likely. On the contrary, a decreased electron density of the haloaromatic nucleus should enhance anaerobic enzymatic attack by a reductive (nucleophilic) mechanism (Sahm et al. 1986). Beland et al. (1976) generalized accordingly that haloaromatic compounds with a first reduction potential (E_{2d}) more positive than -1.52 V versus a saturated calomel electrode will be reductively dehalogenated and those compounds witn an $E_{2d} < -1.76$ V will not. Significant reductive dehalogenation is reported to occur only when the redox potential (E_h) of the environment is 0.35 V and lower, the exact requirements appear to depend upon the compound involved (Kobayashi and Rittmann 1982). The proximity of free electrons derived from the decaying organic matter in organic-rich reduced sludges and the chlorinated compounds adsorbed onto the organic matter make conditions ideal for energy-efficient reductive dechlorination, nitroreduction and reduction of sulphoxides to occur (Matsumura and Esaac 1979).

Examples of biological reductive dechlorination are well documented, especially for pesticides and chlorinated one- and two-carbon aliphatic compounds. Reductive dehalogenation appears to be the initial process of these transformations that may ultimately lead to complete mineralization (which often warrants, however, aerobic microorganisms), which converts, e.g. lindane to benzene (Beland et al. 1976), several halobenzoates even to methane and carbon dioxide (Suflita et al. 1982, Horowitz et al. 1983) and chlorinated biphenyls (Brown et al. 1984; Brownawell and Farrington 1986) by anaerobic microorganisms in highly contaminated anoxic sediments. Microbial degradation of halobenzoates was not observed in the absence of methanogenesis, and it was inhibited by the addition of oxygen. Microbe acclimation to halobenzoates in sediments, e.g. were observed after a lag period which lasted from 1 week to several months, depending on the chemical compound (Horowitz et al. 1983). Analyses of sediments from the Wadden Sea (The Netherlands) showed concentrations of tetrachlorophenol (TCP) that were 2 to 20 times higher than the original contaminant, pentachlorophenol (PCP) (Butte et al. 1985). Because anaerobic dehalogenation occurs with PCP but not with TCP, it was suggested that microbial activity in the anoxic muds may be responsible for the ratio convertion of TCP to PCP noted in this area. Vogel and McCarty (1985) reported that tetrachloroethylene biodegraded anaerobically through a series of reductive dehalogenation and eventual conversion to CO_2. Trichloroethylene (TCE), dichloroethylene and vinyl chloride were all detected as intermediate transformation products. Similarly, Kleopfer et al. (1985) detected TCE biotransformation by reductive dehalogenation in anoxic soil layers to dichloroethylene. As an important technical application of these results, Wilson and Wilson (1985) demonstrated that TCE could be ultimately mineralized aerobically to CO_2 in soils exposed to a mixture of natural methane-rich gas and air.

Conclusions for Contaminated Anoxic Sludge Management to Prevent Secondary Environmental Pollution

There have been many considerations of the various alternatives for the disposal of dredged material (Kester et al. 1983, and references therein) and mine tailings. There is controversy, however, concerning whether the containment disposal for highly contaminated, unstabilized material should be on land (resulting in negative oxidation effects) or in the marine environment (with long-term containment under anoxic conditions). In a review of various marine disposal options Kester et al. (1983) suggested that the best strategy for disposing the sludges is to isolate them in a permanently reducing environment, which represents a highly efficient, economic method for immobilization of heavy metals once introduced into sediment. Accordingly, it has been shown in the foregoing sections of this chapter that the latter option provides advantages for all four pollutant groups considered valuable. They are summarized in Table 4. A "regression" of biochemical diagenetic succession from a freshwater methanic to a marine sulphidic environment affects pollutant mobility in three ways: (1) many of the trace metals and metalloids (e.g. Cd, Hg, As) are effectively immobilized by the formation of insoluble sulphides or capture by iron sulphides; (2) the process of reduction of sulphate to sulphide, together with abundant chloride ligands, is highly efficient in degrading methyl mercury, and repressing its further formation; and (3) reducing conditions are most favourable to initiate organochlorine chemical biodegradation by reductive dehalogenation processes.

It has been shown that knowledge of the behaviour of organochlorines in dredged material not only offers appropriate measures to immobilize them, but also to deteriorate them, which is in contrast to trace metal pollution managing. The evaluation of organochlorine biodegradation conducted to date indicates that the use of properly selected populations of microbes, and the maintenance of environmental conditions most conductive to their metabolism, can be an important means of improving biological treatment of contaminated sludges. The emphasis in the search for microorganisms suitable for biodegradation of compounds arising from human activity, however, has been largely on aerobic ones, because oxidative processes are best known, and because aerobic techniques are relatively simple compared to anaerobic culture methods. As a remarkable example, mineralization rates of 100 radiolabelled chemicals in aerobic-activated sludge were reported

Table 4. Summary of the advantages and disadvantages of anoxic (sulphidic) conditions as to trace metal and metalloid, methyl mercury and organochlorine immobilization from sludges

Compound	Advantages	Disadvantages
Trace metals (cadmium)	Sulphide precipitation	Formation of mobile polysulphide and organic complexes under certain conditions with low Fe-oxide concentrations warrants modelling of free-sulphide generation potential, highly mobile under post-oxic and acidic conditions
Metalloids (arsenic)	Capture by sulphides	Highly mobility under post-oxic and neutral to slightly alkaline conditions warrants Fe/Mn oxide-bearing sediment barriers
Methyl mercury	Degradation and inhibition of CH_3Hg^+ formation by HgS-precipitate formation	Formation of mobile polysulphide complexes under certain conditions, especially low Fe concentrations
Organochlorines	Initiation of biodegradation by reductive dechlorination (methanic environment more favourable)	Formation of harmful terminal residues with certain compounds, especially in sulphidic environments Remobilization through colloidal matter suspended in pore-water warrants capping with inert barrier layers

recently as part of an "Environmental Hazard Profile" (Freitag et al. 1985). Such studies are warranted also for anoxic sludges. Interest in the anaerobic biodegradation of xenobiotics will increase with the improvement of the appropriate laboratory culture and field-screening methods. This attempt seems to be timely since substantial progress in this field indicates that nature offers remarkable biodegradative capabilities even to halogenated aromatic compounds, leading to newly emerging techniques like in situ microbiological treatment to control secondary pollution potential of waste deposits.

In order to achieve higher rates and better yields than those found in nature, it is necessary to gather more in-depth information on the organisms involved and to learn more about their environmental requirements and metabolic processes. The rates of biodegradation may be enhanced by manipulating the physicochemical characteristics of the dredged material to favour the proliferation of pollutant-degrading organisms (microbes and especially fungi), as well as by application of modern genetics (Omenn 1986). The ability of anaerobes to dehalogenate highly toxic and otherwise persistent haloaromatic compounds in anoxic sediment-water systems should be especially helpful. Natural environments, however, present to date severe challenges for survival and effectiveness of genetically engineered microorganisms, or those acclimated in the laboratory and re-inoculated into contaminated soils. This may be due to low concentrations of specific substrates, toxins, predators, abiotic mediates, physical and mineralogical barriers, and to seasonal or other fluctuations in environmental conditions (Omenn 1986). Thus, to quote Kilbane (1986),

"it is difficult to think of something better than mimicking nature's solution of having microorganisms degrade and recycle wastes. In order to degrade toxic synthetic compounds, the biochemical abilities of microorganisms will have to be expanded; thus we must ask if microorganisms display sufficient metabolic plasticity to make it likely that this can be accomplished. For all the sophistication of current molecular genetics, scientists are unable to do anything fundamentally different than to apply knowledge learned from studying biochemical processes basic to nearly all forms of life".

However, the most promising targets for genetic control of environmental organic pollutants appear to involve the incorporation of microbial conversion, mineralization or toxic by-product degradation into industrial processes rather than relying on end-of-the-pipe treatments of relatively dilute, complex waste materials deposited in the natural ecosystem.

Acknowledgements. I am indepted to Professors Ulrich Förstner and Wim Salomons for encouragement in writing this chapter and for many enlivened discussions of environmental biogeochemistry. Many thanks also to my colleague Reinhard Weinberg, who kindly provided literature and critical comments to the section on xenobiotics biogeochemistry.

References

Aggett J, Roberts LS (1986) Insight into the mechanism of accumulation of arsenate and phosphate in Hydro Lake sediments by measuring the rate of dissolution with ethylendiaminetetra-acetic acid. Environ Sci Technol 20:182–186

Alexander M (1979) In: Bourquin AW, Pritchard PH (eds) Microbial degradation of pollutants in marine environments. EPA, Gulf Breeze, pp 67–75

Andreae MO (1979) Arsenic speciation in seawater and interstitial waters: the influence of biological-chemical interactions on the chemistry of a trace element. Limnol Oceanogr 24:440–452

Andreae MO, Byrd JT, Froelich PN, Jr (1983) Arsenic, antimony, germanium and tin in the Tejo Estuary, Portugal: Modelling a polluted estuary. Environ Sci Technol 17:731–737

Andrews RI (1977) Tailings: Environmental consequences and controls Proc Int Conf Heavy metals in the environment, Toronto 1975, vol 2, pp 645–675

Baker MD, Inniss WE, Mayfield CI (1983) Effect of pH on the methylation of mercury and arsenic by sediment microorganisms. Environ Technol Lett 4:89–100

Beland FA, Farwell SO, Robocker AE, Geer RD (1976) Electrochemical reduction and anaerobic degradation of lindane. J Agric Food Chem 24:753–756

Belzile N, Lebel J (1986) Capture of arsenic by pyrite in near-shore marine sediments. Chem Geol 54:279–281

Berman M, Bartha R (1986) Control of the methylation process in a mercury-polluted aquatic sediment. Environ Pollut Ser B 11:41–53

Berner RA (1980) Early Diagenesis. A theoretical approach. Univ Press, Princeton NJ, 241 pp

Berner RA (1981) A new classification of sedimentary environment. J Sediment Petrol 51:359–365

Blum JM, Bartha R (1980) Effect of salinity on methylation of mercury. Bull Environ Contam Toxicol 25:404–408

Bothner MH, Carpenter R (1973) Proc 1st Annu NSF Contaminants Conf, August 1973, Oak Ridge Natl Lab, p 198

Brannon JM (1984) Transformation, fixation, and mobilization of arsenic and antimony in contaminated sediments. Tech Rep D-84-1, US Army Eng Waterways Exp Stn, Vicksburg, MS

Brannon JM, Engler RM, Rose, JR, Hunt PG, Smith I (1976) Distribution of toxic heavy metals in marine and freshwater sediments. In: Krenkel PA, Harrison J, Burdick III JC (eds) Proc Spec Conf Dredging and its environmental effects. ASCE, New York, pp 455–495

Brannon JM, Plumb RH Jr, Smith I (1980) Long-term release of heavy metals from sediments. In: Baker RA (ed) Contaminants and sediments, vol 2. Ann Arbor Science, Ann Arbor, MI, pp 221–266

Brannon JM, Hoeppel RE, Gunnison D (1984) Efficiency of capping contaminated dredged material. In: Montgomery RL, Leach JW (eds) Dredging and dredged material disposal, vol 2. Proc Conf Dredging '84, 14–16 November 1984, Clearwater Beach, ASCE, New York, pp 664–673

Briggs GG (1981) Theoretical and experimental relationships between soil adsorption, octanol-water partition coefficients, water solubilities, bioconcentration factors, and the parachor. J Agric Food Chem 29:1050–1059

Brookins DG (1986) Geochemical behaviour of antimony, arsenic, cadmium and thallium: Eh-pH diagrams for 25°C, 1-bar pressure. Chem Geol 54:271–278

Brooks RR, Fergusson JE, Holzbecher J, Ryan DE, Zhang HF, Dale JM, Freedman B (1982) Pollution by arsenic in a gold-mining district in Nova Scotia. Environ Pollut Ser B 4:109–117

Brown JF, Wagner RE, Bedard DL, Brennan MJ, Carnahan JC, May RJ, Tofflemire TJ (1984) PCB transformations in the upper Hudson sediments. Northeast Environ Sci 3:167–179

Brownawell BJ, Farrington JW (1986) Biogeochemistry of PCBs in interstitial waters of a coastal marine sediment. Geochim Cosmochim Acta 50:157–169

Bulte W, Denker J, Kirsch M, Höpner Th (1985) Pentachlorophenol and tetrachlorophenols in wadden sediment and clams *Mya arenaria* of the Jadebusen after a 14-year period of waste water discharge containing pentachlorophenol. Environ Pollut B 9:29–39

Callender E, Bowser CJ, Rossmann R (1974) Geochemistry of ferro-manganese and manganese carbonate crusts from Green Bay, Lake Michigan. Trans Am Geophys Un 54:340

Calmano W, Ahlf A, Förstner U (1983) Heavy metal removal from contaminated sludges with dissolved sulfur dioxide in combination with bacterial leaching. Proc Int Conf Heavy metals in the environment, September 1983, Heidelberg, CEP Consultants, Edinburgh (GB), pp 952–955

Calmano W, Förstner U, Kersten M, Krause D (1986) Behaviour of dredged mud after stabilization with different additives. In: Assink JW, Brink WJ van den (eds) Contaminated soil, 1st Int TNO Conf Contaminated soil, Utrecht, 11–15 Nov 1985. Nijhoff, Dordrecht, pp 737–741

Carter CW, Suffet IH (1983) Interactions between dissolved humic and fulvic acids and pollutants in aquatic environments. In: Swann R, Eschenroeder A (eds) Fate of chemicals in the environment. ASC Symp Ser 225

Chapman RA, Cole CM (1982) Observations on the influence of water and soil pH on the persistence of insecticides. J Environ Sci Health B17:487–504

Chau YK, Wong PTS (1978) Occurrence of biological methylation of elements in the environment. ACS Symp Ser 92:39–53

Cheng CN, Focht DD (1979) Production of arsine and methylarsine in soil and in culture. Appl Environ Microbiol 38:494–498

Compeau GC, Bartha R (1983) Effects of sea salt anions on the formation and stability of methyl mercury. Bull Environ Contamin Toxicol 31:486–493

Compeau GC, Bartha R (1984) Methylation and demethylation of mercury under controlled redox, pH, and salinity conditions. Appl Environ Microbiol 48:1203–1207

Compeau GC, Bartha R (1986) Sulfate-reducing bacteria: principal mechylators of mercury in anoxic estuarine sediment. Appl Environ Microbiol 50:498–502

Craig PJ, Moreton PA (1984) The role of sulphide in the formation of dimethyl mercury in river and estuary sediments. Mar Pollut Bull 15:406–408

Craig PJ, Moreton PA (1985) The role of speciation in mercury methylation in sediments and water. Environ Pollut B 10:141–158

Crecelius EA, Bothner MH, Carpenter R (1975) Geochemistries of arsenic, antimony, mercury, and related elements in sediments of Puget Sound. Environ Sci Technol 9:325–333

Dickson AG, Riley JP (1976) The distribution of short chain halogenated aliphatic hydrocarbons in some marine organisms. Mar Pollut Bull 7:167–169

Dolphin D (1982) B_{12}, chemistry, biochemistry and medicine, vol 2. John Wiley, New York

Dymond J, Lyle M, Finney B, Piper DZ, Murphy K, Conrad R, Pisias N (1984) Ferromanganese nodules from MANOP sites H, S, and R – control of mineralogical and chemical composition by multiple accretionary processes. Geochim Cosmochim Acta 48:931–949

Eganhouse RP, Young DR, Johnson JN (1978) Geochemistry of mercury in Palos Verdes sediments. Environ Sci Technol 12:1151–1157

Emerson S, Jahnke R, Heggie D (1984) Sediment-water exchange in shallow water estuarine sediments. J Mar Res 42:709–730

Fagerström T, Jernelöv A (1971) Formation of methyl mercury from pure mercuric sulfide in aerobic organic sediment. Water Res 5:121–122

Farmer JG, Lovell MA (1986) Natural enrichment of arsenic in Loch Lemond sediments. Geochim Cosmochim Acta 50:2059–2067

Farrington JW, Westall J (1986) Organic chemical pollutants in the oceans and groundwater: a review of fundamental chemical properties and biochemistry. In: Kullenberg G (ed) The role of the oceans as a waste disposal option. Reidel, Dortrecht, pp 361–425

Farwell SO, Beland FA, Geer RD (1973) Voltametric identification of organochlorine insecticides, polychlorinated biphenyls, polychlorinated naphtalenes and polychlorinated benzenes. Bull Environ Contamin Toxicol 10:157–165

Feick G, Johanson EE, Yeaple DS (1972) Control of mercury contamination in freshwater sediments. Tech Rep R2-72-077, US Environ Protect Ag, Washington, DC

Ferguson JF, Gavis J (1972) A review of the arsenic cycle in natural waters. Water Res 6:1259–1274

Förstner U (1987) An overview of scientific basis for developing remedial options. In: Thomas RL (ed) The ecological effects of in-situ sediment contaminants. Hydrobiologia 149:221–246

Freeman MC, Aggett J, O'Brien G (1986) Microbial transformation of arsenic in lake Ohakuri, New Zealand. Water Res 20:283–294

Freitag D, Ballhorm L, Geyer H, Korte F (1985) Environmental hazard profile of organic chemicals. Chemosphere 14:1589–1616

Froelich PN, Wlinkhammer GP, Bender ML, Luedtke NA, Hartmann B, Maynard V (1979) Early oxidation of organic matter in pelagic sediments of the eastern equatorial Atlantic: Suboxic diagenesis. Geochim Cosmochim Acta 43:1075–1090

Froelich PN, Kaul LW, Byrd JT, Andreae MO, Roe KK (1985) Arsenic, barium, germanium, tin, dimethylsulfide and nutrient biogeochemistry in Charlotte Harbor, Florida, a phosphorous-enriched estuary. Estuar Coastal Shelf Sci 20:239–264

Fujiki M, Hirota R Yamaguchi S (1977) The mechanism of methyl mercury accumulation in fish. In: Peterson SA, Randolph KK (eds) Management of bottom sediments containing toxic substances. Tech Rep 600/3-77-083, US Environ Protect Ag, Corvallis, Or, pp 89–95

Gambrell RP, Khalid RA, Verloo MG, Patrick WH, Jr (1977) Transformations of heavy metals and plant nutrients in dredged sediments as affected by oxidation-reduction potential and pH, vol 2: materials and methods/results and discussion. Tech Rep D-77-4, US Army Eng Waterways Exp Stn, CE, Vicksburg, MI

Gambrell RP, Khalid RA, Patrick WH Jr (1978) Disposal alternatives for contaminated dredged material as a management tool to minimize adverse environmental effects. Tech Rep DS-78-8, US Army Eng Waterways Exp Stn, Vicksburg, MS

Gambrell RP, Khalid RA, Patrick WH, Jr (1980) Chemical availability of mercury, lead, and zinc in mobile bay sediment suspensions as affected by pH and oxidation-reduction conditions. Environ Sci Technol 14:431–436

Gendron A, Silverberg N, Sundby B, Lebel J (1986) Early diagenesis of cadmium and cobalt in sediments of the Laurentian Trough. Geochim Cosmochim Acta 50:741–747

Glass BL (1972) Relation between the degradation of DDT and the iron redox system in soils. J Agric Food Chem 20:324–327

Goldman P (1972) Enzymology of carbon-hydrated bonds. In: Natl Acad Sci (ed) Degradation of synthetic organic molecules in the biosphere. Washington DC, pp 147–161

Gordon AS, Millero FJ (1985) Adsorption mediated decrease in the biodegradation rate of organic compounds. Microb Ecol 11:289–298

Grossmann D, Kersten M, Niecke M, Puskeppel A, Voigt R (1985) Determination of trace elements in membrane filter samples of suspended matter by the Hamburg proton microprobe. In: Degens ET, Kempe S, Herrera R (eds) Transport of carbon and minerals in major world rivers, pt 3. Mitt Geol Paläontol Inst Univ Hamburg, SCOPE/UNEP Sonderbd 58:619–630

Hakanson L (1980) An ecological risk index for aquatic pollution control. A sedimentological approach. Water Res 14:975–1001

Hearn PP, Parkhurst DL, Callender E (1983) Authigenic vivianite in Potomac river sediments: control by ferric oxy-hydroxides. J Sediment Petrol 53:165–177

Herms U, Brümmer G (1979) Einflup der Redoxbedingungen auf die Löslichkeit von Schwermetallen in Böden und Sedimenten. Mitt Dtsch Bodenkundl Ges 27:533–544

Higgins TJ, Gilbert PD (1978) The biodegradation of hydrocarbons. In: Chater KWA, Somervill HJ (eds) The oil industry and microbial ecosystems. Academic Press, London, New York, pp 80–114

Hines ME, Berry-Lyons WM, Armstrong PB, Orem WH, Spencer MJ, Gaudette HE (1984) Seasonal metal remobilization in the sediments of Great Bay, New Hampshire. Mar Chem 15:173–187

Hoeppel RE, Meyers TE, Engler RM (1978) Physical and chemical characterization of dredged material influents and effluents in confined land disposal areas. Tech Rep D-78-24, US Army Eng Waterways Exp Stn, CE, Vicksburg, MI

Hoff JT, Thompson JAJ, Wong CS (1982) Heavy metal release from mine tailings into sea water—a laboratory study. Mar Pollut Bull 13:283–286

Holdren GR, Bricher OP, Matisoff G (1975) A model for the control of dissolved manganese in the interstitial waters of Chesapeake Bay. In: Church TM (ed) Marine chemistry in the coastal environment. Am Chem Soc Symp Ser 18:364–381

Holm TR, Anderson MA, Stanforth RR, Iverson DG (1980) The influence of adsorption on the rates of microbial degradation of arsenic species in sediments. Limnol Oceanogr 25:23–30

Holmes CW, Slade EA, McLerran CJ (1974) Migration and redistribution of zinc and cadmium in marine estuarine systems. Environ Sci Technol 8:255–259

Horowitz A, Suflita JM, Tiedje JM (1983) Reductive dehalogenations of halobenzoates by anaerobic lake sediment microorganisms. Appl Environ Microbiol 45:1459–1465

Horzema LM, Ditoro DM (1983) The extent of reversibility of poly-chlorinated biphenyl adsorption. Water Res 17:851–859

Huang PM, Liaw WK (1979) Adsorption of arsenite by lake sediments. Int Revue ges Hydrobiol 64:263–271

Huang PM, Oscarson DW, Liaw WK, Hammer UT (1982) Dynamics and mechanisms of arsenite oxidation by freshwater lake sediments. Hydrobiologia 91:315–322

Hunt CD, Smith DL (1983) Remobilization of metals from polluted marine sediments. Can J Fish Aquat Sci 40(2):132–142

Hutzinger O, Veerkamp W (1981) Xenobiotic chemicals with pollution potential. In: Leisinger T, Hutter R, Cook AM, Nuesch J (eds) Microbial degradation of xenobiotics and recalcitrant compounds. Academic Press, London New York, pp 3–45

Jeffries TW (1982) The microbiology of mercury. Progr Ind Microbiol 16:23–75

Jenne EA (1977) Trace element sorption by sediments and soils—sites and processes. In: Chappel W, Petersen (eds) Symp Molybdenum. Dekker, New York, pp 425–553

Jenne EA, Luoma SN (1977) Forms of trace elements in soils, sediments, and associated waters: an overview of their determination and biological availability. In: Wildung R, Drucker H (eds) Biological Implications of metals in the environment. ERDA Symp Ser 42, Oak Ridge, pp 110–143

Jensen S, Jernelöv A (1969) Biological methylation in aquatic organisms. Nature (London) 223-753-754

Jones BF, Bowser CJ (1978) The mineralogy and related chemistry of lake sediments. In: Lerman A (ed) Lakes—chemistry, geology, physics. Springer, Berlin, Heidelberg, New York pp 179–235

Jörgensen BB (1977) The sulfur cycle of a coastal marine sediment (Limfjorden, Denmark). Limnol Oceanogr 22:814–832

Karickhoff SW, Morris KR (1986) Impact of tubificid oligochaetes on pollutant transport in bottom sediments. Environ Sci Technol 19:51–56

Karickhoff SW, Brown DS, Scott TA (1979) Sorption of hydrophobic pollutants on natural sediments. Water Res 13:241–248

Kersten M (1988) Geobiological effects on the mobility of contaminants in marine sediments. In: Salomons W, Bayne B, Duursma E, Förstner U (eds) Pollution of the North Sea: an assessment. Springer, Berlin Heidelberg New York Tokyo (in press)

Kersten M, Förstner U (1985) Trace metal partitioning in suspended matter with special reference to pollution in the southeastern North Sea. In: Degens ET, Kempe S, Herrera R (eds) Transport of carbon and minerals in major world rivers, vol 3. Mitt Geol Paläontol Inst Univ Hamburg, SCOPE/UNEP Sonderbd 58, pp 631–645

Kersten M, Förstner U (1986) Chemical fractionation of heavy metals in anoxic estuarine and coastal sediments. Wat Sci Technol 18:121–130

Kester DR, Ketchum BH, Duedall IW, Park PK (1983) Have the questions concerning dredged-material disposal been answered? In: Kester DR (ed) Wastes in the ocean, vol 2. John Wiley, New York, pp 275–287

Khalid RA (1980) Chemical mobility of cadmium in sediment-water systems. In: Nriagu JO (ed) Cadmium in the environment, pt 1: ecological cycling. AEST Ser. John Wiley, New York, pp 257–304

Khan SU (1980) Role of humic substances in predicting fate and transfer of pollutants in the environment. In: Haque R (ed) Dynamics, exposure, and hazard assessment of toxic chemicals. Ann Arbor, Mich, pp 215–230

Kilbane JJ (1986) Genetic aspects of toxic chemical degradation. Microb Ecol 12:135–145

Kloepfer RD (1985) Anaerobic degradation of trichloroethylene in soil. Environ Sci Technol 19:277

Kobayashi H, Rittmann BE (1982) Microbial removal of hazardous organic compounds. Environ Sci Technol 16:170A–183A

Krom MD, Sholkovitz ER (1977) Nature and reactions of dissolved organic matter in the interstitial waters of marine sediments. Geochim Cosmochim Acta 41:1565–1573

Larsson P (1983) Transport of C-14 labelled PCB compounds from sediment to water and from water to air in laboratory model systems. Water Res 17:1317–1326

Lee FY, Kittrick JA (1984) Elements associated with the cadmium phase in a harbor sediment as determined with the electron beam microprobe. J Environ Qual 13:337–340

Lindberg SE, Harriss RC (1974) Mercury-organic matter associations in estuarine sediments and interstitial water. Environ Sci Technol 8:459–462

Lindsay WL (1979) chemical equilibria in soils. John Wiley, New York
Liu Ching-I, Tan Hongxiao (1985) Chemical studies of aquatic pollution by heavy metals in China. In: Irgolic KJ, Martell AE (eds) Proc Worksh Environmental inorganic chemistry, 2–7 June 1983, San Miniato. VCH, Deerfield Beach, Fl, pp 359–371
Lockwood RA, Chen KY (1973) Adsorption of Hg(II) by hydrous manganese oxides. Environ Sci Technol 7:1028–1034
Lovell MA, Farmer JG (1983) The geochemistry of arsenic in the freshwater sediments of Loch Lomond. Proc Int Conf Heavy metals in the environment, vol 2, September 1983, Heidelberg. CEP Consultants, Edinburgh, pp 776–779
Lovley DR, Klug MJ (1986) Model for the distribution of sulfate reduction and methanogenesis in freshwater sediments. Geochim Cosmochim Acta 50:11–18
Lu CSJ, Chen KY (1977) Migration of trace metals in interface of seawater and polluted surficial sediments. Environ Sci Technol 11:174–182
Lynn DC, Bonatti E (1965) Mobility of manganese in diagenesis of deep-sea sediments. Mar Geol 3:457–474
Mabey WM, Mill T (1978) Critical review of hydrolysis of organic compounds in water under environmental conditions. J Phys Chem Ref Data 7:383–415
Macalady DL, Wolfe NL (1985) Effects of sediment sorption and abiotic hydrolysis 1. organophosphorothioate Esters. J Agric Food Chem 33:167–178
Matsumura F (1974) Microbial degradation of pesticides with particular reference to the formation of terminal residues. In: Coulston, Korte F (eds) Proc 3rd Int Symp Chemical and toxicological aspects of environmental quality, Tokyo, November 19–22, 1973. EQS, vol 4, pp 160–174
Matsumura F, Esaac EG (1979) Degradation of pesticides by algae and microorganisms. In: Khan NAQ, Lech JJ, Menn JJ (eds) Pesticides and xenobiotic metabolism in aquatic organisms. Am Chem Soc, Washington DC, pp 371–387
McBride BC, Merilees H, Cullen WR, Pickett W (1978) Anaerobic and aerobic alkylation of arsenic. In: Brinckman FE, Bellama JM (eds) Organometals and organometalloids, occurrence and fate in the environment. ACS Symp Ser 82:94–115
Means JC, Wijayaratne R (1982) Role of natural colloids in the transport of hydrophobic pollutants. Science 215:968–970
Mill T (1980) Chemical and photo oxidation. In: Hutzinger O (ed) the handbook of environmental chemistry, vol 2A. Springer, Berlin, Heidelberg, New York
Moore JW, Ramamoorthy S (1984) Heavy metals in natural waters. Springer, Berlin Heidelberg New York
Neff JW, Foster RS, Slowey JF (1978) Availability of sediment-adsorbed heavy metals to benthos with particular emphasis on deposit-feeding in fauna. DMRP Tech Rep D-78-42, US Army Eng Waterways Exp Stn, CE, Vicksburg, MI
Nelson LA (1981) Mercury in the Thames estuary. Environ Technol Lett 2:225–232
Nriagu JD (ed) (1980) Cadmium in the aquatic environment, pt 1: ecological cycling. John Wiley, New York
Nriagu JD (ed) (1987) Cadmium in the aquatic environment, pt 2.7 John Wiley, New York (in press)
Omenn GS (1986) Genetic control on environmental pollutants: a conference review. Microb Ecol 12:129–134
Oscarson DW, Huang PM, Defosse C, Herbillon A (1981) Oxidative power of Mn(IV) and Fe(III) oxides with respect to As(III) in terrestrial and aquatic environments. Nature London 291:50–51
Pavlon SP, Dexter RN (1980) Thermodynamic aspects of equilibrium sorption of persistent organic molecules at the sediment-sea-water interface: a framework for predicting distribution in the aquatic environment. In: Baker RA (ed) Contaminants and sediments, vol 2. Ann Arbor Sci Publ, Mich, pp 323–329
Peng An, Wang Zijian (1985) Mercury in river sediments. In: Irgolic JK, Martell AE (eds) Proc Worksh Environmental inorganic chemistry, 2–7 June 1983, San Miniato. VCH, Deerfield Beach, Fl, pp 393–400
Peterson ML, Carpenter R (1983) Biogeochemical processes affecting total arsenic and arsenic species distributions in an intermittently anoxic fjord. Mar Chem 12:295–321
Peterson ML, Carpenter R (1986) Arsenic distributions in pore waters and sediments of Puget Sound, Lake Washington, the Washington coast and Seanich Inlet, B.C. Geochim Cosmochim Acta 50:353–369

Pierce ML, Moore CB (1982) Adsorption of arsenite and arsenate on amorphous iron hydroxide. Water Res 16:1247–1253

Prause B, Rehm E, Schulz-Baldes M (1985) The remobilization of Pb and Cd from contaminated dredge spoil after dumping in the marine environment. Environ Technol Letts 6:261–266

Questel JH, Scholefield PG (1953) Arsenite oxidation in soil. Soil Sci 75:279–285

Ramamoorthy S, Rust BR (1976) Mercury sorption and desorption characteristics of some Ottawa river sediments. Can J Earth Sci 13:530–536

Reimers RS, Krenkel PA (1974) Kinetics of mercury adsorption and desorption in sediments. J Water P C 46:352–365

Reimers RS, Krenkel PA, Eagle M, Tragift G (1975) Sorption phenomenon in the organics of bottom sediments. In: Krenkel PA (ed) Heavy metals in the aquatic environment. Pergamon, Oxford, pp 117–136

Rontani JF, Rambelaorisoa E, Bertrand JC, Giusti G (1985) Favourable interaction between photooxidation and bacterial degradation of anthracene in seawater. Chemosphere 14:1909–1912

Sahm H, Brunner M, Schoberth SM (1986) Anaerobic degradation of halogenated aromatic compounds, Microb Ecol 12:147–153

Salomons W (1980) Adsorption processes and hydrodynamic conditions in estuaries. Environ Technol Letts 1:356–365

Salomons W (1985) Sediments and water quality. Environ Technol Lett 6:315–326

Salomons W, Förstner U (1984) Metals in the hydrocycle. Springer, Berlin Heidelberg New York

Salomons W, Geritse RG (1981) Some observations on the occurrence of phosphorus in recent sediments from western Europe. Sci Total Environ 17:37–49

Salomons W, Kerdijk H (1983) Trace metal cycling in lakes (Ijsselmeer, The Netherlands). Proc Int Conf Heavy metals in the environment, vol 2, September 1983, Heidelberg. CEP Consultants, Edinburgh, pp 880–883

Schwarzenbach RP, Giger W, Schaffner C, Wanner O (1986) Groundwater contamination by volatile halogenated alkanes: abiotic formation of volatile sulfur compounds under anaerobic conditions. Environ Sci Technol 19:322–327

Scudlark JR, Johnson DL (1982) Biological oxidation of arsenite in seawater. Estuar Coastal Shelf Sci 14:693–706

Serne RJ, Mercer BW (1975) Dredge disposal study, San Francisco Bay and Estuary. Appendix F: crystalline matrix study. US Army Waterways Eng Distr, San Francisco, Cal

Singer PC, Stumm W (1970) Acidic mine drainage: the rate determining step. Science 167:1121–1123

Street JJ, Sabey BR, Lindsey WL (1978) Influence of pH, phosphorus, cadmium sewage sludge, and incubation time on the solubility and plant uptake of cadmium. J Environ Qual 7:286–290

Stumm W, Baccini P (1978) Man-made perturbation of lakes. In: Lerman A (ed) Lakes – chemistry, geology, physics. Springer, Berlin, Heidelberg, New York, pp 91–126

Stumm W, Morgan JJ (1981) Aquatic chemistry. John Wiley, New York, 780 pp

Suflita JM, Horowitz A, Shelton DR, Tiedje JM (1982) Dehalogenation: a novel pathway for the anaerobic biodegradation of haloaromatic compounds. Science 218:1115–1117

Vidal FV, Vidal VMV (1980) Arsenic metabolism in marine bacteria and yeast. Mar Biol 60:1–7

Vogel TM, McCarty PL (1985) Biotransformation of tetrachloro ethylene to trichloro ethylene, dichloro ethylene, vinyl chloride, and carbon dioxide under methanogenic conditions. Appl Environ Microbiol 49:1080–1089

Wagemann R (1978) Some theoretical aspects of stability and solubility of inorganic arsenic in the freshwater environment. Water Res 12:139–145

Wauchope RD, McDowell LL (1984) Adsorption of phosphate, arsenate, methanoarsenate, and cacodylate by lake and stream sediments: comparison with soils. J Environ Qual 13:499–504

Wienberg R, Heinze E, Förstner U (1986) Experiments on specific retardation of some organic contaminants by slurry trench materials. In: Assink JW, Brink WJ van den (eds) Contaminated soil. Nijhoff, Dordrecht, pp 849–857

Williams RR, Bidleman TF (1978) Toxaphene degradation in estuarine sediments. J Agric Food Chem 26:280–282

Williamson KJ, Bella DA (1980) Estuarine sediments: successional model. J Environ Eng, ASCE 106(EE4), Proc Pap 15595, August 1980, pp 695–709

Wilson JT, Wilson BH (1985) Biotransformation of trichloro-ethylene in soil. Appl Environ Microbiol 49:242–243

Windom HL, Gardner WS, Dusnten WM, Pfaffenhofer GA (1976) Cadmium and mercury transfer in a coastal marine ecosystem. In: Windom HL, Duce RA (eds) Marine pollutant transfer. Heath, Lexington, pp 135–157

Wollast R, Billen G, Mackenzie FT (1975) Behavior of mercury in natural systems and its global cycle. In: McIntyre AD, Mills CF (eds) Ecological toxicological research. Plenum, New York, pp 97–107

Wright DR, Hamilton RD (1982) Release of methyl mercury from sediments: effects of mercury concentration, low temperature, and nutrient addition. Can J Fish Aquat Sci 39:1459–1466

Yoshida I, Kobayashi H, Ueno K (1976) Selective adsorption of arsenic ions on silica gel impregnated with ferric hydroxide. Anal Lett 9:1125–1133

Yoshida I, Ueno K, Kobayashi H (1978) Selective separation of arsenic (III) and (V) ions with ferric complex of chelating ion-exchange resin. Sep Sci 13:173–184

Assessment of Metal Mobility in Dredged Material and Mine Waste by Pore Water Chemistry and Solid Speciation

U. FÖRSTNER and M. KERSTEN[1]

1 Introduction

The general experience that the environmental behaviour and toxicity of an element can only be understood in terms of its actual molecular form led to the introduction of the term "speciation", which is used in a vague manner both for the operational procedure for determining typical metal species in environmental samples and for describing the distribution and transformation of such species in various media (Leppard 1983; Bernhard et al. 1986; Landner 1987). The two major categories in applied research are "analyte species" and "matrix species" (Bernhard et al. 1986): During chemical analysis the species present in a given volume will often be transformed into a single species for which the analytical instrument is sensitive; species exposed to different matrices change their reactivity, solubility, mobility as well as their bioavailability and toxicity.

Among the criteria to assess which element or elemental species may be of major concern, two questions deserve primary attention (Nriagu 1984):

1. Is the element *mobile in geochemical processes* because of either its volatility or its solubility in natural water, so that the effect of geochemical perturbations can propagate through the environment?
2. What are the critical pathways by which the most *toxic species* of the element can reach the most sensitive organs? Characterizing mobility of a certain element species at the molecular, organelle and cellular level, with respect to its *bioavailability* and *toxicity*, requires further insight into the interactions of different metals with complexing ligands at physiological concentrations.

Problems of speciation become particularly complex in *heterogeneous systems*, e.g. in soils, sediments and aerosol particles; thermodynamic models may give suggestions as to the possible species to expect, but due to the important role of kinetically controlled processes in biogeochemistry, the actual speciation is often different from what can be expected (Andreae et al. 1984). In *polluted ("stressed") systems* entropy increases and there is a concomitant increase in instability in both the physical and biological context (Wood et al. 1986): The greater the stress in the environment, the more difficulty in *sample handling* and storage prior to analysis. Many of the analytical techniques are handicapped by disruptive preparation

[1]Technological University of Hamburg-Harburg, Department for Environmental Science and Technology, P.O. Box 901403, D-2100 Hamburg 90, FRG

techniques which may alter the chemical speciation of inorganic components or lead to loss of analyte before analysis, e.g. freezing, lyophilization, evaporation, oxidation, changes in pH, light-catalyzed reactions, reactions with the sample container, time delays before analysis with biologically active samples, sample contamination, statistically invalid sampling, extraction with close to 100% yield; validation of analytical methodology at least needs comparison of authentic samples in the same matrix (Wood et al. 1986).

On the other hand, it is just the "stressed" system, where action is immediately needed and where for an assessment or prognosis of possible adverse effects the species and their transformations of pollutants have to be evaluated. The following questions have been raised with respect to the mobility and bioavailability of potentially toxic metals in contaminated systems (Förstner 1987):

1. How reactive are the metals introduced with *solid materials* from anthropogenic activities (hazardous waste, sewage sludge, atmospheric fallout) in comparison to the natural compounds?
2. Are the *interactions* of critical metals between solution and solid phases comparable for natural and contaminated systems?
3. What are the factors and *processes of remobilization* to become particularly effective, when either the solid inputs or the solid/solution interactions lead to weaker bonding of certain metal species in contaminated compared to natural systems?

Once the impact ot toxic elements or elemental species has been measured or predicted by direct or indirect methods (e.g. bioassays), a management plan can be formulated which usually includes engineering activities (see contribution by Lee and the chapters on treatment, this Vol.).

2 Assessment of Critical Pools for Remobilization of Metals

2.1 Metal Mobilization from Pore Water

The composition of interstitial waters in sediments is perhaps the most sensitive indicator of the types and the extent of reactions that take place between sediment particles containing trace metal and the aqueous phase which contacts them. The large surface area of fine-grained sediment with respect to the small volume of its entrapped interstitial water ensures that minor reactions with the solid phases will be indicated by major changes in the composition of the aqueous phase. Thus, the theoretical assessment of the nature of trace metal phases via the *equilibrium solution* composition has often been used in soil science. Data obtained from this type of analysis refer to a closed steady state system whose chemistry is controlled by the solution geochemistry of metals and by their adsorption-desorption and complexation reactions on interfaces. Though the results are impressive for well-characterized systems, the value of the predicted chemical speciation in natural sediment-water systems obtained from these models is limited (Florence and Batley 1980). In particular, the environmental impact of polluted dredged material and mine tailings is characterized by frequent exchange of pore water by drainage and resuspension. Since aquatic systems are physically dynamic, they are open to a variety of chemical and biological dynamics; even when such dynamics directly affect only selected chemical equilibria, these equilibria are likely to be coupled with others, which will then be disturbed in turn.

2.1.1 Pore Water Sampling

Extraction of interstitial waters from sediments is usually performed by either *centrifugation* or *squeezing* (Salomons and Förstner 1984); oxidation must be prevented during these procedures (Bray et al. 1973; Troup et al. 1974; Lyons et al. 1979). Watson et al. (1985) showed that sediments stored prior to the separation of interstitial water yield significant changes in chemical composition compared to samples processed within 24 h of collection; in addition, the authors described spatial and seasonal variations in the chemistry of estuarine sediment interstitial waters, which need to be carefully examined before defining a proper sampling strategy. Kinniburgh and Miles (1983) described a new centrifugation technique using immiscible displacement with a dense, inert fluoro-carbon liquid; yields of interstitial water from soils at field capacity are typically 20–50% of the total water present. The authors showed that the difference in the chemical composition of the extracted and the remaining water in their samples was negligible.

In-situ methods are considered more promising because of their inherent simplicity, and appear to be well adapted to the study of trace metals at the sediment-water interface under field conditions. A technique described by Mayer (1976) consists of a dialysis bag filled with distilled water, which is displaced into the sediment allowing equilibrium to take place over a period of some days to weeks. An improved sampler of this type has been described by Bottomley and Bayly (1984). Another in-situ sampler for close-interval, pore water studies as presented by Hesslein (1976) can be made from a clear acrylic plastic panel with small compartments pre-drilled in 1-cm steps or less. This panel can be covered by a non-degradable dialysis membrane or by a polysulphone membrane filter sheet (Carignan 1984). Yoshida (1984) reported that the measured values for nutrients from centrifugal separation are much larger than those obtained by the dialysis sampler; he stressed that nutrients are released dynamically during the centrifugal separation, and thus do not indicate the true values which are obtained statically by the dialysis technique. Recently, Carignan et al. (1985) compared both techniques: Centrifugation of the sediment at 5,000 rpm followed by filtration through a 0.45–μm membrane was found equivalent to dialysis for Co, Ni, Cr, Fe and Mn but gave higher and more variable concentrations of Cu, Zn and organic carbon. However, concentrations comparable to those obtained by dialysis were found when centrifugation speed was increased 11,000 rpm and when 0.2- or 0.03–μm membranes were used to filter the supernatant; this procedure was also found equivalent to dialysis for Cr, Co, Ni, Cd and organic carbon.

2.1.2 Factors Affecting Metal Species in Pore Water

Significant enrichment of trace metals in pore waters has been found in (anoxic) sediment samples from Southern California basin, Saanich Inlet (British Columbia) and Loch Fyne (Scotland), and has been explained by effects of *complexation* by organic substances (Brooks et al. 1968; Presley et al. 1972; Duchart et al. 1973). Nissenbaum and Swaine (1976) found that with the exception of iron, nickel and cobalt (which mostly occur as sulphides under anaerobic conditions) the elements concentrated in the interstitial solution are those which are also enriched in sedimentary humates. In this respect Jonasson (1977) established a probable order of binding strength for a number of metal ions onto humic or fulvic acids: $Hg^{2+} > Cu^{2+} > Pb^{2+} > Zn^{2+} > Ni^{2+} > Co^{2+}$. The transport of dissolved metals in the pore waters is, therefore, strongly influenced by the vertical gradient of dissolved humic substances (Krom and Sholkovitz 1978). On the other hand, iron in pore solutions shows a typical temporal evolution, which seems to be controlled by *precipitation* reactions rather than by complexation. In a large-scale experiment on $80 \times 30 \times 6$ m pits dug in the Rhine estuary below the water table and filled with both marine and freshwater dredged sediments (Kerdijk and Salomons; cit. Förstner and Salomons 1983) iron concentrations in the pore water reached a maximum (approx. 50 mg l^{-1}) after 40–50 days; subsequently the Fe content decreased to levels less than 10 mg l^{-1} in the 1-year experiment as a result of precipitation of iron sulphide (mainly marine conditions) or iron carbonate.

It has been stressed by Salomons (1985) that from an *impact* point of view it is important to know whether the concentrations in the pore waters are determined by *adsorption/desorption* processes or by *precipitation/dissolution* processes (see contribution by Bourg, chap. 1, this vol.). If the latter is the case the concentrations in the pore waters of pollutants are independent of the concentrations in the solid phase. There is strong, direct (Luther et al. 1980; Lee and Kittrick 1984) and indirect (Lu and Chen 1977) evidence that the concentrations of copper, cadmium and zinc in sulphidic pore waters are determined by precipitation-dissolution processes; the concentrations of arsenic and chromium in pore waters are probably controlled by adsorption-desorption processes, and mainly depend on the concentrations in the solid phase (Salomons 1985).

There are as yet only few data on the effects of higher concentrations of organic ligands on metal speciation in pore waters. Results from a sediment sample of an urban creek in Melbourne, Australia (Hart and Davies 1977) indicate that the major portion of cadmium was associated with colloidal labile species and more strongly bound complexes; the major proportion of zinc and lead was associated with the colloidal phase, 20% in labile forms and 10% in more tightly bound (organic) forms. Comparison of metal species in surface and pore waters (from mangrove sediment) of the Woronora River, New South Wales (Australia) was made by Batley and Giles (1980). Table 1 indicates an increased solubility of metals in the anoxic interstitial water compared to the surface water; for lead and copper this increase is related both to the labile and bound (organic) fraction, while the difference for cadmium in the organically complexed form is less significant. Most

Table 1. Trace metal species in interstitial and surface waters from the Woronora River, New South Wales, Australia (Batley and Giles 1980)

	Cadmium		Lead		Copper	
	Labile[a] $\mu g\ l^{-1}$	Bound[b] $\mu g\ l^{-1}$	Labile $\mu g\ l^{-1}$	Bound $\mu g\ l^{-1}$	Labile $\mu g\ l^{-1}$	Bound $\mu g\ l^{-1}$
Surface water (oxic)	0.11	0.09	0.28	0.32	0.74	1.15
Interstitial water (mangrove sediment)	0.29	0.11	0.50	0.60	1.35	1.80

[a]Labile metal measured at pH 4.7, acetate buffer.
[b]Bound = total minus labile.

of the organically-associated metals are "colloidal", equivalent to the high molecular weight organics, and it has been proposed by Elderfield (1981) that these substances can be formed from dissolved organic matter during diagenesis.

2.2 Solid Materials Controlling Metal Mobilization

Solid surfaces play an important role in mediating the chemical behaviour of heavy metals; such experience was set in relation to environmental problems by Farmer and Linton (1984): "Accessibility to the environment (via washout, rainout, groundwater leaching, lung fluids, etc.) is governed by both metal *surface accessibility* (extent of surface enrichment) and metal *surface solubility* (surface speciation)". Since adsorption of pollutants onto airborne and waterborne particles is a primary factor in determining the transport, deposition, reactivity and potential toxicity of these materials, analytical methods should be related to the chemistry of the particle's surface and/or to the metal species highly enriched on the surface.

Surface-related studies on solid materials have been performed or are aimed for:

1. Assessment of Sources by Characterization of Typical Inputs. There are as yet only few studies in this field. We have performed a regional survey on the distribution of thallium in soils in an area, where two point sources were mainly effective (Schoer 1984). One was an abandoned lead-zinc mine, the other was the chimney of a cement factory which had used sulphidic roasting residues as additives to special cement. Results from leaching experiments with 1 N ammonium acetate (with 0.01 M EDTA) showed statistically highly significant differences. In the mine area, the extractable portion was in the range of 0.5–5%, whereas in the soils affected by the cement plant emissions up to 40% could be extracted. The lower absolute concentrations in the latter area were therefore more available, in fact, 75–100% of the thallium contents of flue gas dust from the cement industry is soluble in water, leading also in some cases to increased toxicity and accumulation of thallium in plants (compilation by Schoer 1984).

2. Estimation of Biological Availability of Metal Pollutants. Initial applications, about 30 years ago, have been performed in soil science. There is a vast amount of literature (Luoma 1983), and the discussion of this aspect is out of the scope of this presentation. However, two recent developments should be mentioned. One is the combination of chemical and biological tests. [In such an experiment Diks and Allen (1983) found a high correlation between the uptake of copper and the amount of copper present in the manganese/easily reducible phase, and it is suggested that the redox potential and pH in the gut of the studied worm is such that manganese coatings are dissolved.] Another is the application of body fluids, for example, for studying the effect of leaching of contaminants from airborne particles in the lung. It is suggested that biological chelators, possibly cysteine and other serum proteins are important leaching agents, particularly to remove Zn, V, Cu and Fe from fly ash particles in vivo (Harris and Silberman 1983).

3. Differentiation of Geochemical Environments. Diagenesis involves processes in the interstital water and gases, which are strongly affected by changes of redox conditions. The sequence of "redox titration" comprises the major reactions respiration, manganese reduction, nitrate reduction, iron reduction and sulphate reduction mediated by bacteria. In ancient sedimentary deposits, these zones can be identified from characteristic mineral assemblages (Berner 1981): The oxic environment contains oxyhydrates of Mn and Fe at low contents of organic matter (which is mostly degraded); the post-oxic environment is characterized by the presence of manganese carbonate, both oxides and carbonates of iron, and low organic matter as well. Under anoxic conditions two branches can be distinguished: In the marine milieu reduction of sulphate provides sufficient sulphide ions to form iron sulphide, whereas in the freshwater environment there is a tendency to form carbonate of iron, when the sulphide ions are consumed; the general tendency of diagenesis in anoxic freshwater sediments is the formation of methane. Sediments collected on four campaigns represent different early diagenetic environments according to the aforementioned classification (Kersten and Förstner 1987). Cadmium is associated in the oxic environment with reducible phases, carbonates and to some extent with exchangeable forms. Higher percentages of sulphides are found already in the post-oxic environment and in both anoxic milieus. These findings reflect the strong affinity of cadmium to the sulphidic phase, even in such environments, where the concentration of sulphide ions is limited.

4. Evaluation of Diagenetic Effects. The typical effects of the earliest stages of "diagenesis" (involving transformations of organic matter, "ageing" of mineral components and formation of new equilibria between solid and dissolved species) have been demonstrated by Salomons (1980) with respect to the behaviour of trace metals at the sediment/seawater interface. Desorption was studied by adding cadmium and zinc to suspended matter in river water; after adsorbing periods of 1, 3, 8, 24 and 60 days, NaCl was added to the suspension to increase the chloride concentration to 1.9% (approx. seawater composition). After an adsorption period of only 1 day, 24% of the adsorbed cadmium and 60% of the adsorbed zinc remain bound to the sediment; after 60 days 40% of the cadmium and 88% of the zinc bound to the sediment are not released after NaCl treatment.

In the present contribution major emphasis is given to another application, namely the characterization of solid waste material with respect to the potential environmental impact following a release of metal contaminants into the water phase.

2.2.1 Methods for Solid Speciation

Basically there are *three methodological concepts* for determining the distribution of an element within or among small particles (Keyser et al. 1978; Förstner 1985):

1. *Analysis of single particles* by X-ray fluorescence using either a scanning electron microscope (SEM) or an electron microprobe can identify differences in the matrix composition between individual particles. The total concentration of the element can be determined as a function of particle size. Other physical fractionation and pre-concentration methods include density and magnetic separations.
2. The *surface of the particles* can be studied directly by the use of electron microprobe X-ray emission spectrometry (EMP), electron spectroscopy for chemical analysis (ESCA), Auger electron spectroscopy (AES) and secondary ion-mass spectrometry. Depth-profile analysis determines the variation of chemical composition below the original surface.
3. *Solvent leaching,* apart from the characterization of the reactivity of specific metals, can provide information on the behaviour of metal pollutants under typical environmental conditions. Common single reagent leachate tests, e.g. the US EPA, ASTM, IAEA, ICES and the German Water Chemistry Group (Deutsche Einheitsverfahren) use either distilled water or acetic acid. A large number of test procedures have been designed particularly for soil studies; these partly used organic chelators such as EDTA, DTPA, both as single extractants or in sequential procedures. A single-step method of the US EPA (Ham et al. 1979) designed for studies on the leachability of waste products consists of a mixture of sodium acetate, acetic acid, glycine, pyrogallol and iron sulphate.

2.2.2 Application of Chemical Extraction Sequences

One of the potential advantages of leaching methods is in obtaining relevant information from a small number of samples, one possibly being sufficient. A number of test protocols in *soil science* have been designed initially for the assessment of plant-available soil nutrients and speciation of trace metals in sewage sludge-amended soils (Jackson 1958; Lake et al. 1984). In *geochemical exploration* the major applications of selective extraction procedures are (Chao 1984): (1) to elucidate the mode of occurrence of trace metals in soils and sediments; (2) to enhance the "geochemical contrast" between mineralized and background areas above that which can be accomplished by the bulk analysis; and (3) to differentiate between effects on metal distribution caused by anomalous mineralization, and those resulting from lithological and environmental factors. In the *sediment-petrographic field* interest was focussed initially on the differentiation between authigenic and detrital phases in ferromanganese concretions from deep-sea deposits (e.g. Chester and Hughes 1967; Lyle et al. 1984). In general, these studies serve the determination of the major accumulative phases for metals in sedimentary ore deposits and the diagenetic accumulation mechanism (Förstner 1982a). According to Horowitz (1984), two approaches are used in chemical partitioning of sediments: The first is to determine *how* metals are retained on or by sediments, the so-called mechanistic approach; the second determines *where* inorganic constituents are retained on or by sediments (phase or site), the so-called phase approach. The scheme in Table 2 (after Förstner 1982b) combines the contributions (and their estimated significance) of both the mechanistic and the phase approach.

Table 2. Mechanisms and substrates of metal enrichment in aquatic systems (Förstner 1982b)

	Rock debris, solid waste material + organics	Metal -hydroxide -carbonate -sulphide	Reactive organic substance	Hydrous Fe/Mn-oxides (authigenic)	Calcium carbonate
Inert bonding	xxx			x	x
Precipitation		xxx		xx	xx
Adsorption	x	x	x	x	x
Co-precipitation		xx		xxx	xx
Coatings			xxx	xxx	x
Flocculation			x	xx	

In connection with the problems arising from the disposal of *solid wastes,* particularly of dredged materials, *extraction sequences* have been applied which are designed to differentiate between the exchangeable, carbonatic, reducible (hydrous Fe/Mn oxides), oxidizable (sulphides and organic phases) and residual fractions (Engler et al. 1977). One of the more widely applied extraction sequences of Tessier and co-workers (1979) is presented in Table 3.

This sequence has been used to examine the different "pools" of Cd and Pb, and to estimate their reactivity in various types of metal-rich particulates (Fig. 1).

With respect to the different substrates, the extreme leachability of both cadmium and lead in *urban particulate matter* (Lum et al. 1982) and *street dust* (Harrison et al. 1981) is particularly relevant for subsequent interactions with acid, complexing or salty solutions (the high cation and chloride concentration used in the exchange solution may reflect conditions in soils contaminated with de-icing salt.) Despite the relatively low concentration of cadium in the *sewage sludge* sample (activated sludge; Förstner et al. 1981), there is a significant shift to higher percentages in the carbonate fraction, whereas lead is typically enriched in the organic/sulphidic and residual forms. The general experience that the (anthropogenically) more enriched elements are also the more reactive ones seems

Table 3. Sequential extraction scheme for partitioning sediment samples (Kersten and Förstner 1986)

Fraction	Extractant	Extracted component
Exchangeable	1 M NH_4OAc, pH 7	Exchangeable ions
Carbonatic fraction	1 M NaOAc, HOAc pH 5	Carbonates
Easily reducible	0.01 M NH_2OH HCl + 0.01 $MHNO_3$	Mn-oxides
Moderately reducible	0.1 M oxalate buffer pH 3	Amorphous Fe-oxides
Sulphidic/organic	30% H_2H_2 + 0.02 HNO_3 pH 2 extr. with 1 M NH_4OAc-6% HNO_3	Sulphides together with organic matter
Residual fraction	Hot HNO_3 conc.	Lithogenic crystallites

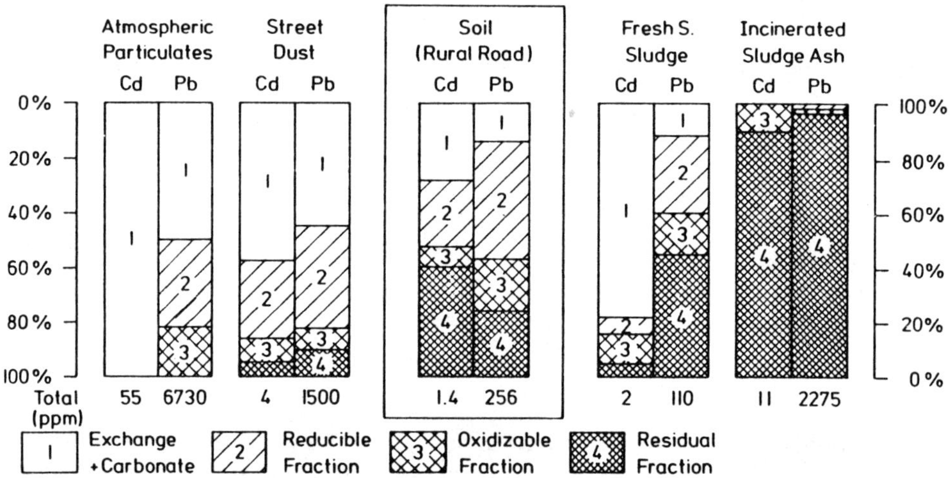

Fig. 1. Chemical fractionation of cadmium and lead in solid waste particles (urban particulate matter = standard reference material SRM 1648; Lum et al. 1982), street dust (Harrison et al. 1981), fresh sewage sludge (Förstner et al. 1981) and incinerated sewage sludge ash (Fraser and Lum 1982)

to be valid only for waste material, which has not been treated by high temperature processes; this is exemplified by the data from *incinerated sludge ash* (Hamilton municipal incinerator), where the remaining metals are highly enriched but rather stably bound in the "residual fraction" (Fraser and Lum 1982). (Better efficiency of emission control, however, of stack gases from combustion processes will recycle higher percentages of atmospheric particulate matter into the residues to be deposited; these materials are expected to contain higher fractions of leachable metals, as shown in the first two columns of Fig. 1.)

In relation to the species distribution in the *soil* example (Fig. 1 represents a moderately polluted soil near a rural road studied by Harrison et al. 1981), the inputs of cadmium and lead from atmospheric particulates, street dust and fresh sewage sludge are bound in more labile forms and should, therefore, more easily be affected by changes of the chemical environment. *Diagenetic effects*, which have been mentioned above, can be extrapolated to *geologic time scales* by a comparison of the bonding intensity of stable metal isotopes and their unstable counterparts, the latter supplied from radioactive emissions of nuclear power and reprocessing plants. In Fig. 2 the effects are shown of sequential leaching of a sediment sample from the lower Rhone River in France. The reducing agents hydroxylamine (pH 2) and oxalate buffer (pH 3) only extract 15% of the natural stable manganese, while the artificial isotope Mn-54 from the reprocessing plant is mobilized at more than 80% by these treatments (Förstner and Schoer 1984).

2.2.3 Problems with Sample Treatment and Individual Extractants

Despite clear advantages of a differentiated analysis over investigations of total sample, sequential chemical extraction is probably the most useful tool for *predicting long-term adverse effects* from contamined solid material. It has become obvious that there are many problems associated with these procedures (e.g. Calmano and Förstner 1983; Nirel et al. 1986; Rapin et al. 1986; Kersten and Förstner 1986):

1. *Reactions are not selective* and are influenced by the duration of the experiment and by the ratio of solid matter to the volume of extractants. A too high solid content, together with an increased buffer capacity may cause the system to overload; such an effect is reflected, for example, by changes of pH values in time-dependent tests (Pfeiffer et al. 1982).
2. Processes of *readsorption and precipitation* have to be considered, particularly during extraction with ammonium acetate.
3. Most important: labile phases could be *transformed during sample preparation*, which can occur especially for samples from reducing environments.

In this respect, earlier warnings have been made by various authors, not to forget changes of the sample matrix during recovery and treatment of the material. The first relates to the anoxic sediment material, where changes are quite obvious: "The integrity of the samples must be maintained throughout manipulation and

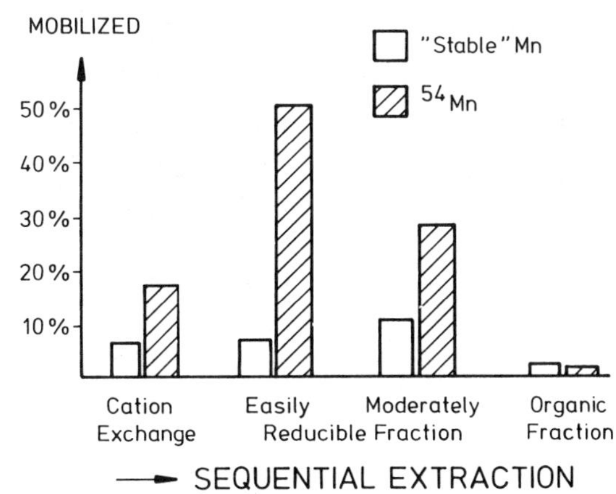

Fig. 2. Comparison of chemical extractability of artificial and stable isotopes of manganese from a sediment sample of the Rhone River (After Förstner and Schoer 1984)

extraction" (Engler et al. 1977; Khalid et al. 1978). The second indicates that even oxic materials are not safe for changes during treatment (Thomson et al. 1980). Although these problems, particularly for anoxic sediments, have been well known for many years, we have clearly underestimated the effects for a long period. This was realized when we tried to separate iron forms in dredged sediments. Our results were totally different from the model calculations, which proposed a very considerable percentage of iron carbonate. When we tried to produce artificial iron carbonate, we failed as the originally white material disintegrated to form red iron oxide within short time periods. It became obvious that these materials were highly sensitive to aeration.

A simple but impressive experiment on the effect of oxidation in regulating the chemical form of cadmium and other trace metals has been performed on an anoxic sediment sample from Hamburg harbour. The sample was divided into four series under an argon flushed glove box in order to study the effect of various sample pre-treatments including aeration and dehydration on the chemical forms of cadmium (Fig. 3; Kersten and Förstner 1986):

1. Manipulations of the first series were all done under inert atmosphere to serve as a control.
2. The second series was treated by the elutriate test modified for air bubbling (Lee et al. 1976). This test was initially designed by the US Environmental Protection Agency to detect any short-term release of chemical contaminants from polluted material during dredging manipulations and disposal. This test involves the mixing of 1 vol. harbour sludge with 4 vol. dredging or disposal site water for a 30-min shaking period. If the soluble chemical constituent in the water exceeds 1.5 times the ambient concentration in the original water, special conditions will

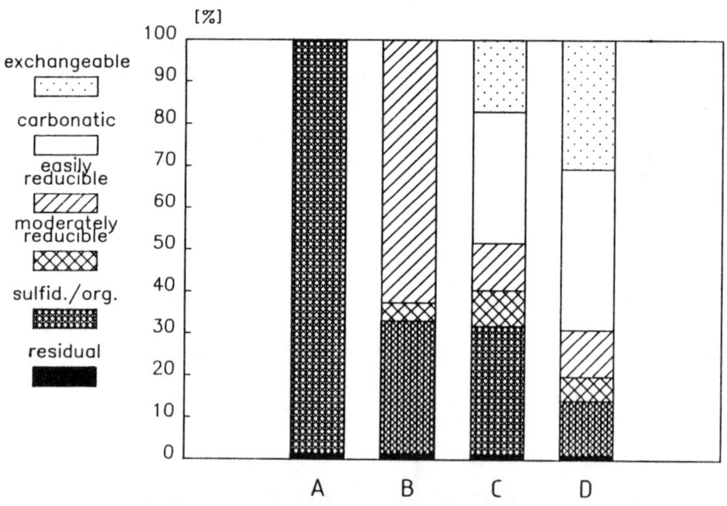

Fig. 3. Partition of cadmium in anoxic mud from Hamburg harbour in relation to the pre-treatment procedures: *A* control sample extracted as received under oxygen-free conditions; *B* after treatment with the US EPA/CE elutriate test; *C* freeze-dried; *D* oven-dried (60°C) (After Kersten and Förstner 1986)

govern the remedial measures to be undertaken (Engler 1980). A study conducted on the factors influencing the results of the elutriate tests has shown that this test as originally developed cannot yield a reliable estimate of the potential release, especially of Cd, since it did not define the conditions of mixing to enable a well-defined, reproducible oxygen status existing during the test period (Lee et al. 1976). The result was that a modified elutriate test has been proposed in which compressed air agitation is utilized during the mixing period.

3. The third subsample series was freeze dried; and
4. The fourth series was dried under air at 60°C.

Subsequent to the preservation and pre-treatment measures respectively, the subsamples were extracted by a six-step sequential leaching technique (modification of Tessier's method). The significant differences as shown in Fig. 3 can be ascribed to the contact of the sediment with air and by dehydration rather than to experimental artefacts such as inhomogeneity of the sediments or variations in the extraction protocol. Indeed, no differences were obtained for oxic-suspended matter (Kersten and Förstner 1985), and the sum of the metal concentrations in the individual fractions of all four series of each sample agreed within 10%.

In the pre-treatment scheme, which was developed from this experience, anaerobic dredged samples were taken immediately from the centre of the material with a polyethylene spoon, filled into a polyethylene bottle up to the surface. Immediately after arriving at the laboratory, sediments were inserted into a glove box prepared with an inert argon atmosphere. Oxygen-free conditions in the glove box were maintained by purging continuously with argon under slight positive pressure. Extractants were deaerated prior to the treatment procedure.

2.2.4 Extent and Velocity of Species Changes During Oxidation

One of the most obvious features in the foregoing oxidation sequences is the transformation of the oxidizable (sulphidic/organic) fraction. Unfortunately, this fraction is still less well defined because it comprises both labile and more refractory organic substances in combination with sulphidic metal associations. With respect to the practical uses of these investigations, there is much information available from the percentage of oxidizable phases. In Fig. 4 bar graphs show the average percentages of the peroxide-oxidizable metal fraction of the respective non-residual bound metals as obtained from the sequential extraction of the Hamburg harbour mud samples. It is shown that the concentration of metal in this fraction increases in the order Mn < Fe < Cr = Ni < Zn < Pb < Cd < Cu, which widely corresponds to the Irving-Williams order for the sequence of complex stability, but also to the stability of sulphidic metal compounds. Following application of the elutriate test, decreased portions of the sulphidic/organic fractions of Pb (42%) and Cu (21%) are chiefly found in the moderately reducible fraction. Transformation of oxidizable Ni (11%), Zn (62%) and Cd (67%) mainly results in increased percentages of easily reducible fractions.

At present there is a study in progress on the efficiency of phase transformations under varying environmental conditions, but also on the velocity of such

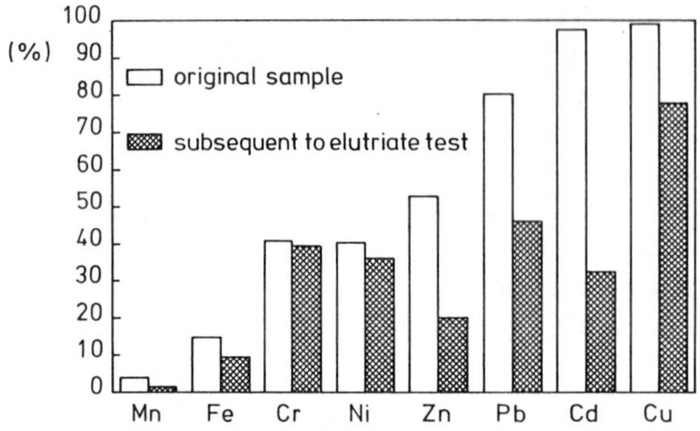

Fig. 4. Changes of oxidizable metal fractions in a sediment sample from Hamburg harbour during elutriate test procedure

effects (Recke and Förstner 1986). The example of cadmium in a very strongly polluted sediment of Hamburg harbour indicates that aeration is significantly affecting phase distributions even after 10 min. Here, one can observe first a transformation of the sulphidic compounds from moderately reducible to easily reducible (1–10 h), after which time the phases occur which have been found earlier after application of drying methods.

3 Remobilization of Metals from Solid Waste Materials

Solubility, mobility and bioavailability of particle-bound metals can be increased by *four major factors* in terrestrial and aquatic environments:

1. *Lowering of pH*, i.e. acidity imposes problems in all aspects of metal mobilization in the environment, e.g. toxicity of drinking water, growth and reproduction of aquatic organisms, increased leaching of nutrients from the soil and the ensuing reduction of soil fertility, increased availability and toxicity of metals and the undesirable acceleration of mercury methylation in sediments (Fagerström and Jernelöv 1972). On a regional scale, acid precipitation is probably the prime factor affecting metal mobility in surface waters.
2. Increasing occurrence of natural and synthetic *complexing agents*, which can form soluble metal complexes with trace metals that are otherwise adsorbed to solid matter.
3. *Increasing salt concentrations*, by the effect of competition on sorption sites on solid surfaces and by the formation of soluble chloro-complexes with some trace metals.
4. *Changing redox conditions*, e.g. after land deposition of polluted anoxic dredged materials.

Here, particular attention will be given to the effects of *pH and redox changes* in surface waters and soils.

3.1 Interactions of Mobilizing Factors

It has been stressed by Plant and Raiswell (1983) that the predominance of simple mineral solution equilibria explains the concentration of *major elements* in the surface environment, but the behaviour of many *trace elements* is more complex and is also determined by co-precipitation, surface effects and interactions with organic phases. Many interactions, however, can be estimated from the simple scheme in Table 4 (Förstner et al. 1986), which includes data from soil studies and dredged material (Gambrell et al. 1977; Mattigod and Page 1983; Plant and Raiswell 1983), and from recent studies on the effects of acid precipitation (Campbell et al. 1983).

Table 4. Relative mobilities of elements in sediments and soils as a function of Eh and pH (Förstner 1985, cf. references therein)

	Electron activity		Proton activity	
Relative mobility	Reducing	Oxidizing	Neutral-alkaline	Acid
Very low mobility	Al, Cr, Mo, V, U, Se, S, B, Hg, Cu, Cd, Pb	Al, Cr, Fe, Mn	Al, Cr, Hg, Cu, Ni, Co	Si
Low mobility	Si, K, P, Ni, Zn, Co, Fe	Si, K, P, Pb	Si, K, P, Pb, Fe, Zn, Cd	K, Fe (III)
Medium mobility	Mn	Co, Ni, Hg, Cu, Zn, Cd	Mn	Al, Pb, Cu, Cr, V
High mobility	Ca, Na, Mg, Sr	Ca, Na, Mg, Sr, Mo, V, U, Se	Ca, Na, Mg, Cr	Ca, Na, Mg, Zn, Cd, Hg, Co, (Mn)
Very high mobility	Cl, I, Br	Cl, I, Br, B	Cl, I, Br, S, B, Mo, V, U, Se	Cl, I, Br, B

It can be expected that changes from reducing to oxidizing conditions, which involve transformations of sulphides and a shift to more acid conditions, will increase the mobility of typical "chalcophilic" elements, such as Hg, Zn, Pb, Cu and Cd. On the other hand, the mobility is characteristically lowered for Mn and Fe under oxidizing conditions. At lower pH values, one can observe a typical antagonistic behaviour for cadmium and iron under different redox conditions. An example of solid waste *compost* in Fig. 5 has been studied by Herms and Brümmer (1978). At pH 5, iron is most mobile at redox values below zero, and concentrations in solution were up to 600 mg Fe l^{-1}; cadmium, on the other hand, has the highest dissolved concentrations, up to 1 mg Cd l^{-1} at redox conditions of +500 mV. It seems that the study of such *coupled cycles* will be particularly promising with respect to the interactions of large redox-controlled systems like iron and manganese with critical trace metals such as cadmium, mercury and arsenic.

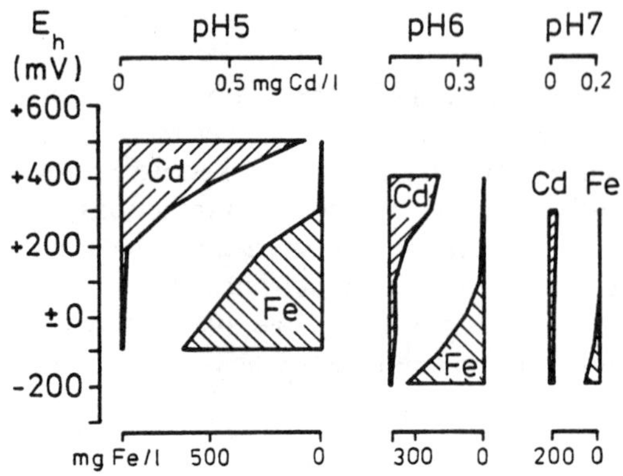

Fig. 5. Solubility of cadmium and iron in waste compost at different pH and redox conditions (After Herms and Brümmer 1978)

3.2 Metal Release from Pore Water During Dredging

Early sediment changes and element mobilization from pore water in a man-made estuarine marsh have been investigated by Darby et al. (1986). This study exemplifies both the mechanisms of release of metals via pore water extraction and subsequent changes by the effect of oxidation (Table 5).

Compared to the river water concentration, the channel sediment pore water is enriched by a factor of 200 for Fe and Mn, 30–50 for Ni and Pb, approx. 10 for Cd and Hg and 2–3 for Cu and Zn. When the expected concentrations of metals following hydraulic dredging, which were calculated from the rate of pore water to river water to be about 1:4, were compared with the actual measurements at the pipe exiting from the dredging device, characteristic differences were observed (Table 5). When the actual concentrations at the pipe's exit were less than expected, that element was removed from the solution presumably by scavenging or precipitation. This was valid for iron (approx. half of the expected concentration) and, to a lesser extent, for manganese. When the exiting solution was greater than the expected concentration in Table 5, mobilization from sediment phases was

Table 5. Mobilization of metals and nutrients during dredging (After Darby et al. 1986)[a]

Metal	Channel sediment pore water (a)	River water concentration (b)	a/b	Effluent at man-made marsh: Expected concentration	Measured concentration	Change (%)
Mn	6.94	0.03	230	1.34	1.19	− 11
Fe	57.3	0.26	220	11.12	6.01	− 46
Ni	0.054	0.001	54	0.011	0.035	+ 218
Pb	0.077	0.002	38	0.016	0.142	+ 788
Hg ($\mu g\ l^{-1}$)	3.2	0.26	12	0.82	2.0	+ 144
Cd	0.009	0.001	9	0.0025	0.019	+ 660
Cu	0.012	0.004	3	0.0055	0.051	+ 827
Zn	0.12	0.052	2	0.065	5.30	+8069

[a] All concentrations are in mg l^{-1} except Hg.

assumed; in this respect, highest rates of release were found for zinc, followed by copper, lead and cadmium.

According to Darby et al. (1986), the levels of heavy metal mobilization were higher at this time than at any time in the subsequent 2 years of marsh maturation. The amount of metal mobilization detected at the effluent pipe of the disposal area during dredging could be accounted for by the release of relatively small amounts of those elements bound to labile sediment phases (see Sect. 3); while only 3–5% of the labile phase Cu and Ni would account for the measured increases reported in Table 2, up to 36% of the labile phase Pb and Zn was required to account for the higher than expected concentrations of these elements (see also Adams and Darby 1980).

3.3 Changes of Cadmium Forms During Oxidation of Dredged Spoil

Hoeppel et al. (1978) were among the first to observe changes in element forms during dredged spoil disposal on land containment areas. They compared five influent and effluent samples of suspended matter in such ponding systems (Table 6). From their findings it is obvious that cadmium concentrations increased significantly in the carbonate phase of the effluent samples, presumably as a direct result of the transfer from organic/sulphidic associations present in the influent slurries.

This findings are confirmed by our data from sequential leaching procedures on Hamburg harbour sediments (Fig. 3; Kersten and Förstner 1986). Following the application of the elutriate test, the oxidizable sulphidic/organic portion of Cd decreases drastically and is now found in the easily reducible fraction. Coprecipitation and adsorption of Cd with the precipitated oxyhydrates may have removed the liberated metal from the solution. Freshly precipitated oxyhydrates are much more effective in scavenging high concentrations of trace metals because of a greater reactive surface area than aged crystalline materials (Jenne 1977). After freeze- and oven-drying of the initially anoxic samples, cadmium proportions were found even in the most mobile, operationally defined carbonatic and exchangeable fractions. The high concentration of cadmium present in these fractions may have a hazardous impact on water quality during dredging and disposal operations as well as upland disposal of these sediments (Gambrell et al. 1977; Khalid 1980).

Table 6. Cadmium forms in solids from confined land disposal (Hoeppel et al. 1978)

Chemical fraction	Percent of total cadmium content	
	Influent	Effluent
Exchangeable fraction[a]	21.0%	18.0%
Carbonatic fraction[b]	21.4%	56.7%
Easily reducible fraction[c]	9.2%	11.8%
Remaining phases	49.3%	13.5%

[a] Ammonium acetate extractable.
[b] 1 M acetic acid extractable.
[c] 0.1 M hydroxylamine hydrochloride in 0.01 M nitric acid.

3.4 Species Transformations in Tidal Sediments

An example of oxidative remobilization of cadmium and other heavy metals has been studied in a *tidal freshwater flat* in the upper Elbe estuary near Hamburg (Kerner et al. 1986). This mud flat (diurnal tidal water fluctuations in the range of 3 m are affecting this productive site) is colonized by dense monodominant reed stands providing an effective trap for suspended matter containing heavy metal from upstream. Elevated cadmium contents in the rhizomes of the emerged macrophytes indicate a high proportion of bioavailable Cd species in the root zone. Examination of sediment cores taken at this site showed a distinct pattern of redox potential and heavy metal *fractionation profiles* (Fig. 6). Below a marked redox potential discontinuity the Fe fractionation profile is characterized by increasing

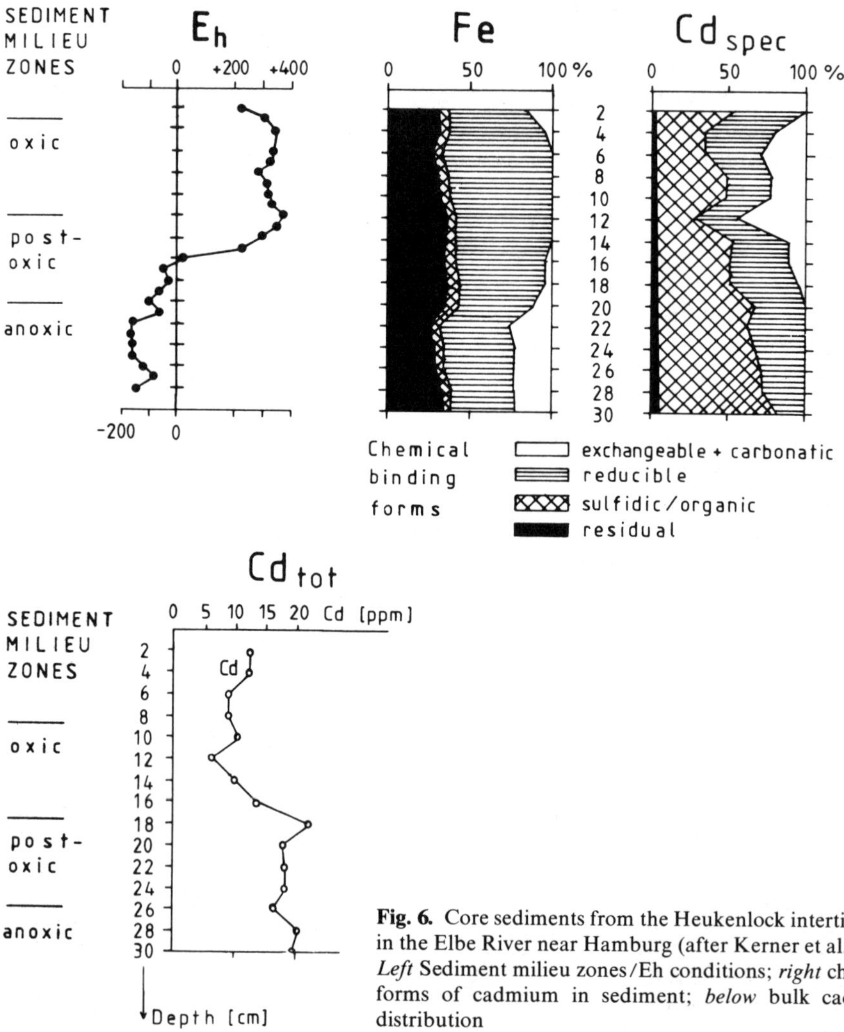

Fig. 6. Core sediments from the Heukenlock intertidal flat in the Elbe River near Hamburg (after Kerner et al. 1986). *Left* Sediment milieu zones/Eh conditions; *right* chemical forms of cadmium in sediment; *below* bulk cadmium distribution

Na-acetate extractable proportions, which presumably represents carbonatic Fe (II) forms. The particulate cadmium-binding forms, on the other hand, reveal a behaviour inverse to that of iron. While in the anoxic zone approximately 60 to 80% of Cd is found in the oxidizable fraction, high percentages of Na-acetate extractable forms are found in the oxic and post-oxic zones of the sediment cores. The higher amounts of labile cadmium forms are accompanied with a marked depletion in the total content of the toxic metal compared to that in the anoxic sediment zone. Comparison of the fractionation patterns and total contents of other diagenetically less mobile metal examples indicates that a significant proportion of cadmium is leached from the surface sediment by a process of "*oxidative pumping*" by tidal water drainage in this high-energetic environment. This could result in migration of the remobilized metal into either the deeper anoxic zone, where it can precipitate again to contribute to the enhanced oxidizable sulphidic/organic fraction, or to the surface water, from where it can be exported into the outer estuary. It could, however, also contribute to bioavailable cadmium portions such as indicated by the enhanced macrophyte cadmium concentrations.

These data demonstrate the problematic effect of dispersing *anoxic waste materials* in ecologically productive, high-energy, nearshore, estuarine and inlet zones (Khalid 1980). This may also pertain to procedures such as "sludge harrowing" as it is occasionally performed in the cold season in some sections of Hamburg harbour. Most critical is the dispersion of anoxic mud on the land surface, rather than on a heap deposit, due to stronger interactions of the former material with the atmosphere.

4 Predicting Potential Metal Mobilization from Solid Speciation

From these findings it seems that one of the major applications of solid speciation data includes an estimation of the long-term changes of metal mobilities under variable environmental conditions by analyzing characteristic rearrangements of chemical forms in these materials. This will be particularly interesting for assessing the effects of changing redox conditions, and these effects have practical implications for the disposal alternatives for dredged materials.

Although the individual "fraction" received from the *chemical leaching experiments* rarely reflect specific metal "phases", but rather are defined by the selection of the extracting medium and by the experimental conditions ("operational phases"), the elution medium is designed to simulate certain, mostly extreme, environmental conditions, such as the interaction with saline waters in estuaries or reducing conditions during land disposal of sludge materials. The overall most significant effects on the remobilization of heavy metals can be expected from acidification, either from atmospheric emissions or from oxidation of sulphidic compounds in anoxic waste materials. Experiments on the long-term behaviour of sludge-induced metals should focus, therefore, particular attention on changes of redox and pH conditions. In addition, the effects of organic substances have to be studied, since the mobility of released metals can be affected by complexation processes.

Species differentiations can be used for the estimation on the remobilization of metals under *changing environmental conditions:*

1. In the *estuarine environment* the "exchangeable fraction" might be particularly affected; however, changes of pH and redox potential could also influence other easily extractable phases, e.g. carbonate and manganese oxides. The high proportion of Cd in exchangeable fractions of polluted sediments is reflected in the typical mobilization effect of this element in the estuarine environment.
2. Among the factors enhancing metal mobility, *acid interactions,* both from acid precipitation and oxidation of sulphide minerals in mine wastes and dredged sediments, deserve particular attention due to the fact that ionic species predominate and are readily available for biological uptake. Lowering of pH will affect, according to its strength, the "exchangeable", then the "easily reducible" and in some cases part of the "moderately reducible" fraction, the latter consisting of Fe-oxyhydrates in less crystallized forms. In strongly reducing environments, e.g. in highly polluted sediments, the moderately reducible fraction can be affected too, especially when the iron is present in the form of "coatings".
3. The effects on *organically bound metals* are more complex; however, it has been argued that this fraction is highly susceptible to environmental changes, especially during early diagenetic reactions, where recycling of mineralized organic matter and pore fluid transfer processes are controlling the dynamics of pollutants and nutrients in solid waste materials.

Estimations on the long-term behaviour of metals in solid waste materials should focus particular attention on *changes of redox and pH conditions.* For determining both effects, short-term release and long-term rearrangement, we have modified an experimental scheme, which was originally used by Patrick et al. (1973) and Herms and Brümmer (1978) for the study of soil suspensions and municipal waste materials. Chemical fractions are analyzed with a six-step extraction sequence before and after treatment and the released metals are recovered with an ion-exchanger system from solutions. The variables applied in the *self-regulating system* are "pH", "redox" and "temperature" (the latter for simulating the effect of time). The system can be applied for different intensities of contact between solid materials and solution, e.g. by using shakers, flow-through columns and dialysis bags.

The present state of knowledge on the interrelation between solid matter speciation and the quantitative extent of *bioavailable element concentration* is still unsatisfactory, since the leachable fraction does not necessarily correspond to the amount available to biota (Pickering 1981). In most cases information is lacking about the specific mechanism by which the organisms actively participate in the removal of nutrients from the solid substrate. Plant root activities, for example, include redox changes, pH alterations and organic complexing processes. Competition exists between adsorption sites on the solid substrate (Fe/Mn-oxides, organic matter) and selective mechanisms of metal uptake by the different organisms (Luoma and Bryan 1977).

5 Conclusions

Reactivity, mobility and availability of trace metals for metabolic processes are closely related to their chemical species both in solution and in particulate matter. Because of its inherent simplicity, in situ dialysis appears particularly well adapted to the study of trace constituents in sediment pore waters under field conditions (Carignan et al. 1985). For the differentiation of the solid metal species, e.g. cation exchangeable forms, carbonate phases, reducible fractions, associations with organic substances and sulphides and the inert "residual" fractions, chemical extraction sequences have been developed, which can be used, among others, for the estimation on the potential remobilization of metals under changing environmental conditions.

There is a tendency that elements introduced with solid waste material are less stably bound than those in natural systems. Even with relative small proportions of these materials, therefore, mobilization (and subsequent transfer to biota) of potentially toxic elements by acidity, complexing agents or redox changes, may be significantly increased. From these findings it seems that one of the major applications of solid speciation data includes an estimation of the long-term changes of metal mobilities under variable environmental conditions by analyzing characteristics rearrangements of chemical forms in these materials. This will be particularly interesting for assessing the effects of changing redox conditions, and these effects have practical implications for the disposal alternatives for dredged materials.

References

Adams DD, Darby DA (1980) A dilution-mixing model for dredged sediments in freshwater systems. In: Baker RA (ed) Contaminants and sediments, vol 1. Ann Arbor Sci Publ, pp 373–392

Andreae MO, Asami T, Bertine KK, Buat-Menard PE, Duce RA, Filip Z, Forstner U, Goldberg ED, Heinrichs H, Jernelov AB, Pacyna JM, Thornton I, Tobschall HJ, Zoller WH (1984) Changing biogeochemical cycles (Group report). In: Nriagu JO (ed) Changing metal cycles and human health. Dahlem Konferenzen. Life Sci Res Rep 28: pp 359–374

Batley GE, Giles MS (1980) A solvent displacement technique for the separation of sediment interstitial waters. In: Baker RA (ed) Contaminants and sediments, vol 2. Ann Arbor Sci Publ, pp 101–117

Berner RA (1981) A new geochemical classification of sedimentary environment. J Sediment Petrol 51:359–365

Bernhard M, Brinckman FE, Sadler PS (eds) (1986) The importance of chemical speciation in environmental processes. Dahlem Konferenzen, Berlin. Springer, Berlin Heidelberg New York Tokyo

Bottomley EZ, Bayly IL (1984) A sediment pore water sampler used in root zone studies of the submerged macrophyte, Myriophyllum Spicatum. Limnol Oceanogr 29:671–673

Bray JT, Bricker OP, Troup BN (1973) Phosphate in interstitial waters of anoxic sediments: oxidation effects during sampling procedure. Science 180:1362–1364

Brooks RR, Presley BJ, Kaplan IR (1968) Trace elements in the interstitial waters of marine sediments. Geochim Cosmochim Acta 32:397–414

Calmano W, Förstner U (1983) Chemical extraction of heavy metals in polluted river sediments in Central Europe. Sci Total Environ 28:77–90

Campbell PGC, Stokes PM, Galloway JN (1983) Effects of atmospheric deposition on the geochemical cycling and biological availability of metals. In: Proc Int Conf Heavy metals in the environment. CEP Consultants, Edinburgh, pp 760–763

Carignan R (1984) Interstitial water sampling by dialysis. Methodological notes. Limnol Oceanogr 29:667–670

Carignan R, Rapin F, Tessier A (1985) Sediment pore water sampling for metal analysis: a comparison of techniques. Geochim Cosmochim Acta 49:2493–2497

Chao TT (1984) Use of partial dissolution techniques in geochemical exploration. J Geochem Explor 20:101–135

Chester R, Hughes MJ (1967) A chemical technique for the separation of ferromanganese minerals, carbonate minerals and adsorbed trace elements from pelagic sediments. Chem Geol 2:249–262

Darby DA, Adams DD, Nivens WT (1986) Early sediment changes and element mobilization in a man-made estuarine marsh. In: Sly PG (ed) Sediment and water interactions. Springer, Berlin Heidelberg New York Tokyo, pp 343–351

Diks DM, Allen HE (1983) Correlation of copper distribution in a freshwater-sediment system to bioavailability. Bull Environ Contam Toxicol 30:37–43

Duchart P, Calvert SE, Price NB (1973) Distribution of trace metals in the pore waters of shallow water marine sediments. Limnol Oceanogr 18:605–610

Elderfield H (1981) Metal-organic associations in interstitial waters of Narragansett Bay sediments. Am J Sci 281:1184–1196

Engler RM (1980) Prediction of pollution potential through geochemical and biological procedures: development of regulatory guidelines and criteria for the discharge of dredged fill material. In: Baker RA (ed) Contaminants and sediments, vol 1, Ann Arbor Sci Publ, pp 143–169

Engler RM, Brannon JM, Rose J, Bigham, G (1977) A practical selective extraction procedure for sediment characterization. In: Yen TF (ed) Chemistry of marine sediments. Ann Arbor Sci Publ, pp 163–171

Fagerström T, Jernelöv A (1972) Aspects of the quantitative ecology of mercury. Water Res 6:1193–1202

Farmer ME, Linton RW (1984) Correlative surface analysis studies of environmental particles. Environ Sci Technol 18:319–326

Florence TM, Batley GE (1980) Chemical speciation in natural waters. Crit Rev Anal Chem 9:219–296

Förstner U (1982a) Accumulative phases for heavy metals in limnic sediments. Hydrobiologia 91:269–284

Förstner U (1982b) Chemical forms of metal enrichment in recent sediments. In: Amstutz GC El Goresy A, Frenzel G et al. (eds) Ore genesis – the state of the art. Springer, Berlin Heidelberg New York Tokyo pp 191–199

Förstner U (1985) Chemical forms and reactivities of metals in sediments. In: Leschber R, Davis RD, L'Hermite P (eds) Chemical methods for assessing bioavailable metals in sludges and soils. Elsevier, London, pp 1–30

Förstner U (1987) Metal speciation in solid wastes – factors affecting mobility. In: Landner L (ed) Speciation of metals in water, sediment and soil systems. Lecture Notes Earth Sci 11:13–41. Springer, Berlin-Heidelberg-New York-Tokyo

Förstner U, Salomons W (1983) Trace element speciation in surface waters: interactions with particulate matter. In: Leppard GG (ed) (1983) Trace element speciation in surface waters and its ecological implications. Proc NATO Adv Res Worksh, Nervi/Italy. Plenum, New York, pp 245–273

Förstner U, Schoer J (1984) Diagenesis of chemical associations of Cs-137 and other artificial radionuclides in river sediments. Environ Technol Letts 5:295–306

Förstner U, Calmano W, Conradt K, Jaksch H, Schimkus C, Schoer J (1981) Chemical speciation of heavy metals in solid waste materials (sewage sludge, mining wastes, dredged materials, polluted sediments) by sequential extraction. In: Proc Int Conf Heavy metals in the environment, Amsterdam. CEP Consultants, Edinburgh, pp 698–704

Förstner U, Ahlf W, Calmano W, Kersten M, Salomons W (1986) Mobility of heavy metals in dredged harbor sediments. In: Sly PG (ed) Sediments and water interactions. Springer, Berlin Heidelberg New York Tokyo, pp 371–380

Fraser JL, Lum KR (1982) Availability of elements of environmental importance in incinerated sludge ash. Environ Sci Technol 17:52–54

Gambrell RP, Khalid RA, Verloo MG, Patrick WH, Jr (1977) Trace and toxic metal uptake by marsh plants as affected by Eh, pH, and salinity. US Army Eng Waterways Exp Stn, Tech Rep D-77-4, Corps of engineers, Vicksburg/MS

Ham RK, Anderson MA, Stanforth R, Stegmann R (1979) Background study on the development of a standard leaching test. US EPA-600/2-79-109

Harris WR, Silberman D (1983) Time-dependent leaching of coal fly ash by chelating agents. Environ Sci Technol 17:139-145

Harrison RM, Laxen DPH, Wilson SJ (1981) Chemical associations of lead, cadmium, copper, and zinc in street dusts and roadside soils. Environ Sci Technol 15:1378-1383

Hart BT, Davies SHR (1977) A new dialysis-ion exchange technique for determining the forms of trace metals in water. Aust J Mar Freshwater Res 28:105-112

Herms U, Brümmer G (1978) Löslichkeit von Schwermetallen in Siedlungsabfällen und Böden in Abhängigkeit von pH-Wert, Redoxbedingungen und Stoffbestand. Mitt Dtsch Bodenkd Ges 27:23-43

Hesslein R (1976) An in-situ sampler for close interval pore water studies. Limnol Oceanogr 21:912-914

Hoeppel RE, Meyers TE, Engler RM (1978) Physical and chemical characterization of dredged material influents and effluents in confined land disposal areas. US Army Eng Waterways Exp Stn, Techn Rep D-78-24, CE Vicksburg/MS

Horowith AJ (1984) A primer on trace metal-sediment chemistry. US Geol Surv, Open-File Rep 84-709. Doraville/GA, 82 pp

Jackson ML (1958) Soil chemical analysis. Prentice-Hall, Englewood Cliffs/NJ

Jenne EA (1977) Treace element sorption by sediments and soils – sites and processes. In: Chappell W, Petersen K (eds) Symp Molybdenum. Dekker, NY, pp 425-553

Jonasson IR (1977) Geochemistry of sediment/water interactions of metals, including observations on availability. In: Shear H, Watson AEP (eds) The fluvial transport of sediment-associated nutrients and contaminants. IJC/PLUARG, Windsor/Ont, pp 255-271

Kerner M, Kausch H, Kersten M (1986) Der Einfluß der Gezeiten auf die Verteilung von Nährstoffen und Schwermetallen in Wattsedimenten des Elbe-Aestuars. Arch Hydrobiol/Suppl 75:118-131

Kersten M, Förstner U (1985) Trace metal partitioning in suspended matter with special reference to pollution in the southeastern North Sea. In: Degens ET, Kempe S, Herrera R (eds) Transport of carbon and minerals in major world rivers, vol 3. Mitt Geol-Paläontol Inst Univ Hamburg, SCOPE/UNEP Spec Vol 58:631-645

Kersten M, Förstner U (1986) Chemical fractionation of heavy metals in anoxic estuarine and coastal sediments. Wat Sci Technol 18:121-130

Kersten M, Förstner U (1987) Cadmium associations in freshwater and marine sediment. In: Nriagu JO (ed) Cadmium in the aquatic environment. John Wiley, New York pp 51-88

Keyser TR, Natusch DFS, Evans CA jr, Linton RW (1978) Characterizing the surface of environmental particles. Environ Sci Technol 12:768-773

Khalid RA (1980) Chemical mobility of cadmium sediment-water systems. In: Nriagu JO (ed) Cadmium in the environment, part I: ecological cycling. John Wiley, New York, pp 257-304

Khalid RA, Patrick WH jr, Gambrell RP (1978) Effect of dissolved oxygen on chemical transformations of heavy metals, phosphorus and nitrogen in an estuarine sediment. Estuar Coast Mar Sci 6:21-35

Kinniburgh DG, Miles DL (1983) Extraction and chemical analysis of interstitial water from soils and rocks. Environ Sci Technol 17:362-368

Krom MD, Sholkovitz ER (1978) On the association of iron and manganese with organic matter in anoxic marine pore water. Geochim Cosmochim Acta 42:607-611

Lake DL, Kirk PWW, Lester JN (1984) Fractionation, characterization, and speciation of heavy metals in sewage sludge and sludge-amended soils: a review. J Environ Qual 13:175-183

Landner L (ed) (1987) Metal speciation in water, sediment and soil systems. Lecture notes in earth sciences. Springer Berlin Heidelberg New York Tokyo

Lee FY, Kittrick JA (1984) Elements associated with the cadmium phase in a harbor sediment as determined with the electron beam microprobe. J Environ Qual 13:337-340

Lee GF, Lopez JM, Piwoni, MD (1976) Evaluation of the factors influencing the results of the elutriate test for dredging material disposal criteria. In: Krenkel PA, Harrison J, Burdick JC III. Proc Spec Conf Dredging and its environmental effects. Am Soc Civ Eng, New York, pp 253-288

Leppard GG (ed) (1983) Trace element speciation in surface waters and its ecological implications. Proc NATO Adv Res Worksh, Nov 2–4, 1981 Nervi/Italy. Plenum, New York
Lu CSJ, Chen KY (1977) Migration of trace metals in interfaces of seawater and polluted surficial sediments. Environ Sci Technol 11:174–182
Lum KR, Betteridge JS, Macdonald RR (1982) The potential availability of P, Al, Ca, Co, Cr, Cu, Fe, Mn, Ni, Pb, and Zn in urban particulate matter. Environ Technol Letts 3:57–62
Luoma SN (1983) Bioavailability of trace metals to aquatic organisms: a review. Sci Total Environ 28:1–22
Luoma SN, Bryan GW (1978) Factors controlling the availability of sediment-bound lead to the estuarine bivalve *Scrobicularia plana.* J Mar Biol Assoc UK 58:793–802
Luther GW, Meyerson AL, Krajewski JJ, Hires R (1980) Metal sulfides in estuarine sediments. J Sediment Petrol 50:1117–1120
Lyle M, Heath GR, Robbins JM (1984) Transport and release of transition elements during early diagenesis: Sequential leaching of sediments from MANOP sites M and H. Part I. pH 5 acetic acid leach. Geochim Cosmochim Acta 48:1705–1715
Lyons WB, Gaudette HE, Smith GM (1979) Pore water sampling in anoxic carbonate sediments: oxidation artefacts. Nature (London) 277:48–49
Mattigod SV, Page AL (1983) Assessment of metal pollution in soils. In: Thornton I (ed) Applied environmental geochemistry. Academic Press, London New York, pp 355–394
Mayer LM (1976) Chemical water sampling in lakes and sediments with dialysis bags. Limnol Oceanogr 21:909–911
Nirel P, Thomas AJ, Martin JM (1986) A critical evaluation of sequential extraction techniques. In : Bulman RA, Cooper JR (eds) Speciation of fission and activation products in the environment. Elsevier, London, pp 19–26
Nissenbaum A, Swaine DJ (1976) Organic matter-metal interactions in recent sediments. The role of humic substances. Geochim Cosmochim Acta 40:809–816
Nriagu JO (ed) (1984) Changing metal cycles and human health. Dahlem Konferenzen, Life Sci Res Rep 28. Springer, Berlin Heidelberg New York
Patrick WH, Williams BG, Moraghan JT (1973) A simple system for controlling redox potential and pH in soil suspensions. Soil Sci Soc Am Proc 37:331–332
Pfeiffer G, Förstner U, Stoffers P (1982) Speciation of reducible metal compounds in pelagic sediments by chemical extraction. Senckenbergiana Mar 14:23–38
Pickering WF (1981) Selective chemical extraction of soil components and bound metal species. CRC Crit Rev Anal Chem Nov: 233–266
Plant JA, Raiswell R (1983) Principles of environmental geochemistry. In: Thornton I (ed) Applied environmental geochemistry. Academic Press, London New York, pp 1–39
Presley BJ, Kolodny Y, Nissenbaum IR (1972) Early diagenesis in a reducing fjord, Saanich Inlet, British Columbia. II. Trace element distribution in interstitial water and sediment. Geochim Cosmochim Acta 36:1073–1090
Rapin F, Tessier A, Campbell PGC, Carignan R (1986) Potential artifacts in the determination of metal partitioning in sediments by a sequential extraction procedure. Environ Sci Technol 20:836–840
Recke M, Förstner U (1986) Oxidation increases mobility of toxic metals in dredged sludges. In: Lester JN, Perry R, Sterritt RM (eds) Proc Int Conf Chemicals in the environment. Selper, London, pp 809–818
Salomons W (1980) Adsorption processes and hydrodynamic conditions in estuaries. Environ Technol Letts 1:356–365
Salomons W (1985) Sediments and water quality. Environ Technol Lett 6: 315–368
Salomons W, Förstner U (1984) Metals in the hydrocycle. Springer, Berlin Heidelberg New York
Schoer J (1984) Thallium. In: Hutzinger O (ed) Handbook of environmental chemistry, vol 3, pt C. Springer, Berlin Heidelberg New York
Tessier A, Campbell PGC, Bisson M (1979) Sequential extraction procedure for the speciation of particulate trace metals. Anal Chem 51:844–851
Thomson EA, Luoma SN, Cain DJ, Johansson C (1980) The effect of sample storage on the extraction of Cu, Zn, Fe, Mn, and organic material from oxidized estuarine sediments. Water Air Soil Pollut 74:215–233
Troup BN, Bricker OP, Bray JT (1974) Oxidation effect on the analysis of iron in the interstitial water of recent anoxic sediments. Nature (London) 249:237–239

Watson PG, Frickers PE, Goodchild CM (1985) Spatial and seasonal variations in the chemistry of sediment interstitial waters in the Tamar estuary. Estuar Coast Shelf Sci 21: 105–119

Wood JM, Chakrabarty AM, Craig PJ, Förstner U, Fowler BA, Herms U, Krull IS, Mackay D, Olson GJ, Russell DH, Salomons W, Silver S (1986) Speciation in systems under stress (Group report). In: Bernhard M, Brinckman FE, Sadler PS (eds) The importance of chemical speciation in environmental processes. Dahlem Konferenzen, September 2–7, 1984 Berlin. Springer, Berlin Heidelberg New York Tokyo

Yoshida T (1984) On the behaviour of nutrients in sediment pore water and on sediment surface. Proc Int Conf Environmental Contamination, London 1984, pp. 172–175. CEP Consultants, Edinburgh

Diagenetic Processes in Aquatic Mine Tailings Deposits in British Columbia

T. F. PEDERSEN[1] and A. J. LOSHER[2]

1 Introduction

Mine wastes have been deposited in coastal or terrestrial aquatic systems for at least two millenia, historically without regard for the biological, chemical or physical consequences of such disposal. The rise of industrialization and coincident exponential growth in the rates of metal consumption and exploitation have been associated with a profound increase in the annual volume of mine tailings requiring permanent storage on the earth's surface. A parallel consequence has been the steady shift towards exploitation of lower grade, larger volume orebodies which has radically increased the volume of tailings produced by most contemporary milling operations. This phenomenon has increased the need to find or construct large-volume repositories for tailings storage near mine sites; when coupled with the growing awareness of the serious long-term environmental risks posed by terrestrial disposal, and typical attractive economics, submerged storage in both fresh and saline waters is becoming an increasingly appealing alternative.

Submerged tailings deposits are, of course, not free from often severe environmental impact, as has been made abundantly clear elsewhere in this volume. However, until recently, research on the environmental effects of such discharge to marine or lacustrine environments has focussed primarily on physical dispersion and biological impact, with less emphasis being placed on the nature of post-depositional (i.e. diagenetic) chemical reactivity. In this chapter we will review this latter and somewhat neglected area of concern, using examples from British Columbian waters to demonstrate the extent and form of early diagenesis in sulphide-bearing aquatic tailings deposits. The approach we use in attempting to understand post-depositional reactivity is to study interstitial water chemistry and the composition of the solid phase in concert which allows significantly more information to be derived than studying either factor in isolation.

We will address several questions in this review:

1. What processes are controlling the mobility of trace metals in deposited tailings, and to what extent are such metals mobile?
2. How is benthic regeneration of nutrients in coastal sediments affected by the rapid deposition of large volumes of tailings over large surface areas?

[1]Department of Oceanography, University of British Columbia, Vancouver B.C., V6T 1W5, Canada
[2]EAWAG–The Swiss Federal Institute for Water Resources and Water Pollution Control, CH-8600 Dubendorf, Switzerland

3. In what direction and how significant to the environment are benthic fluxes of trace metals in contrasting tailings deposits?
4. How do diagenetic reactions in the tailings compare to diagenesis in natural sediments native to the receiving depositional basin?

These questions, where applicable, will be considered with reference to the three submerged tailings deposits, one aquatic (Buttle Lake) and two marine (Rupert Inlet and Alice Arm), which we have studied to date.

2 History and Environmental Settings

Buttle Lake is a large (30 km long by 1.5 km wide) water body which occupies a U-shaped valley in an area of high relief on Vancouver Island (Fig. 1). The south basin of the lake reaches a maximum depth of 87 m and received tailings, via a slightly submerged outfall, from 1967 to 1984. The tailings were derived from the milling of high-grade zinc-, copper- and lead-bearing ore at the mine site 6 km west of the south basin, adjacent to Myra Creek (Fig. 1). By 1985, nearly 5.5×10^6 t of mill tailings had been released to the lake; this discharge ceased in 1984 and land disposal is now being practiced.

The lacustrine deposit is confined areally to the south basin by a relatively shallow sill 5 km north of the discharge site which effectively limits benthic physical dispersion of the material. The tailings consist largely of sand- and silt-sized silicate gangue minerals from the host metavolcanic rocks and iron, zinc, copper and lead sulphides. Concentrations of Zn, Cu and Pb in the tailings range widely but average about 7000, 1300 and 900 mg kg^{-1} respectively.

Rupert and Holberg inlets, respectively 10 and 34 km long, are relatively shallow fjords located in part of a glacially-scoured trough that lies obliquely across the northern tip of Vancouver Island (Fig. 1). Open communication with the Pacific Ocean through Quatsino Sound is restricted by an 18-m deep sill at Quatsino Narrows; on the fjord side of the sill, at the confluence of the inlets, a maximum depth of 170 m is reached. The average depths are considerably shallower: 110 m in Rupert and only 80 m in Holberg Inlet. Vigorous mixing, resulting from confinement of the tidal prism through Quatsino Narrows, occurs in the area of the confluence; the mixing produces a homogeneous water body in the Rupert-Holberg Basin which is characterized by a uniformly high dissolved oxygen concentration (Drinkwater and Osborn 1975).

The discharge of tailings to Rupert Inlet differs substantially in degree from that in Buttle Lake. Currently, approximately 40×10^3 t day^{-1} are released via a submerged outfall at 50 m depth. This disposal rate has increased slightly since high-volume mining of the low-grade chalcopyrite- and molybdenite-bearing deposit commenced in 1971. Tailings now cover much of the floor of the inlet and extend a significant distance up the adjoining Holberg Inlet (Fig. 1). Near the outfall, the tailings are now more than 25 m thick; the deposit thins progressively toward the margins, although intermittent slumping and channelized down-inlet turbidity current flows have produced a complex morphology (Hay 1982).

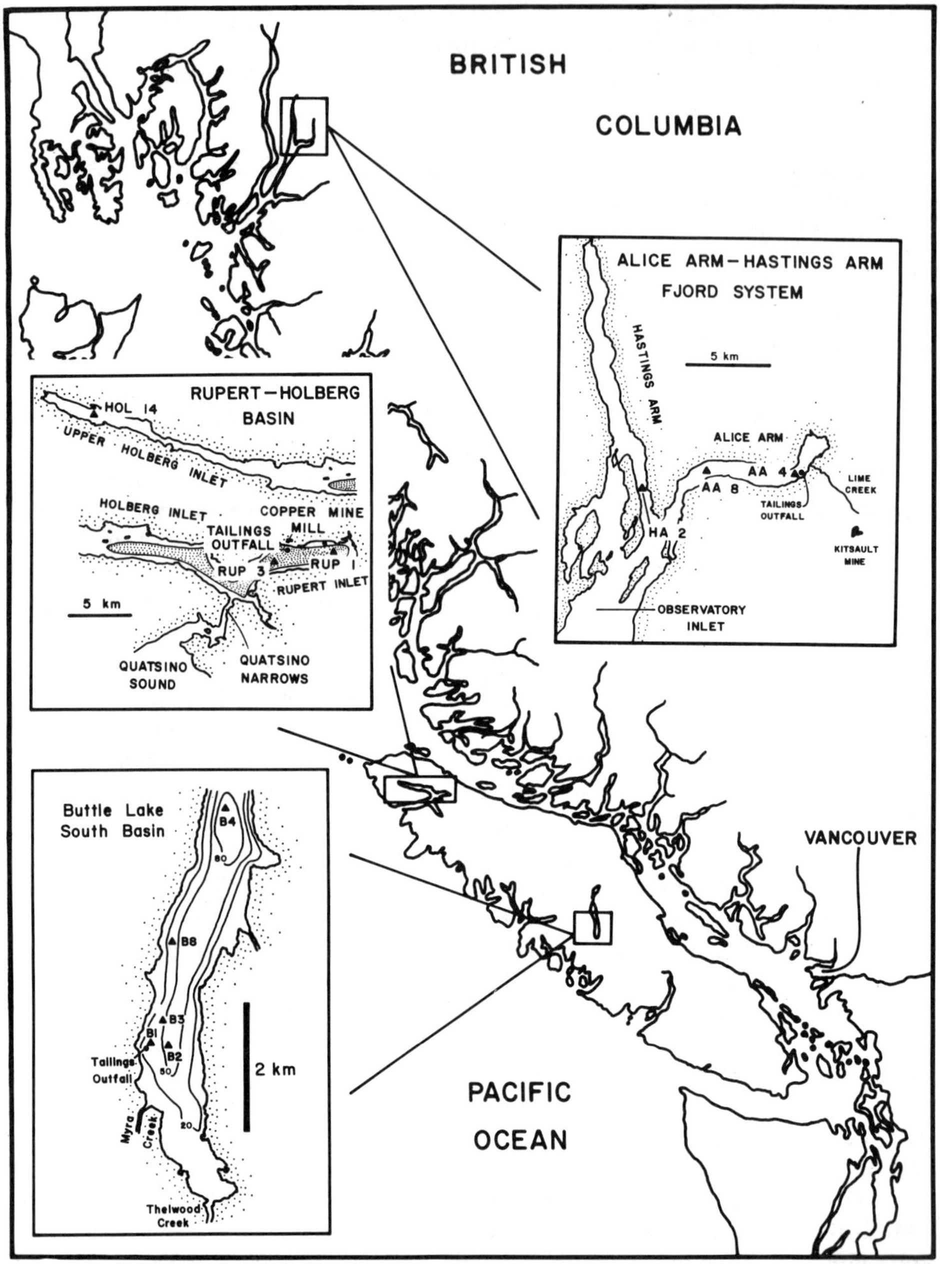

Fig. 1. Location map showing sites of tailings deposition and core locations (*insets*). The *shaded zone* in the Rupert-Holberg Basin inset shows the approximate distribution of tailings in the area in 1980

The Rupert Inlet tailings are typically silt-sized and are dominated mineralogically by quartz (50–70%), biotite and chlorite (5–10%) and feldspar (2–20%). Magnetite (2–4%), pyrite (2–4%), calcite (2.5%), chalcopyrite (0.2%), sphalerite (0.02%) and molybdenite (0.01%) occur in much smaller concentrations (Poling 1982).

The 19-km-long Alice Arm and 22-km-long Hastings Arm in northwestern British Columbia are the landward termini of the Portland Canal–Observatory Inlet fjord system, which stretches some 110 km inland from the Pacific Ocean (Fig. 1). A complex of shallow sills ranging in depth from 21 to 51 m separates the arms from Observatory Inlet, but does not seriously restrict the tidal exchange of water within the inlet system; as a result, the water in both inlets is relatively well oxygenated year-round. Maximum depths reach 307 and 386 m respectively.

Tailings derived from mining of the Kitsault molybdenum deposit have been discharged to Alice Arm during two recent and relatively short periods. During the initial mining phase (1967–1972), about 6×10^3 t day^{-1} of mill tailings were simply discharged into Lime Creek (Fig. 1) which carried them into Alice Arm near its eastern end. Mining recommenced in April, 1982 at double the previous throughput, and until the mine shut down 18 months later, tailings were discharged directly into the inlet through a submerged outfall at a depth of 50 m.

Unlike the Buttle Lake or Rupert Inlet orebodies, the low-grade Kitsault ore contains essentially only one metal of economic interest, molybdenum, which is hosted by a quartz monzonite porphyry. The mineralogic composition of the tailings reflects this association: K-feldspar and quartz are the most common constituents in the mill discharge, while molybdenite particles are rare. As in the other two examples considered in this chapter, the tailings are typically silt-sized.

The only quantitatively significant reagent added to all three tailings deposits is lime (CaO), which is used to raise the pH in the milling circuits and to enhance coagulation in the thickening tanks. At the Rupert Inlet mill, for example, about 0.6 kg t^{-1} ore is added; the presence of such relatively large concentrations of lime appears to have a considerable influence on the geochemistry of pore waters in the Rupert Inlet and Alice Arm deposits, as is discussed in the following section.

3 Case Studies

3.1 Buttle Lake

Pedersen (1983) discussed the distribution of dissolved Zn, Cd, Cu, Mn and Fe in pore waters extracted by centrifugation under nitrogen from four cores (Table 1) collected from both tailings and natural sediments in the south basin of Buttle Lake. At the time this investigation was carried out, dissolved Zn and Cu levels in the overlying lake water were very high (Table 1) and it was speculated that two factors were responsible for these enrichments. First, it had been discovered that water flowing into the lake from Myra Creek (Fig. 1) contained extraordinarily high concentrations of dissolved metals, which were traced to active leaching of a sulphide-bearing waste rock dump by *Thiobacilli,* 6 km upstream from the lake;

Table 1. Dissolved interstitial metal concentrations in Buttle Lake cores, measured by graphite furnace AAS (Data from Pedersen 1983). Core locations are shown in Fig. 1

Core	Sample depth (cm)	Dissolved metal concentration ($\mu g\ l^{-1}$)				
		Zn	Cd	Cu	Mn	Fe
B1	0–3	35	< 0.5	15.6	170	24
	3–6	20	< 0.5	8.0,8.3	130	18
	6–10	27	< 0.5	13.4,13.1	140	16
	10–14	31,31	< 0.5	3.4	170	7.2
	14–19	34	< 0.5	2.2	210	3.4
	25–30	36,37	< 0.5	5.0	550,560	23
	35–40	26	< 0.5	0.5	114	13
	45–50	23	< 0.5	1.9	90	29
B2	0–3	34	< 0.5	5.6	650	42
	3–6	34,34	< 0.5	6.3	800	62
	6–10	13	< 0.5	4.2	630	30
	10–14	19	< 0.5	5.9	370,350	15
	20–25	23	< 0.5	0.7,0.9	650	36
	35–40	26	< 0.5	6.2	350	13
	55–60	29	< 0.5	6.5	290,300	15
	78–83	30	< 0.5	4.4	280	39
B3	0–3	33,31	2.7,2.1	13.2,13.1	3190,3120	42
	3–6	6.4	< 0.5	2.0	2740	32
	6–10	78,79	< 0.5	2.0	2240	23
	10–14	35	< 0.5	4.7	2390	38
	20–25	28	< 0.5	1.1	1610	58
	30–35	5.8	< 0.5	< 0.5	580	481
	45–50	42	< 0.5	0.5	520	111
	60–65	< 3.0	< 0.5	< 0.5	7400,7330	1550
B4	0–5	22,20	< 0.5	< 0.5	12100	8700
	5–10	21	< 0.5	< 0.5	26900	13900
	10–15	7.4	< 0.5	< 0.5	24900,25000	9300
	30–35	< 3.0	< 0.5	< 0.5	44800	12900
Lake bottom water, 54-m depth, near B2		170	1	11	–	–
Lake bottom water, 87-m depth, near B4		230	0.4	15	–	–

metal levels in the creek water in July, 1981 averaged 600 $\mu g\ l^{-1}$ Zn, 40 $\mu g\ l^{-1}$ Cu and 1.4 $\mu g\ l^{-1}$ Cd (B.C. Research, cited in Pedersen 1983). Second, remobilization of metals from the tailings on the floor of the lake was assumed to be occurring and contributing to the high metal loads in the overlying water. Although the pore water metal concentrations range quite widely, it is clear from the data in Table 1 that such diagenetic remobilization of metals could not have been significant; indeed, metal concentrations in the tailings pore waters were much *less* in 1982 than in the overlying Buttle Lake water. In interstitial waters from the natural sediments (core B4) the Zn concentration decreased rather sharply and steadily with depth, indicating that the metal was being incorporated into an authigenic precipitate.

Pedersen (1983) concluded that several factors controlled these distributions. First, oxidation of the tailings on the lake bottom was not occurring, presumably because the rate of discharge was sufficient to bury the deposits continuously and quickly, thus minimizing the time of exposure to dissolved oxygen in the bottom water. Because Zn and Cu oxyhydroxides are much more soluble than their sulphide counterparts, oxidation of the detrital sulphides in the tailings would release metals to interstitial solution. An example of this phenomenon is shown in Table 2. Water which was allowed to accumulate by dewatering on the tops of two of the Buttle Lake cores, after exposure to air for 8–11 h, showed significantly higher Zn and Cu concentrations. The absence of any similar increase in dissolved Zn, Cu or Cd concentrations with depth in the tailings cores (Table 1) confirmed that the tailings were relatively unreactive on the lake bottom; there was no evidence that progressive oxidation with concomitant metal release was occurring. Second, the absence of detectable dissolved Zn or Cu in the natural sediments underlying the tailings in core B3 and the decreasing Zn gradient in core B4 indicated that these metals were diffusing *into* the natural sediments from the overlying metal-rich lake water and were being removed from solution, presumably by incorporation into authigenic sulphide minerals. Support for this contention is given by the associated *increase* of dissolved Mn and to a lesser extent, Fe (Table 1). Because manganese and iron oxyhydroxides serve as preferred electron acceptors after oxygen and nitrate have been depleted (Stumm and Morgan 1970), their presence at high concentrations in solution in the uppermost sample in core B4 is evidence that anoxic conditions are established at shallow depths in the natural sediments. Iron sulphide precipitation in this facies is of course limited by the scarcity of reducible sulphate, but is nevertheless indicated by the decreased Fe concentration at depth in B4 relative to the steadily increasing dissolved Mn level.

The Buttle Lake study demonstrated clearly that the tailings were not diagenetically reactive while deposition was proceeding at a high rate. In this type of example, the key factor which mitigates against release of metals is the high

Table 2. Zn, Cd and Cu concentrations in supernatant (core-top) water immediately after core collection (A) and after 8–11 h of partial exposure of the tailings on the top of the cores to air (B) (Data from Pedersen 1983)

Core		Dissolved metal concentrations ($\mu g\ l^{-1}$)		
		Zn	Cd	Cu
B1	(A)	22,20,20	$<0.5, <0.5$	0.8
	(B)	56,56	1.0,1.0	14.6,14.8
B2	(A)	44	<0.5	5.2
	(B)	117	<0.5	12.0,13.9

sedimentation rate, which places a strong limit on the amount of oxygen which can diffuse into the tailings from bottom water to support oxidation of detrital sulphide minerals. As the data in Table 2 demonstrate, oxidation with associated metal release can occur quite rapidly in the presence of high oxygen concentrations. Therefore, metal remobilization from the lacustrine tailings could occur in the period following cessation of discharge but preceding burial by subsequent natural sedimentation. In the south basin of Buttle Lake, the natural sedimentation rate measured at site B8 using ^{210}Pb data (Table 3) is about 2 mm yr^{-1}. In the same core the solid-phase Mn concentration decreases sharply between 0 and 3 cm depth (Table 3), indicating solubilization at depths below 1 cm. This distribution is consistent with the pore water data in core B4. Assuming that the data from both cores are representative of the natural sediments in the basin in general, then it can be suggested logically that the tailings would become covered with a veneer of natural sediments, anoxic at a shallow depth, within 15 to 20 years following the cessation of discharge. At that point the metal-rich deposit would, in effect, become chemically sealed with respect to upward diffusion of metals from the previously oxidized, but now buried tailings.

Table 3. ^{210}Pb (supported and unsupported) and Zn concentrations in sediments from core B8. The ^{210}Pb data yield a sedimentation rate for the site of 2 mm yr^{-1}. The Zn analyses indicate that tailings are confined to the top 2 cm of the core. Zn and Mn were measured by X-ray fluorescence and ^{210}Pb by counting the ^{210}Bi daughter in a low-background proportional counter

Depth in core (cm)	210_{Pb} (dpm g^{-1})	Zn ($\mu g\ g^{-1}$)	Mn ($\mu g\ g^{-1}$)
0–1	12.7 + 3.3	3800	10400
1–2		4100	4410
2–3	9.8 + 2.7	740	–
3–4		184	2170
4–5	5.6 + 3.3	142	–
5–8		128	2100
8–11	4.7 + 2.7	–	–
11–15		135	2030
15–20	0.0 + 4.6	–	–
30–35	0.0 + 4.0	–	–

Epilogue: Collection and treatment of the leachate from the waste-rock dump near the mine site commenced in 1983. As a result dissolved metal levels in Myra Creek water and in the lake have been falling steadily and are now near historical background levels of $<5\ \mu g\ l^{-1}$ Zn, $<0.5\ \mu g\ l^{-1}$ Cd and $<1\ \mu g\ l^{-1}$ Cu. Tailings deposition in the lake ceased in 1984, largely for aesthetic reasons; although it had been demonstrated that the tailings were not contributing metals to the lake (Pedersen 1983), ongoing use of Buttle Lake as a major source of domestic drinking water essentially obviated continued tailings discharge. The tailings are now being stored in well-drained ponds near the mine; drainage water from these accumulations is recycled in the mill and dissolved metals are removed by adding lime and settling out the precipitated metal hydroxides.

3.2 Rupert Inlet

Two aspects of early diagenesis of the tailings and natural sediments in the Rupert-Holberg Basin have been considered in detail previously (Pedersen 1984, 1985): i.e. benthic nutrient regeneration and remobilization of metals, and these will be reviewed in turn. Both studies used pore water and solid-phase data obtained from analysis of three cores, each representing a different sediment facies: core HOL 14 was collected from organic-rich sediments in the upper Holberg Inlet, an area unaffected by tailings input; RUP 3 was raised from an area of rapidly accumulating tailings (on the order of 0.5–1 m yr^{-1}) on the (presumably) stable flank of the tailings deposit down-inlet from the outfall; and RUP 1 was obtained from an area of slow tailings accumulation (4–5 cm yr^{-1}) near the eastern end of Rupert Inlet (Fig. 1) and consisted of about 40 cm of tailings overlying natural sediments.

Profiles of dissolved metabolites (NH_4^+ and PO_4^{3-}) in the interstitial waters of all three facies are shown in (Fig. 2). In upper Holberg Inlet sediments, ammonia and phosphate are clearly regenerated from organic matter in stoichiometric proportion; the N:P molar ratio in HOL 14 pore water is about 10, as demonstrated by the slope of the correlation plot in Fig. 3. Such proportional regeneration of N and P is commonly observed in interstitial waters in organic-rich coastal sediments (e.g. Sholkovitz 1973; Martens et al. 1978). A radically different distribution is characteristic of both tailings-bearing cores, however. Relative to the ammonia concentration, dissolved phosphate in both RUP 3 and the tailings portion of RUP 1 is strongly depleted as is evident in Figs. 2 and 3; the dearth of phosphate indicates that the tailings are actively removing nearly all regenerated PO_4^{3-} from the interstitial solution. The explanation for this phenomenon lies in the addition of lime to the tailings during the milling of the ore. The quantity added is sufficient to increase both the dissolved Ca concentration and the pH in the tailings pore water (Fig. 4); these distributions are almost certainly due to continuing dissolution of particulate CaO after the tailings have been deposited on the floor of the inlet. Pedersen (1984) suggested that the relatively high pH and calcium enrichment were responsible for producing supersaturation with respect to carbonate fluorapatite ($Ca(PO_4)_3(CO_3, F)$) and that such a phase, of unknown stoichiometry, was forming authigenically within the deposited tailings. This conclusion was supported by a saturation model based on the composition of offshore Californian phosphorite (Atlas and Pytkowicz 1977), which is known to precipitate authigenically from phosphate-rich pore waters during early diagenesis in nearshore and hemipelagic sediments (Jahnke et al. 1983).

The removal of phosphate from pore water in both the slowly and rapidly accumulating tailings facies in Rupert Inlet appears to eliminate any regenerative flux of phosphorus from the sediments to the overlying water. It is well known that in many relatively shallow aquatic environments, recycling of phosphorus is accomplished by bacterial metabolization of organic matter at or near the sediment-water interface. This process is frequently of major importance to the phosphorus budget in coastal waters (see, for example, Nixon 1981; Elderfield et al. 1981; and Fisher et al. 1982) and in lakes (Emerson and Widmer 1978; Holdren

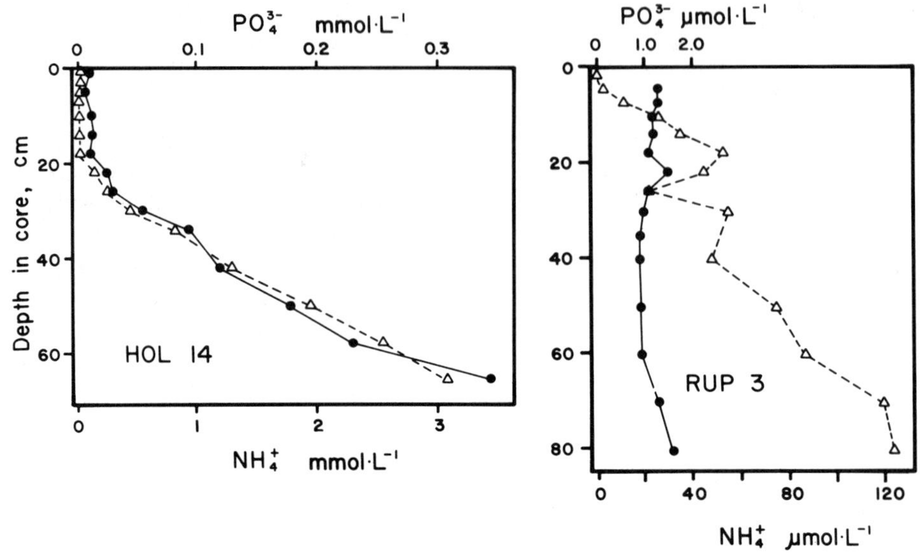

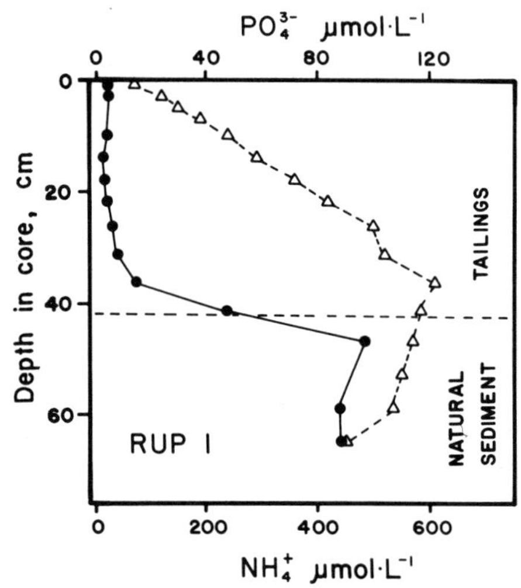

Fig. 2. Profiles of dissolved PO_4^{3-} (•) and NH_4^+ (△) in Rupert and Holberg Inlet pore waters. Note the varying scales on the abscissae (Data after Pedersen 1984)

and Armstrong 1980). Pedersen (1984) showed, however, that the phosphate demand by the tailings and consequent elimination of a benthic P flux over much of the floor of Rupert Inlet do not significantly affect the local phosphorus budget. Even if all the lime added to the tailings (1.8×10^4 kg day^{-1}) combined with phosphorus in the stoichiometric proportions of apatite, less than 2% of the dissolved P inventory in the inlet could be removed per day. Given this extreme estimate of the possible P demand and the relatively short residence time of water in the inlet (13 days, calculated by Pedersen 1984) it is clear that diagenetic reactions in the tailings will have hardly any influence on the local phosphorus balance.

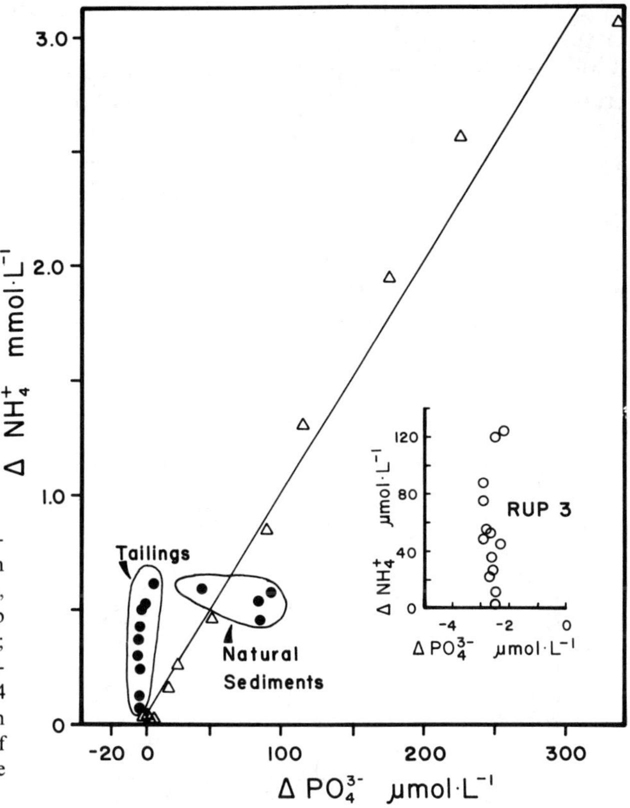

Fig. 3. Dissolved ammonia versus phosphate concentrations in natural sediments and tailings, calculated relative to core-top water. *Open triangles* HOL 14; *filled circles* RUP 1. The regression line through the HOL 14 data yields a slope of 10 which corresponds to the molar ratio of N:P in pore water in the core (Data after Pedersen 1984)

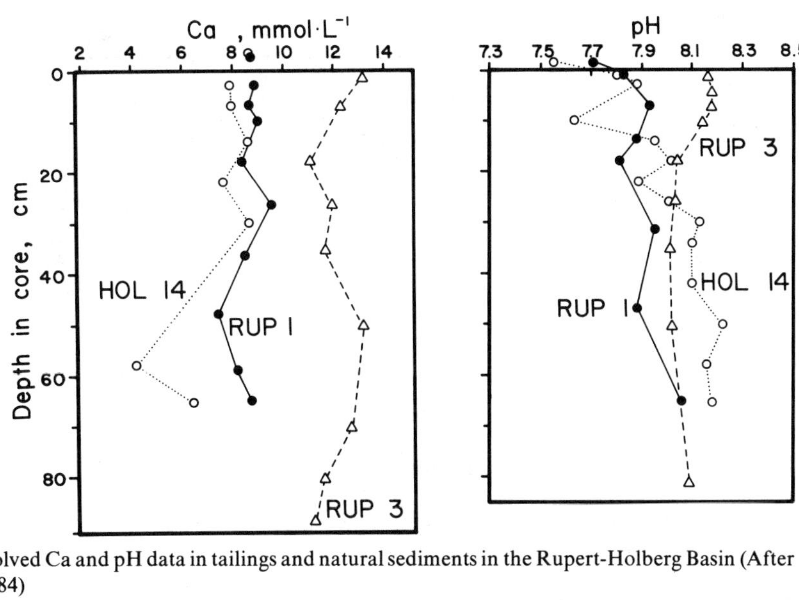

Fig. 4. Dissolved Ca and pH data in tailings and natural sediments in the Rupert-Holberg Basin (After Pedersen 1984)

Unlike the case for phosphate, diagenetic reactions in both the tailings and natural sediments do support benthic fluxes of trace metals into the overlying water, as indicated in the profiles shown in Figs. 5 and 6. Dissolved manganese and iron are clearly released to pore water in the top 2 cm of core HOL 14, due to solution of their respective oxides, which demonstrates that anoxic conditions are established at a very shallow depth in this organic-rich facies. Both metals will reprecipitate as oxides upon diffusing upwards and encountering oxygen at or near the sediment-water interface, only to dissolve again upon burial. This well-known recycling process is typically characterized by near-surface enrichments of Mn and Fe oxides (see, for example, Lynn and Bonatti 1965; Calvert and Price 1972; Pedersen and Price 1982; and Pedersen et al. 1986) and can probably account for

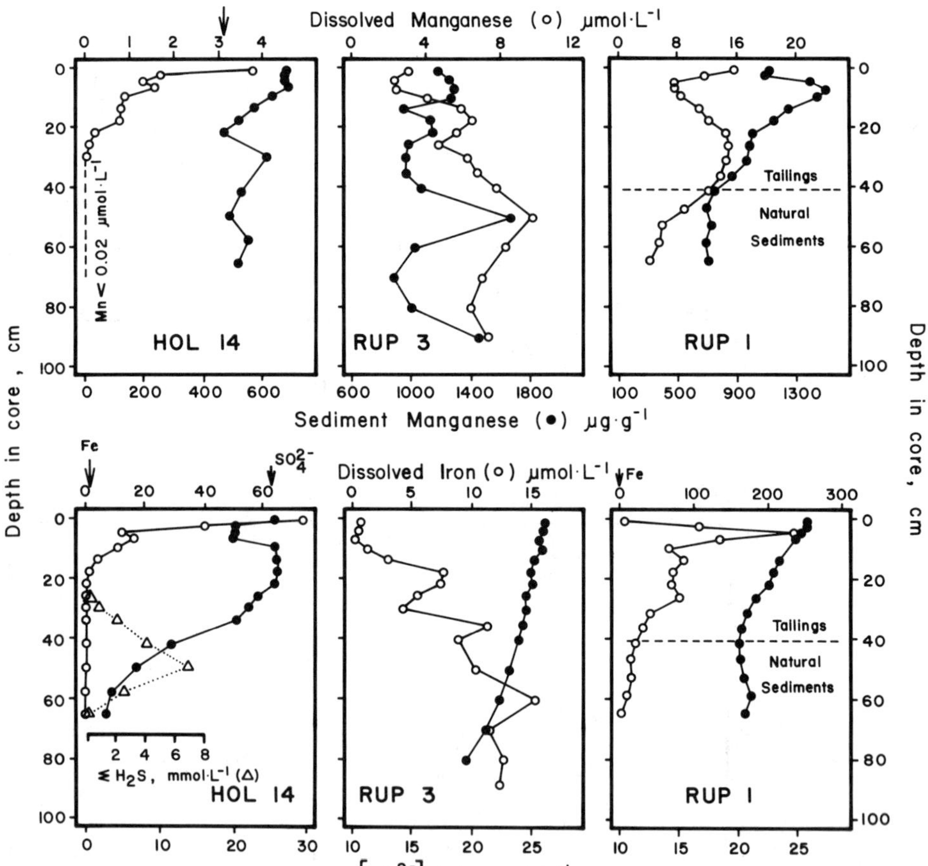

Fig. 5. Profiles of solid-phase manganese and dissolved manganese, iron, H_2S (HOL 14 only) and sulphate in Rupert-Holberg sediments and pore waters (after Pedersen 1985). *Arrows* indicate the concentrations of dissolved species in supernatant (core-top) water. The *vertical dashed line* indicates the analytical detection limit for Mn

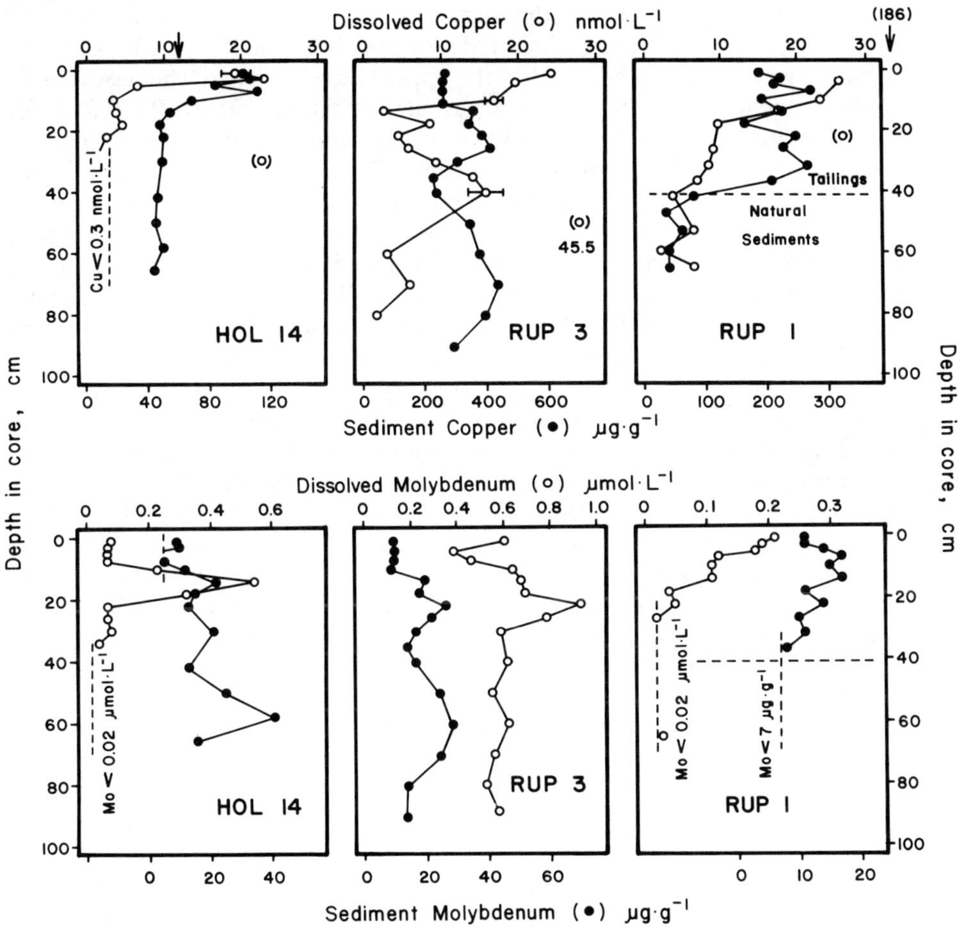

Fig. 6. Profiles of solid-phase and dissolved copper and molybdenum in Rupert-Holberg sediments and pore waters (After Pedersen 1985). *Vertical lines* represent analytical detection limits. Note the varying scales on the abscissae

the 100–200 $\mu g\ g^{-1}$ increase in the concentration of solid-phase Mn seen in the upper several cm in HOL 14 (Fig. 5). Iron and manganese are both quantitatively removed from interstitial water below about 30 cm depth in the natural sediments. The presence of high concentrations of dissolved H_2S in this same depth range (Fig. 5) indicates that the removal of dissolved iron is due to precipitation of authigenic FeS or FeS_2. Because alabandite (MnS) is much more soluble than FeS, manganese removal from solution is probably dominated by co-precipitation of Mn in solid solution with either a magnesium-calcium carbonate or an iron sulphide phase (Pedersen 1985).

Similar but less extreme distributions occur in the slowly accumulating tailings, represented by core RUP 1. As in the natural sediments, the profiles indicate solution of oxides in the upper (tailings) portion of the core and precipitation of authigenic sulphides at depth; the latter assertion is supported by the progressive decrease of SO_4^{2-} below the upper few centimetres. The dissolved Fe and Mn distributions in the rapidly accumulating tailings facies (RUP 3) differ significantly from those in the other two cores: the concentrations of both elements are generally very low and increase with depth. The increases are concurrent with a decrease in the pore water sulphate concentration and probably result from reductive solution of oxides. As noted by Pedersen (1985) the coincidence of higher concentrations of solid-phase and dissolved Mn at 50-cm depth suggests that the extent of release to solution is controlled by variability with depth of the quantity or reactive surface area of the soluble, presumably oxide, Mn phase present. The inconsistent nature of the pore water profile clearly indicates that diffusive steady state has not been reached, which reflects the very rapid accumulation of this facies.

The distribution of dissolved copper in surface sediments is similar in all three facies despite the contrasting modes of sedimentation. Concentrations are at a maximum of 20–25 nmol l^{-1} near the interface and decrease with depth. However, there are differences in the magnitude of the decrease: Cu is undetectable below 30 cm depth in core HOL 14 but is depleted only to levels of several nmol l^{-1} in the lower sections of RUP 1 and RUP 3. Pedersen (1985) suggested that these profiles reflected control by three factors: (1) release to solution from a labile copper-bearing phase at or near the water natural-sediment interface; (2) release to solution by newly deposited tailings; and (3) precipitation of authigenic sulphides at depth. A near-surface maximum of copper in pore waters has been observed in a wide range of sedimentary environments, including oxidizing pelagic clays (Callender and Bowser 1980; Klinkhammer et al. 1982; Sawlan and Murray 1983), suboxic hemipelagic sediments (Klinkhammer 1980) and shallow coastal (Elderfield et al. 1981) and anoxic fjord deposits (Heggie 1983). This widespread surficial enrichment is thought to be maintained by the decomposition of labile copper-bearing organic material at or very near the interface, and it is very probable that this process is responsible for the high interstitial Cu concentration at the top of core HOL 14. Similar reasoning can be applied to the slowly accumulating tailings, but an additional factor may influence copper diagenesis in this facies. As in the Buttle Lake study, Pedersen (1985) found that exposure of the tailings at the surface of RUP 1 to atmospheric oxygen for about 6 h greatly increased the concentration of dissolved Cu in supernatant (core-top) water. This rapid release was ascribed to oxidation of detrital Cu sulphides in the tailings, an observation which was consistent with rapid oxidation kinetics observed in laboratory studies by Hoff et al. (1982). Oxidation may therefore be a significant factor contributing to the benthic flux of copper from the slowly accumulating distal fringes of the tailings deposit. This observation probably also applies to the dissolved Cu enrichment present near the surface of RUP 3. The low dissolved Fe concentrations and the absence of measurable sulphate reduction in the upper 10 cm of the core suggest that oxidizing conditions prevail at or slightly below the sediment surface; oxidation of sulphides may be responsible for the surficial Cu maximum.

It should be noted that although oxidation of the tailings and associated release of Cu to solution has been invoked as a potential contributor to the benthic Cu flux, the actual magnitude of the Cu enrichments in surficial pore waters in the tailings is similar to or less than concentrations observed in natural sediments collected from unpolluted environments (see, for example, Heggie 1983). Pedersen (1985) noted, in this context, that the tailings in Rupert Inlet cannot be supporting a benthic flux of Cu to inlet waters which is greater than the evasion of the element from most natural coastal sediments.

A different conclusion has been drawn for Mo, however. The geochemistry of the element in natural sediments is controlled by familiar processes: oxide dissolution, sulphide precipitation and association with organic material. In anoxic environments, coprecipitation of Mo with iron sulphide phases has been described as an important pathway for the removal of Mo from solution (Korolev 1958; Pilipchuk and Volkov 1968, 1974, Bertine 1972; Malcolm 1985) and scavenging of Mo by suspended particulate or sedimentary manganese oxides is known to be an equally important reaction where oxidizing conditions prevail (Berrang and Grill 1974; Pilipchuk and Volkov 1974). In addition, an association of Mo with organic matter in sediments or pore waters has been observed in a number of studies (Jones 1974; Contreras 1978; Brumsack and Gieskes 1983; Calvert and Price 1983; Malcolm 1985). Pedersen (1985) suggested that the interstitial water and solid phase data for the natural sediments represented by core HOL 14 (Fig. 6) are compatible with the diagenetic framework established by Korolev (1958) and Bertine (1972), in which Mo originally coprecipitated with Fe monosulphide is returned to pore solution upon maturation of FeS to FeS_2, only to be subsequently fixed in the sediments as a discrete Mo sulphide phase (presumably $Mo^{VI}S_3$) or by reduction to Mo(V) and adsorption onto organic matter (Szilagyi 1967).

Markedly different distributions characterize both the slowly and rapidly accumulating tailings facies (Fig. 6). Pedersen (1985) noted that the dissolved Mo enrichment in the upper several centimetres of RUP 1 must be due to release from a dissolving oxide phase, probably MoO_3 (molybdite) in the tailings, which is derived from oxidation of MoS_2 during milling of molybdenite ore (Chander and Fuerstenau 1972; Castro 1979); MoO_3 has a high solubility at slightly alkaline or higher pH levels (Killeffer and Linz 1952). The quantitative removal of Mo from solution at depth in the core reflects sulphide precipitation or adsorption onto organic matter; as in the Holberg Inlet sediments, the consuming phase(s) cannot be distinguished. The high concentration of Mo in RUP 3 was similarly ascribed by Pedersen (1985) to be due to MoO_3 dissolution; this conclusion is consistent with the sympathetic correlation between the concentration of solid-phase Mo in the tailings and the pore water profile (Fig. 6). The correlation indicates that the Mo concentrations measured in solution are controlled directly by the available surface area of the dissolving oxide phase, which in turn must be a function of the concentration of MoS_2 present. Because the tailings are accumulating so rapidly at this site, insufficient time has elapsed to allow diffusion to "smooth" the pore water profile and obliterate the direct association with the solid-phase distribution.

Because the concentrations of dissolved Mo in both the tailings-bearing cores are significantly higher than the concentration of the element in seawater (0.10 $\mu mol\ l^{-1}$), the tailings in Rupert Inlet must be supporting a benthic flux of Mo to the overlying inlet waters, in contrast to the clear absence of an evasion of Mo from the natural sediments in upper Holberg Inlet. Pedersen (1985), however, calculated that, given the relatively rapid, tidally-driven flushing rate for the basin, and the high Mo concentration characteristic of seawater, the benthic Mo flux from the tailings will have no measurable impact on the Mo inventory in the local waters.

3.3 Alice Arm

Losher (1985) has considered in detail the nature of diagenesis in the natural sediments and tailings in, respectively, Hastings and Alice Arms, using a suite of cores collected with the light-weight gravity corer described by Pedersen et al. (1985); interstitial water samples were extracted from the cores by centrifugation under N_2 of samples extruded in a nitrogen-filled glove box. In the tailings-free deposits in core HA-2 (Fig. 1), solid-phase Mn and Fe are slightly enriched in the top few centimetres; the dissolved Fe and Mn profiles for the core (Figs. 7 and 8) indicate that both elements are released to solution in the upper two or three decimetres and consumed at greater depths, presumably by precipitation of authigenic phases. These distributions indicate that anoxic conditions occur within the top 30 cm of the core and they are similar in principle to those in upper Holberg Inlet sediments, differing only in degree; in the latter case, anoxia is developed very near the sediment-water interface reflecting the much higher organic carbon concentration at site HOL 14 (typically 5%; Pedersen 1985) compared to 1 to 1.5% in HA-2 (Losher 1985). Core AA-8, which represents the slowly accumulating tailings facies in Alice Arm, consists of about 18 cm of Mo-rich felsic tailings derived from the Kitsault orebody overlying 35 cm of natural sediments, which are

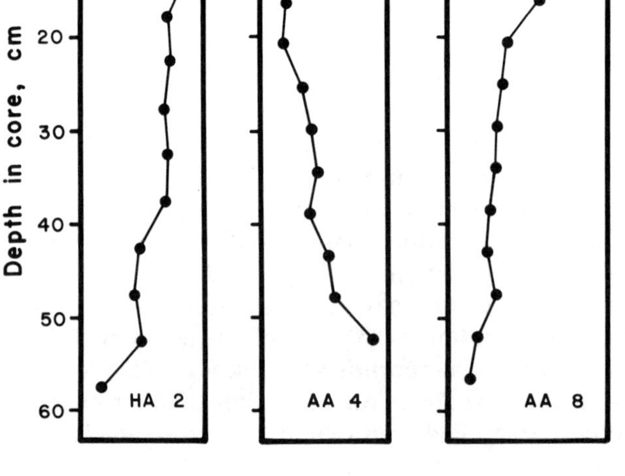

Fig. 7. Dissolved iron in pore waters from the Alice Arm area, measured by graphite furnace AAS. *HA-2* Natural sediments from Hastings Arm; *AA-4* rapidly accumulating tailings from Alice Arm; *AA-8* slowly accumulating tailings from Alice Arm. Note the range of scales on the abscissae

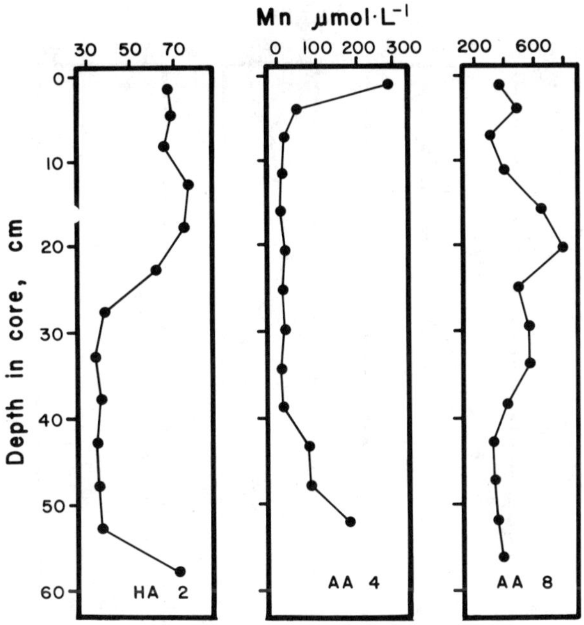

Fig. 8. Dissolved manganese in pore waters from the Alice Arm area, measured by graphite furnace AAS. Facies descriptions as in Fig. 7. Note the range of scales on the abscissae

in turn underlain by several centimetres of barite-bearing, Mo-depleted, riverborne tailings deposited during the mining of a silver deposit nearby during the years 1919 to 1921. As in the Hastings Arm sediments, dissolved Mn and Fe are enriched in pore waters in the upper 30 cm of the core and depleted at greater depths (Figs. 7 and 8), indicating the presence of anoxic conditions below the 30-cm horizon. Core AA-4 consists entirely of rapidly accumulating, modern (Kitsault) tailings; the increasing concentration of dissolved Fe with depth suggests that oxygen is depleted from the pore water below the upper 2 dm at this location. The presence of small but measurable quantities of H_2S below 30-cm depth (Losher 1985) supports this contention. These data when considered in toto indicate quite clearly that both the natural sediments and the tailings deposits in the Alice and Hastings Arm systems become anoxic at depths on the order of 1 to 3 dm.

The distribution of Mo in the pore waters of all three cores is shown in Fig. 9. The most striking aspect of these profiles is the extremely highly dissolved Mo levels measured in the interstitial waters of both tailings cores. To our knowledge, the extraordinary concentrations in core AA-4 (up to 30 μmol l^{-1}) are the highest ever encountered in marine sediments. In contrast, the concentrations in natural sediment (core HA-2) pore waters are more than two orders of magnitude lower. These distributions appear to be controlled by the same diagenetic processes described earlier for the sediments in the Rupert-Holberg Basin. The slight increase over the seawater concentration (0.10 μmol l^{-1}) in the top decimetre of HA-2, followed by the decrease at greater depth indicates release of Mo near the surface, probably from dissolving oxides, and consumption at depth, either by coprecipitation with sulphides or adsorption of Mo(V) onto organic matter. These

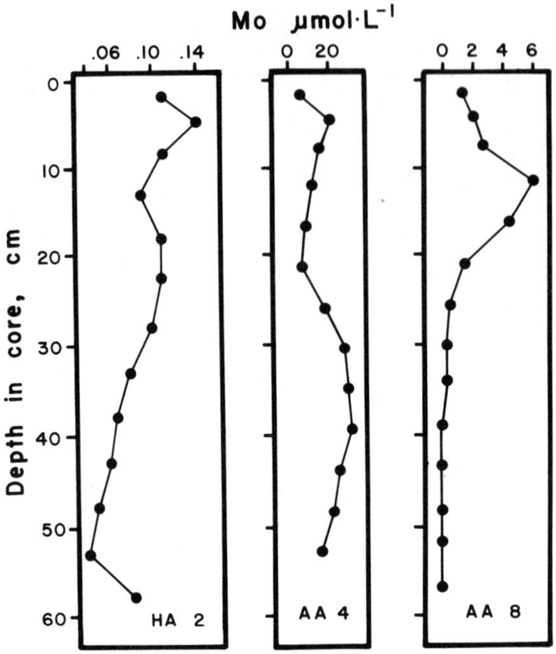

Fig. 9. Dissolved molybdenum in pore waters from the Alice Arm area, measured by direct injection graphite furnace AAS. Facies descriptions as in Fig. 7. Note the range of scales on the abscissae

observations are consistent with the diagenetic framework established by the Mn and Fe profiles.

The primary source of the extremely high concentrations of Mo in the tailings pore waters is very probably dissolution of MoO_3 formed as a surficial oxidation product on molybdenite grains during ore milling; as noted earlier in this chapter, this alteration product is highly soluble in water at pH values >6. In addition, kinetic data from Titley and Anthony (1961) suggest that MoS_2 is not very stable in aqueous solutions and dissolves at seawater pH even under mildly reducing conditions; thus, direct dissolution of molybdenite in the tailings may also contribute to the high dissolved Mo levels.

Further evidence to support this suggestion that the pore water Mo levels in the tailings are related directly to the availability of dissolving solid Mo phases is provided by the comparison between the distributions of both the solid and dissolved Mo shown in Fig. 10 for AA-4, AA-8 and a third tailings-bearing core (AA-1). The latter was collected from the centre of the inlet, immediately to the north of the mouth of Lime Creek (Fig. 1). It is quite clear from this plot that the highest concentrations of Mo in the solid phase correlate with Mo enrichments in pore water. The covariance is particularly striking in AA-1 and in the bottom half of core AA-4. This unique relationship is not only a function of dissolution from tailings particles but also reflects the extremely high sedimentation rate of the tailings. The length of time for diffusion to operate over length scales of a few decimetres is on the order of 10–20 years; because the Alice Arm tailings accumulated recently at a rate an order of magnitude faster, insufficient time has elapsed since deposition to allow diffusion to smooth the dissolved Mo profiles and possibly obscure the solid-phase pore water relationship evident in Fig. 10.

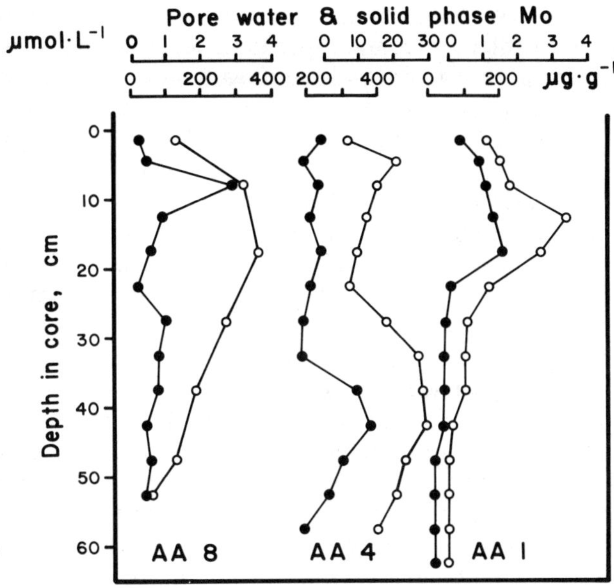

Fig. 10. Solid-phase and dissolved molybdenum in three molybdenum in three molybdenite-bearing tailings cores from Alice Arm. *Open circles* Dissolved Mo; *filled circles* solid-phase Mo. Note the varying scales for the pore water profiles

The enriched Mo levels in tailings pore waters in Alice Arm will certainly support an unnatural flux of the element to the inlet waters. Losher (1985) used a flux estimate based on Fick's First Law and the data shown in Fig. 10 to suggest that the 7 km^2 area of the inlet covered by tailings could contribute about 1500 kg yr^{-1} of Mo to the overlying water. This estimate is very probably an upper limit and amounts to about 4% of the inventory of dissolved Mo in Alice Arm waters. The calculations suggest that the diagenetic release of Mo to the inlet deep waters from deposited tailings is minor but could be measurable.

4 Summary and Overview

Our approach in diagenetic studies of mine tailings deposits is essentially two-pronged: our first concern is to study element distributions in both the solid and dissolved states in order to extract as much information on post-depositional reactivity as possible, and our second concern is to compare the results from the tailings samples with distributions in representative, proximal natural sediments. The three examples discussed in this chapter illustrate the utility of this philosophy.

It is clear that all three of the submerged mine tailings deposits that we have studied are not associated currently with any serious chemical perturbation of the host natural waters. It has been demonstrated that the slightly alkaline tailings in Rupert Inlet inhibit benthic regeneration of phosphate on the inlet floor, probably because the addition of relatively large quantities of lime to the tailings during milling leads to the post-depositional precipitation of an apatite phase. However, because benthic regeneration of PO_4 is not a major contributor to the phosphate budget in the Rupert-Holberg Basin, the consumption of phosphorus by the tailings has no significant effect. Similarly, benthic fluxes of Cu from the tailings in Rupert Inlet are comparable to or less than fluxes measured in natural coastal sediments elsewhere and are substantially less than the benthic flux of copper to oceanic deep waters from pelagic and hemipelagic sediments. However, Mo is being released from the tailings in both Rupert Inlet and, to a much greater degree, in Alice Arm, in contrast to the absence of significant evasion from proximal natural sediments in both areas. In both cases, the fluxes contribute only a minor proportion of the calculated inventory of the element in the fjord waters, which reflects the relatively rapid flushing rate of both inlets and the naturally high concentration of dissolved molybdenum in seawater. The situation in Buttle Lake, prior to the recent cessation of tailings discharge, was anomalous in that the extremely high dissolved metal concentrations in the lake water (derived from bacteriological leaching of a subaerial waste dump near the lake) were supporting diffusive fluxes of trace metals into both the tailings and natural sediments on the lake floor, where the metals were presumably being fixed in authigenic sulphides.

One of the major reasons for the absence of significant remobilization of metals other than Mo from the tailings in the deposits discussed here must be the high rate of deposition associated with high-volume discharge. The rapid burial of tailings particles limits their exposure time to oxygenated bottom water, thus inhibiting oxidation of residual sulphides in the finely-ground mill waste. Pedersen's (1983, 1985) observations of the high metal concentrations in water which had accumulated on the tops of tailings cores exposed to atmospheric oxygen suggest that oxidation of submerged sulphide-bearing tailings deposits may become a problem in the interim period following cessation of tailings discharge, prior to the covering of the metal-rich wastes by a veneer of anoxic natural sediments. We intend to address this problem in future investigations of the diagenetic reactivity of aquatic mine tailings deposits.

Acknowledgements. Our work on mine tailings diagenesis has been supported financially by Fisheries and Oceans Canada, the Natural Sciences and Engineering Research Council of Canada and Westmin Resources Limited; we are grateful for their assistance. The data in Tables 1 and 2 are published with the permission of Pergamon Press Limited; Figs. 2 to 6 are published with the permission of the National Research Council of Canada.

References

Atlas E, Pytkowicz RM (1977) Solubility behaviour of apatites in seawater. Limnol Oceanogr 22:290–300

Berrang PG, Grill EV (1974) The effect of manganese oxide scavenging on molybdenum in Saanich Inlet, British Columbia. Mar Chem 2:125–148

Bertine KK (1972) The deposition of molybdenum in anoxic waters. Mar Chem 1:43–53

Brumsack HJ, Gieskes JM (1983) Interstitial water trace-metal chemistry of laminated sediments from the Gulf of California. Mar Chem 14:89–106

Callender E, Bowser CJ (1980) Manganese and copper geochemistry of interstitial fluids from manganese nodule-rich pelagic sediments of the northeastern equatorial Pacific Ocean. Am J Sci 280:1063–1096

Calvert SE, Price NB (1972) Diffusion and reaction profiles of dissolved manganese in the pore waters of marine sediments. Earth Planet Sci Lett 16:245–249

Calvert SE, Price NB (1983) Geochemistry of Namibian Shelf sediments. In: Suess E, Thiede J (eds) Coastal upwelling. Plenum, New York, London, pp 337–375
Castro S (1979) Flotation of molybdenum. In: Sutulov A (ed) International molybdenum encyclopedia. Intermediate, Santiago, pp 164–181
Chander S, Fuerstenau DW (1972) On the natural floatability of molybdenite. Trans Am Inst Min Eng 252:62–69
Contreras R, Fogg TR, Chastee ND, Gaudette HE, Lyons WB (1978) Molybdenum in pore waters of marine sediments by electron paramagnetic resonance spectroscopy. Mar Chem 6:365–373
Drinkwater KF, Osborn TR (1975) The role of tidal mixing in Rupert and Holberg Inlets, Vancouver Island. Limnol Oceanogr 20:518–528
Elderfield H, Luedtke N, McCaffrey RJ, Bender M (1981) Benthic flux studies in Narragansett Bay. Am J Sci 281:768–787
Emerson S, Widmer G (1978) Early diagenesis in anaerobic lake sediments – II. Thermodynamic and kinetic factors controlling the formation of iron phosphate. Geochim Cosmochim Acta 42:1307–1316
Fisher TR, Carlson PR, Barber RT (1982) Sediment nutrient regeneration in three North Carolina estuaries. Est Coast Shelf Sci 14:101–116
Hay AE (1982) The effects of submarine channels on mine tailing disposal in Rupert Inlet, British Columbia. In: Ellis DV (ed) Marine tailings disposal. Ann Arbor Sci Publ, pp 139–181
Heggie DT (1983) Copper in the Resurrection Fjord, Alaska. Est Coast Shelf Sci 17:613–635
Hoff JT, Thompson JAJ, Wong CS (1982) Heavy metal release from mine tailings into seawater – a laboratory study. Mar Pollut Bull 13:283–286
Holdren GC, Armstrong DE (1980) Factors affecting phosphorus release from intact lake sediment cores. Environ Sci Tech 14:79–87
Jahnke RA, Emerson SR, Roe KK, Burnett WC (1983) The present day formation of apatite in Mexican continental margin sediments. Geochim Cosmochim Acta 47:259–266
Jones GB (1974) Molybdenum in a near-shore and estuarine environment, north Wales. Est Coast Mar Sci 2:185–189
Killefer DH, Linz A (1952) Molybdenum compounds. Wiley Interscience, New York, p 39
Klinkhammer GP (1980) Early diagenesis in sediments from the eastern equatorial Pacific, II. Pore water metal results. Earth Planet Sci Lett 49:81–101
Klinkhammer GP, Heggie DT, Graham DW (1982) Metal diagenesis in oxic marine sediments. Earth Planet Sci Lett 61:211–219
Korolev DF (1958) The role of iron sulphides in the accumulation of molybdenum in sedimentary rocks of the reduced zone. Geochemistry 4:452–463
Losher AJ (1985) The geochemistry of sediments and mine tailings in the Alice Arm area. Thesis, Univ British Columbia, Vancouver
Lynn DC, Bonatti E (1965) Mobility of manganese in diagenesis of deep-sea sediments. Mar Geol 3:457–474
Malcolm SJ (1985) Early diagenesis of molybdenum in estuarine sediments. Mar Chem 16:213–225
Martens CS, Berner RA, Rosenfeld JK (1978) Interstitial water chemistry of anoxic Long Island Sound sediments. 2. Nutrient regeneration and phosphate removal. Limnol Oceanogr 12:605–617
Nixon SW (1981) Remineralization and nutrient cycling in coastal marine ecosystems. In: Neilson BJ, Cronin LE (eds) Estuaries and nutrients. Humana, Clifton, N J, pp 111–138
Pedersen TF (1983) Dissolved heavy metals in a lacustrine mine tailings deposit – Buttle Lake, British Columbia. Mar Poll Bull 14:249–254
Pedersen TF (1984) Interstitial water metabolite chemistry in a marine mine tailings deposit, Rupert Inlet, B.C. Can J Earth Sci 21:1–9
Pedersen TF (1985) Early diagenesis of copper and molybdenum in mine tailings and natural sediments in Rupert and Holberg inlets, British Columbia. Can J Earth Sci 22:1474–1484
Pedersen TF, Price NB (1982) The geochemistry of manganese carbonate in Panama Basin sediments. Geochim Cosmochim Acta 46:59–68
Pedersen TF, Malcolm SJ, Sholkovitz ER (1985) A lightweight gravity corer for undisturbed sampling of soft sediments. Can J Earth Sci 22:133–135
Pedersen TF, Vogel JS, Southon JR (1986) Copper and manganese in hemipelagic sediments at 21° N, East Pacific Rise: diagenetic contrasts. Geochim Cosmochim Acta 2019–2031
Pilipchuk MF, Volkov II (1968) The geochemistry of molybdenum in the Black Sea. Lithol Min Res 4:389–407

Pilipchuk MF, Volkov II (1974) Behaviour of molybdenum in processes of sediment formation and diagenesis in the Black Sea. In: Degens ET, Ross DA (eds) The Black Sea – geology, chemistry and biology. Am Assoc Petrol Geol Mem 20:542–553

Poling GW (1982) Characteristics of mill tailings and their behaviour in marine environments. In: Ellis DV (ed) Marine tailings disposal. Ann Arbor Sci Publ, pp 63–84

Sawlan JJ, Murray JW (1983) Trace metal remobilization in the interstitial waters of red clay and hemipelagic marine sediments. Earth Planet Sci Lett 64:213–230

Sholkovitz ER (1973) Interstitial water chemistry of the Santa Barbara Basin sediments. Geochim Cosmochim Acta 37:2043–2073

Stumm W, Morgan JJ (1970) Aquatic chemistry. John Wiley, London

Szilagyi M (1967) Sorption of molybdenum by humus preparation. Geochem Int 4:1165–1167

Titley SR, Anthony JW (1961) Some preliminary observations on the theoretical geochemistry of molybdenum under supergene conditions. Arizona Geol Soc Digest 4:103–116

Behaviour of Trace Metals in a Tropical River System Affected by Mining

J. JEFFERY[1], N. MARSHMAN[1], and W. SALOMONS[2]

1 Introduction

Modern mining methods have enabled the exploitation of high-volume, low-grade mineral deposits. Mining of these deposits requires safe and efficient disposal of large quantities of mineral wastes that contain low concentrations of extractable minerals.

Disposal of these wastes requires recognition of the environment that surrounds the mine site. At many mine sites topography, net annual precipitation and seismic stability permit the economic containment of mine wastes on land. The disposal system is able to handle all on-site rainfall as well as all liquid and solid wastes. Conditions at some mine sites, however, require that alternative means of waste storage be employed.

At sites where there is no total containment of the wastes careful environmental management is needed to minimize the impact of waste disposal on the environment. This requires an adequate understanding of the environment where the wastes are being disposed and the way in which the wastes behave and could potentially impact the area.

Such an understanding requires knowledge of the physical, chemical and biological elements of the system. It depends upon the collection of data over long periods of time and also on the execution of site-specific research both before and after the disposal system is implemented. Finally, it requires that the information collected on site be integrated with knowledge in the general literature.

This chapter describes part of the geochemistry study programme of the waste disposal system at a mine in the humid tropics. At the mine site net annual precipitation is 3,000 mm yr^{-1}, seismicity is high and topography rugged. Consequently, total containment of rainfall and wastes is not practicable. Particular emphasis is given in this chapter to (1) description of the wastes and the environment into which they are disposed; (2) the chemical processes which take place throughout the disposal system; and (3) the way that the available information has been used to develop a geochemical numerical model. This model has enabled an integrated approach to be taken in the environmental management of the waste disposal system.

[1]Bougainville Copper Limited, Panguna, North Solomons Province, Papua New Guinea
[2]Delft Hydraulics Laboratory, Institute of Soil Fertility, 9750 RA Haren (Gr), The Netherlands

2 The Mine Environment

Bougainville Copper Limited operates a large open pit mine on Bougainville Island, Papua New Guinea. The mine is located at Panguna, 700 m above sea level in the central mountain range of the island in the catchment of the Kawerong/Jaba/Pangara river system (Fig. 1). The river flows into Empress Augusta Bay on the west coast of Bougainville. River slopes range from 7% in the narrow, steep-sided Kawerong Valley close to Panguna to 0.2% in the Jaba close to the coast where the river flows across open swampland.

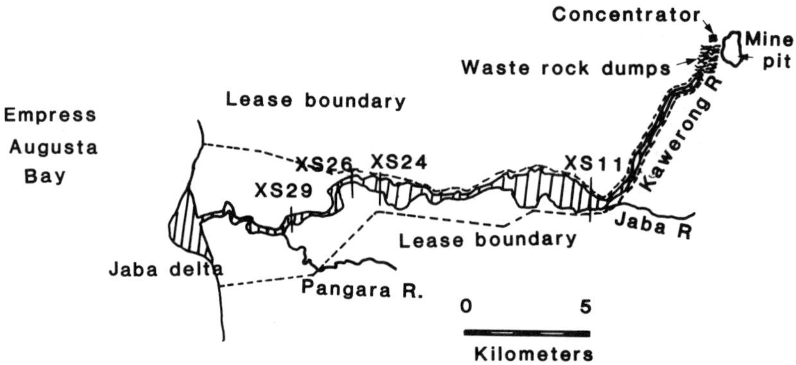

Fig. 1. General layout of the Panguna mine site and waste disposal area. The *hatched areas* represent regions covered with tailings. Four river cross-sections are shown on the diagram

2.1 Mine Waste Disposal at Bougainville

The orebody is a low-grade porphyry copper deposit that contains both some gold and silver. Since the start of mine operations, over 1,000,000,000 t of material have been removed from the pit. Three forms of material are removed from the pit, ore, waste rock and overburden (Table 1).

The ore is taken to the concentrator where it is crushed and ground, lime and flotation chemicals added and the Cu-containing sulphide minerals recovered. Tailings from the concentrator (98.5% of the ore) are discharged via a pipeline directly into the Kawerong River immediately downstream of the toe of the dumps. Sixty percent of the tailings that enter the river system reach Empress Augusta Bay.

The waste rock contains sub-economic concentrations of extractable minerals. It is stacked as stable batters on the sides of the Upper Kawerong Valley adjacent to the pit. An acidic leachate stream emerges from the toe of the dumps and enters directly into the Kawerong River. The overburden was, up until 1981, dumped into the Kawerong River below the dump toe. Currently, it is being placed into the waste dumps.

Table 1. Summary of production statistics and chemical composition of the three forms of mine waste

	Waste type		
	Tailings	Waste rock	Overburden
Total production (tonnes)	500×10^6	280×10^6	160×10^6
Daily production[a] (tonnes)	130,000	90,000	Variable
Surface Area[b] (km^2)	18	2.8	
Cu (mg kg^{-1})	800	1700	[c]
S (g kg^{-1})	3	10–20	[c]

[a]Average for 1985.
[b]Surface area of on-land deposits.
[c]Not regularly assayed.

2.2 Aim of Geochemistry Study Programmes

Overall the mine disposes of large volumes of waste in the headwaters of the Kawerong/Jaba river system. These wastes contain S and trace elements which may enter the environment with potentially adverse impacts. Consequently, the aim of the geochemistry programme is to quantify the major types, locations and forms of acid and trace metals in the disposal area, to ascertain their mobility and if necessary to develop management options to ameliorate adverse effects. In view of the changes that will occur in the geochemistry of the system after river disposal of tailings ceases, it is necessary to examine both during and after mine life conditions. Also, it is necessary to look at both the terrestrial and marine environments. In this chapter only the terrestrial component is discussed.

3 Geochemical and Associated Data Bases

Data bases on different aspects of the waste disposal and river systems have been established since the commencement of mining. These relate to the climate, the hydrology of the catchment, the wastes and the chemistry of streams within the catchment. The information contained in them is fundamental to understanding the current and future behaviour of trace metals. These aspects are described below.

3.1 Composition of Mine Waste

Knowledge of the composition of the three major types of wastes is important to assess the potential for release of trace elements into the environment. Each is briefly discussed after a description of the chemistry of the orebody is given.

The main rock types in the orebody are andesite and granodiorite. Mineralization of Cu (mainly as chalcopyrite) in these rocks occurs along fractures, on joints or is associated with quartz veins. A lower proportion is disseminated within the body of the rock. The chemical composition of ore is presented elsewhere in this book (Chap. 26) and apart from high concentrations of Cu (Table 1), concentrations of most trace metals are low and less than or similar to published average data for the lithosphere (Lindsay, 1979). The Bougainville orebody is free of the more notorious heavy metals (e.g. As, Pb, Hg and Cd). The orebody also contains raised levels of sulphide minerals, mainly pyrite.

The waste rock and tailings are derived from the same rock types as the ore and the mineralogical and chemical composition of the three are similar except for Cu and S (Table 1). Waste rock is fractured rock with a large average particle size (>50 mm) and up to 20% of particles less than 2 mm. As a consequence of this particle size distribution the sulphide minerals located along fracture planes or in the fines are available for rapid oxidation, whereas the sulphides disseminated throughout the waste rock are virtually unreactive. Tailings are composed of sands and silts, and virtually all sulphide minerals in the tailings are available for oxidation.

The overburden material is primarily composed of weathered volcanic ash. It has a sandy to silt consistency and contains low concentrations of Cu and S. Overburden is not routinely chemically analyzed. It is, however, the major parent material of the soils in the Panguna region and is not expected to have unusual chemical properties. Occasional analyses of samples support this conclusion.

3.2 Climate and River Hydrology

An understanding of both the climate and the hydrology of the catchment into which the wastes are disposed is necessary to know how trace elements from the wastes may be leached into the river system and then diluted and transported. The hydrology and climate have been studied as a separate study through a network of stations (Meynink, unpublished data). The climate of Bougainville is typical of the humid tropics and additional information is described in another chapter of this book (Chap 26).

Natural catchment river flow is typified by a comparatively high and slowly changing (i.e. daily to weekly) base flow due to interflow and groundwater discharge with short duration flashy flood peaks superimposed. Overall catchment yield is high (greater than 70%) and regular due to comparatively low evapotranspiration rates and uniform rainfall pattern. Mean discharge rates in the Kawerong-Jaba river system are given in Table 2. There is good consistency between the tributaries of the Jaba River in their discharge per unit catchment surface area.

There are a number of features of the climate and hydrology of the system of relevance to the chemistry of the river system. The high and uniform rainfall and catchment yields means (1) that the waste deposits are rapidly and uniformly (through time) leached with little accumulation within the wastes of soluble forms of heavy metals and acid; (2) that total containment of rainfall and wastes is not practical; (3) that the range of river discharges (i.e. the ratio of peak to low flow discharges) is comparatively small when compared to rivers in regions with more seasonal rainfall; (4) that at any time the ratio of discharges from each of the

Table 2. Mean discharge and composition of the Kawerong/Jaba river system stream

		Stream						
		KR2[a]	KWl[b]	Kawerong XS5	Jaba XS16	Jaba XS35	Jaba[c]	Pangara
Discharge ($m^3 s^{-1}$)		0.75	0.46	5.7	15.2	41.1	7.0	2.5
pH		5.0	4.0	9.2	8.2	8.2	8.0	7.8
Ca	($mg\ l^{-1}$)	200	270	131	46	20	2.1	1.9
Mg	($mg\ l^{-1}$)	150	270	12	4.6	3.6	2	1
Cu	($mg\ l^{-1}$)	70	235	0.004	0.004	0.006	0.002	0.001
Hg	($\mu g\ l^{-1}$)	<0.008		<0.04	<0.04	<0.04	<0.04	<0.04
Cd	($\mu g\ l^{-1}$)	28		<1	<1	<1	<1	<1
Pb	($\mu g\ l^{-1}$)	3		<1	<1	1	<1	1
Zn	($\mu g\ l^{-1}$)	3000		1	1	2	2	2
Al	($mg\ l^{-1}$)	2	25					
Na	($mg\ l^{-1}$)	20	38					
K	($mg\ l^{-1}$)	20	27					
Mn	($mg\ l^{-1}$)	12	32					
Fe	($mg\ l^{-1}$)	0.13	0.2					

[a]Dump leachate stream sampling point immediately upstream of confluence with Kawerong. The data was collected 1983–1985.
[b]Dump leachate stream sampling point at the toe of the dumps (1981–1982).
[c]Immediately above confluence with Kawerong.

tributaries is comparatively small as most rainfall events affect each of the sub-catchments in a similar fashion; (5) that there is high dilution of leachate from waste deposits as it enters the river system.

3.3 Sediment Discharge to the River

Since the commencement of mine construction 600 million tonnes of tailings, overburden and natural catchment erosion have entered the river system. Natural erosion around the mine site is less than 100,000 t per annum (tpa), and is mainly caused by high intensity rainfall.

Sediment transport and bed level aggradation throughout the river system have been monitored since mine start-up through the use of a number of survey lines (cross-sections; XS's) along the river system. These survey points are numbered from XS5 at the confluence of the Kawerong and the dump leachate stream, to XS35 at the coastline. To date, 40% of the material disposed into the river system has deposited on land. The largest areas covered are two depositional basins (Basin 1 and Basin 2; Figs. 1 and 2) of approximately equal surface area in the middle section of the river system.

A numerical sediment transport model has been developed to describe sediment deposition, degradation and transport throughout the river system during and post-mine life (Meynink 1983). The model has assisted our understanding of the geochemistry of the system by quantifying the processes involved in sediment transport. Model predictions indicate that the river will continue to

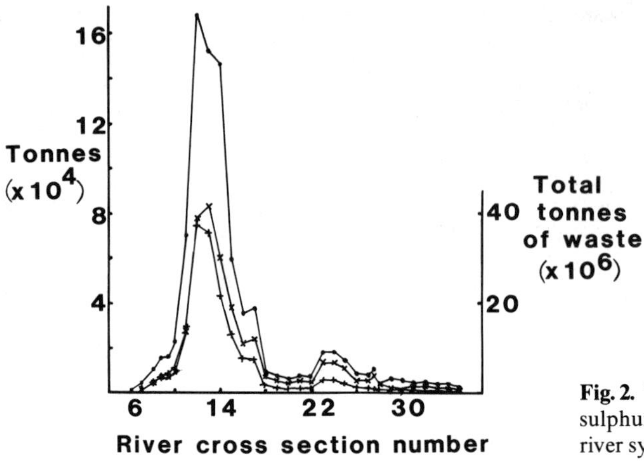

Fig. 2. Total tonnage of mine waste (X), sulphur (•) and copper (+) stored in the river system

transport tailings material well past the end of mine life and thus river water will continue to interact chemically with the deposits.

The tailings deposits in the river system have been hydraulically sorted and they contain a low amount of clay- and silt-sized material. Sulphide and Cu concentrations are higher in tailings deposited in the upper reaches of the river and decrease more or less uniformly throughout the river system from 5 to 2 g kg^{-1} for S and from 1600 mg kg^{-1} to 500 mg kg^{-1} for Cu. The majority of Cu and S present in the river system is immediately downstream of the Jaba/Kawerong confluence (Fig. 2).

3.4 Waste Rock Dump Leachate Stream

Drainage from the waste dumps is not currently treated and flows a short distance from the toe of the dumps before entering the Kawerong River. Before joining the Kawerong it is joined by a number of small side streams.

The flow characteristics of the dump leachate stream are similar to those of streams in natural catchments. Low flows are largely due to groundwater flow from the dump, whereas peak flows include a significant amount of runoff from both natural catchment and dump faces. This runoff has little contact with the waste rock material itself.

Leachate from the waste dump is a major input of acidity, soluble Cu and other metals to the river system. It is primarily an acidic solution containing sulphates of Cu, Ca and Mg (Table 2), and at present 1,500 tpa Cu are released into the river system. Some Mn, Al and K are present but virtually no Fe. With the exception of Cu and occasionally Cd all concentrations of trace elements are low.

The release of acid and Cu from the dumps fluctuates through time due to long-, intermediate- and short-term processes. In the long term the concentration of Cu in the leachate at the toe of the dump progressively increased in the decade after mine start-up to 250 mg l^{-1}. This increase was due to progressive enlargement of the dumps and to an overall increase in the sulphide oxidation rate. Since then concentrations have declined, due to both dilution as the dumps have enlarged and covered additional streams, and to adsorption of Cu by the weathered superficial material which has been placed in the dumps since 1982.

In the intermediate term the actual concentration of Cu leached from the dumps varies little and no correlation is observed over periods of several months between Cu concentration (mg l^{-1}) and stream discharge; Cu yield ($week^{-1}$), however, increases with discharge, i.e. during extended periods of high rainfall a greater percentage of dump volume is wetted and leached.

There are considerable fluctuations in the hourly/daily Cu concentration of the leachate due to rainfall. This is because the flow of water through the dumps is damped, the immediate effect of rainfall is a decrease in concentration (mg l^{-1}) with little change in the total release of Cu (kg h^{-1}).

3.5 Composition of River Waters

Various streams in and around the waste disposal area are sampled and analyzed for major constituents and selected trace elements. This information is important in order to know how these waters will chemically react upon mixing with acidic leachate streams.

For streams unaffected by waste disposal the waters contain low concentrations of both total dissolved solids (Ca, Mg, Na and K) and trace elements (Cu, Zn, Pb and Hg; Table 2). The alkalinity of most tributaries to the Kawerong/Jaba river system is around 0.7 mEq l^{-1}, a level which is capable of neutralizing significant quantities of acid.

The chemical composition of river water affected by mine waste differs little from those of other rivers on Bougainville (Table 2). Soluble Cu concentrations in the Jaba/Kawerong are low. There are two main reasons for this. Firstly, the high sediment load in the river adsorbs dissolved metals and secondly, the tailings stream has a high alkalinity due to lime added in the concentrator. This alkalinity is more than sufficient to neutralize the acid from waste dump leachate and buffers pH between 7.8 to 8.2. This is similar to the pH of other streams where natural bicarbonate alkalinity controls pH.

4 Geochemical Processes Affecting Trace Metals

The data described above has enabled a conceptual model of the waste disposal system to be developed. This model allows a qualitative understanding of the important parameters affecting the system (Fig. 3) and is an important step in the development of a quantitative understanding of the system. The major components of the conceptual model are:

1. Acid and trace metal release from the waste rock dumps and tailings deposits.
2. Alkalinity input to the river from concentrator tailings and natural waters.
3. Sorption of trace metals in the river system onto suspended solids.

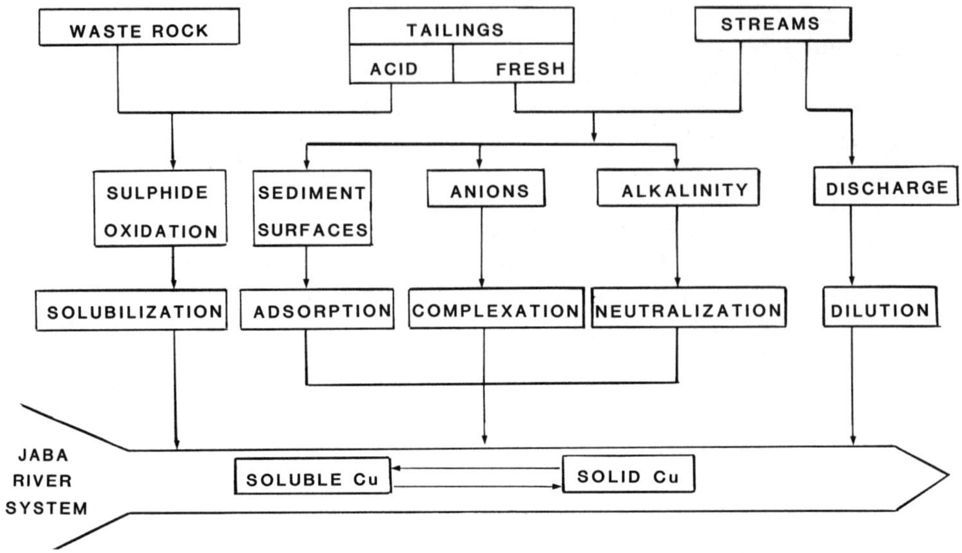

Fig. 3. Conceptual model of the geochemistry of the waste disposal system on Bougainville

4. Trace metal complexation in river water due to organic and inorganic ligands.
5. Temporal variations due to parameters such as flow fluctuations and cessation of dumping of tailings into the river system.

Quantitative data about the major processes highlighted in the conceptual model were required, and a study programme developed. This study was divided into investigations of (1) processes that affect the release of acid and Cu from the wastes and (2) the chemical reactions that take place in the river system itself. Both components are described below.

4.1 Weathering of Mine Wastes

The oxidation of sulphide minerals in mine wastes with the subsequent release of trace metals is an important geochemical process potentially leading to adverse environmental impacts. The overall process has been called *mine waste weathering*. Weathering is divided into three stages; oxidation, neutralization and leaching. Weathering of Bougainville mine wastes has been studied, both in the field and the laboratory.

4.1.1 Sulphide Mineral Oxidation

Aeration is important in sulphide oxidation and sediment acidification and its role has been described elsewhere in this book (Chap. 26). In aerated tailings the pH remains high for the first 2 years due to the lime in the sediment. After this has been depleted pH rapidly falls until there is almost complete oxidation of sulphides. In comparison, the pH of the leachate from the waterlogged tailings remains between 7 and 8. The dissolved oxygen concentrations of the pore water is low due to the low gaseous diffusion in waterlogged sediment. These observations explain why there is currently little oxidation of tailings in the river system. Most of the deposits are waterlogged.

Modelling studies of gaseous diffusion into freely draining tailings deposits indicate that oxygen diffusion also limits sulphide mineral oxidation in sediment more than 500 mm below the surface. Thus, even after the end of mine life when the river runs as an incised channel, oxidation is likely to be confined to the surface layer.

The waste rock dumps are more porous than the tailings deposits and hence gaseous diffusion processes occur much faster. Oxygen is thought to be limiting, however, in the inner parts of the dumps because of the low surface area-to-volume ratio. Despite this, sufficient oxygen does reach the leachate prior to its discharge to enable complete oxidation of Fe^{2+} to Fe^{3+}.

The amount of acid formed in the waste is related to the amount of reactive sulphide present. If one assumes a 2:1 ratio of pyrite to chalcopyrite in Bougainville waste (a value found for tailings) then the overall chemical equation for oxidation can be writen as:

$$2FeS_2 + CuFeS_2 + 47/4\,O_2 + 19/2\,H_2O = Cu^{2+} + 3Fe(OH)_3 + 10H^+ + 6SO_4^{2-}. \quad (1)$$

Thus, for each mole of sulphide sulphur that completely oxidizes, 5/3 mol acid are produced. The total amount of acid that can be produced from the waste is 150–500 mEq kg^{-1} depending upon sulphide content.

The minimum pH during the oxidation phase of both the waste material and leachate water is related to the initial sulphide content of the waste. Sediment pH of acid waste rock (4.0–4.5) is less than that of tailings (4.5–5.0) due to the differences in the initial sulphide content of the material. Also, the tailings in the upper reaches of the river will, when they oxidize, become more acid than those closer to the coast due to their higher sulphide content.

The actual pH of the leachate, however, is also affected by the solubility of various compounds within the wastes. This has been investigated by examination of the solubility constants of a range of Cu and Al compounds. Solution pH is affected by the solubility of Al minerals at values less than 4.5 and by bronchiatite at higher pH values.

An estimate of the sulphide oxidation rate of the waste dumps has been derived from the release of SO_4^{2-} from the dumps. Current release of SO_4^{2-}/S in the dump leachate stream is 24 t day^{-1}. This is equivalent to 3% of the sulphide being added to the dump in fresh waste rock daily. The figure underestimates the actual sulphide oxidation rate because of the formation of hydroxy sulphate precipates.

Field and lysimeter trials indicate that active oxidation occurs in aerated tailings over a 5–8 year period. When the reserves of reactive sulphide are virtually depleted, the pH of the leachate and sediment pH slowly rises. After 10 years the pH of the tailings in lysimeters rose to 5.5. At this stage only 600 mg kg^{-1} S and 400 mg kg^{-1} Cu remained in the waste.

4.1.2 Acid Neutralization

Not all of the acid produced from sulphide oxidation enters the river system. Fresh wastes have considerable capacity to immobilize (or neutralize) sulphide oxidation products. Three different types of neutralization reactions are recognized. They are ion exchange, carbonate mineral dissolution and incongruent dissolution of silicate minerals. They have also been divided into long-term and short-term reactions.

In the ion exchange reactions, acidic cations (e.g. H^+, Cu^{2+} and Al^{3+}) exchange with basic cations (mainly Ca^{2+}) that initially saturate the waste. Exchange reactions are rapid but of limited extent due to the low cation exchange capacity (CEC) of mine waste.

Neutralization of acidic cations by carbonate minerals is also rapid and occurs in tailings as a result of the lime added to the ore in the concentrator. Carbonate compounds are formed in the Ca-saturated alkaline tailings on mixing with air. Some of this lime deposits with the tailings and acts to buffer acid production. This type of reaction does not occur in waste rock due to the absence of added lime.

The actual extent of the above two reactions has been measured in tailings by titration with weak acid and has been called *short-term neutralization capacity*. Short-term neutralization capacity was defined as the amount of acid needed to decrease pH to 4. For river-deposited tailings it was 10 mEq kg^{-1}.

The third type of neutralization reaction involves dissolution of silicate minerals and is slower than short-term neutralization. It has been called *long-term neutralization.* Ultimately it is dissolution of silicate minerals that controls sediment pH. The actual extent of long-term neutralization is difficult both to define and to measure since it is pH- and rate-dependent. Two methods have been used to estimate it. Firstly, samples of tailings have been slowly titrated with metallic solutions (i.e. Fe^{3+}, Al^{3+} and Cu^{2+}) over 4 to 5 weeks. Average acid consumption was 100 mEq kg^{-1}, with a final pH of approximately 4. Secondly, a leaching trial was established in which fresh tailings were leached with waste dump leachate. Between 70–80 mEq kg^{-1} of acid was consumed by the tailings before the effluent pH fell to 4.5. This represents between 15 to 70% of the possible acid production from Bougainville mine wastes. It can therefore be concluded for tailings deposits that the acid generated in the surface 0.5 m would be neutralized if it passed through an additional 0.5–1.0 m of alkaline tailings.

The importance of the above three reactions can also be seen by examining the composition of waste dump leachate. The presence of Ca, Mg, K, Na and Al in dump leachate indicates that there is significant attack of gangue (feldspar) minerals. Currently, the leachate contains 25 mEq^{-1} of basic cations (mainly Ca and Mg) compared with only 5 mEq^{-1} of acidic cations (mostly Cu and Al). Therefore, neutralization reactions are important in reducing the release of acid and Cu into the environment.

4.1.3 Solute Leaching

The third stage in tailings weathering is the leaching of solutes from the waste deposits profile into the river system. Solute movement is related to both water flow and to the chemistry of the material through which the water is moving.

The water table of the waste dumps is close to the original ground surface, consequently water entering the dumps moves vertically until it reaches a layer with low permeability. At this point the water moves sideways until the vertical permeability increases. Regions with low vertical conductivities exist in the waste rock mainly because of compaction zones underneath road surfaces. As a consequence of these regions up to half of the material in the waste rock is poorly leached.

Modelling of flow in tailings deposits has indicated that after disposal of tailings to the river ceases there will be significant leaching of solutes from the banks of incised channels. These banks will be freely draining and will be 300 m wide in the upper part of Basin 1 and less than 100 m wide in the lower portion of Basin 2. Away from the river channel the water table will be close to the surface of the tailings deposit. Groundwater movement in these saturated zones will be low, decreasing approximately with the square of the distance from the channel.

The solutes in the groundwater interact with the mine waste through which the water flows. A range of reactions occur. The most important of which have been described (Sect 4.1.2). These reactions are most marked where there is a pH gradient in the waste material.

4.2 River Processes

After the leachate reaches the river system the dissolved trace metals and acid undergo a number of reactions that affect their speciation. Four main processes are recognized. They are dilution, precipitation, carbon dioxide equilibration and adsorption.

Dilution of trace metals occurs throughout the river system. This dilution is important in reducing the initially high concentrations of cations in leachate. Dilution alone, however, is limited in its ability to decrease cation concentrations especially in cases where the solid and aqueous phases are in equilibrium. In such cases the phases re-equilibrate after dilution so that the decline in metal concentration is less than that of the dilution and the total quantity of metal in solution increases.

Precipitation of compounds can occur when streams of different composition mix and the resultant concentration quotient exceeds the solubility product for that compound. The precipitation of Cu from acid leachate solution upon mixing with natural river water has been studied (Fig. 4). A precipitate that contains Cu, S and Al and traces of Ca, Mg and Si forms. It is amorphous by X-ray diffraction, but its solubility characteristics are similar to those of hydrated bronchiatite ($Cu_4(OH)_6SO_4$ 1.3 H_2O) with a log K_{sp} of –66.4. The precipitate has a minimum Cu solubility of 0.2–0.3 mg l^{-1} which occurs when the pH is raised to 7.5.

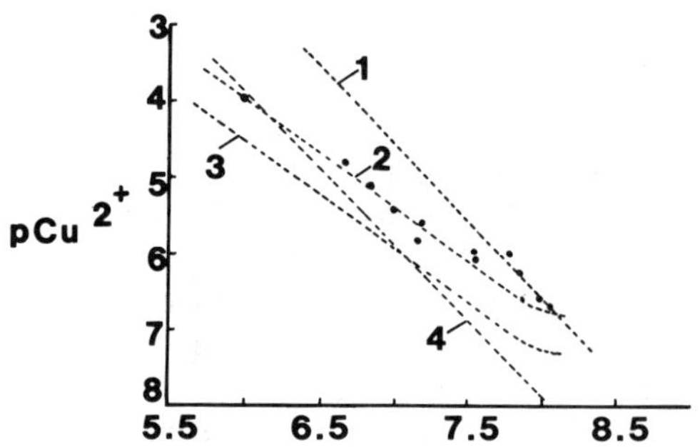

Fig. 4

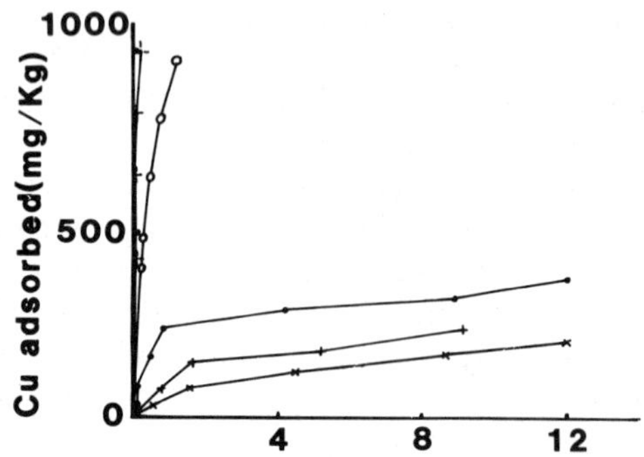

Fig. 5

Fig. 4. Relationship between solution pH and the concentration (-log molar) of cupric ions in leachate-river water mixes. *Dots* represent actual measurements made. *Dashed lines* are the theoretical solubilities of *1* $Cu(OH)_2$; *2* CuO; *3* $Cu_4(OH)_4SO_4$; *4* $Cu_4(OH)_6SO_4$ 1.3 H_2O

Fig. 5. Adsorption isotherms of Cu onto tailings at pH 4.5 (X), 5.0 (+), 6.0(•) 7.4 (O) and 7.8 (- -)

Two environmentally important conclusions can be drawn from these precipitation studies. The first is that the majority of the Cu released into the river system will precipitate downstream of the Jaba/Kawerong confluence, even after cessation of disposal of tailings into the river. The combined alkalinity of the natural tributaries at the confluence is similar in magnitude to the acidity of the waste dump leachate stream. Secondly, there is a chemical barrier below which soluble Cu concentrations cannot fall as a result of precipitation reactions alone. Consequently, dissolved Cu concentrations cannot be reduced, by management procedures, below this level by pH control alone.

Carbon dioxide equilibration between the atmosphere and the river water is an important process affecting solution pH and thus Cu solubility. Equilibration of aqueous and atmospheric CO_2 is slow. When leachate is mixed with natural river water, solution pH is lower and Cu content higher than is predicted by equilibrium reactions. With time CO_2 is driven off and there is a gradual rise in solution pH.

Data about this equilibration effect was collected during a 2-week period when mining operations ceased. During this period, precipitation reactions controlled Cu chemistry in the Kawerong River. Alkalinity and pH measurements were made at selected sites and these were used to calculate pCO_2 values. For water in the Kawerong River just downstream of the dump leachate discharge point, pCO_2 was -2.2. Further downstream pCO_2 decreased slightly although even just above the Jaba confluence it was still five times greater than for air. Below the confluence of the Kawerong and Jaba Rivers, pCO_2 was between -3.1 and -3.4, the shallow and braided river morphology aided gaseous exchange.

Adsorption reactions are important in reducing soluble Cu levels below those possible by precipitation alone. Adsorption of Cu onto tailings has been studied in the pH range 4.5–7.8 (Fig. 5). Adsorption showed Langmuir behaviour and increased with pH. When the pH was less than 6.0 the majority of adsorption occurred whilst dissolved Cu concentrations were undersaturated with respect to bronchiatite. Also most of the Cu was adsorbed as hydrated species of $CuOH^+$ as maximum adsorption was greater than the CEC of the tailings.

At pH 7.8 the shape of the adsorption isotherm was quite different and Cu continued to adsorb to the tailings until the solution Cu concentration was 0.2 mg l^{-1}. At this concentration Cu would have begun to precipitate.

5 Modelling of Copper in the River System

Research and monitoring programmes described above have provided information on many aspects of trace metal behaviour in both mine wastes and the river system. They have enabled quantitative understanding of the processes defined in the conceptual model (Sect. 4) of the waste disposal system. The information covers many different types of chemical reactions that occur over a number of locations. Therefore, a numerical model, RIVCHEM, was developed in order to integrate the information and to assess the most important parameters affecting river Cu chemistry both during and post-mine life.

5.1 Model Description

RIVCHEM quantifies pH, Cu, and Al throughout the river system. It has a hydrological component and a chemical component. The hydrological component divides the river system into a number of reaches, with provisions within each reach for the input of water from the upstream reach (or dump leachate in the case of the first reach), the input of leachate from oxidizing tailings deposits and the addition of waters from tributaries. The chemistry component combines the compositions of all the waters mixing at each reach and then considers the equilibrium speciation of the Ca, Mg, Al and Cu as they are affected by complexation, precipitation, adsorption both onto organic matter and particulate material and the release of excess CO_2 to the atmosphere. It also has a sub-component that derives the composition of leachate from the tailings deposits. The model takes into account only the main elements that influence Cu and Al speciation either directly (e.g. via complexation or pH) or indirectly (i.e. via ionic strength).

The input of Cu from oxidized tailings is based on the sulphide content and the acid-buffering capacity of tailings. The dimensions of the oxidation zone in each reach of the river are specified and it is assumed that each gram of sulphide sulphur in the zone oxidizes to give 60 mEq of acidity. A designated amount of this acidity is buffered by gangue minerals and is released as Ca. The residual acidity is released as Cu (up to the designated amount of leachable Cu in the tailings) and the remainder as Al.

Appropriate reaction constants for complex-ion formation were taken from the literature. Solubility products of amorphous precipitates were derived by experimentation and used in preference to published solubility products for well-crystallized minerals. Good agreement was found between experimentally derived data and calculated Cu and Al speciation values where it was assumed that Al precipitated as gibbsite or alunite and Cu as bronchiatite, tenorite and malachite.

5.2 Model Validity

Various factors affect the accuracy of model predictions. They can be divided into a number of categories, i.e. model realism, chemical logic and input variable accuracy. It is necessary prior to running the model to understand each of these parameters and the overall accuracy of the model.

Any numerical model needs to accurately portray the major processes occurring in the system whilst retaining an overall simplicity. The balance between the two has been termed physical realism. In the case of RIVCHEM it is felt that the study programme prior to model development enabled inclusion of the major processes that occur in the river system. The use of a chemically based equilibrium model for the speciation component of the model appears to be valid. The main areas where simplification has occurred is in the input of minor tributaries to the river and the empirical manner in which tailings leachate is calculated. In the former case sensitivity analysis indicates that there is little effect on the predictions, whereas in the latter case our general understanding of the processes preclude a more fundamental algorithm.

The chemical logic of the model has been compared with a series of predicted speciation results from the more sophisticated CHARON chemical model (Delft Hydraulics Laboratory), and also with a series of laboratory mixing experiments involving dump leachate and river water (Fig. 6). In both cases the agreement was good. The exclusion of minor elements from RIVCHEM did not affect model predictions and the chemical constants used were appropriate.

Selection of appropriate input variables significantly affects model predictions. Two forms of variability in model inputs are recognized. Firstly, some variables have not been properly quantified. These include the actual CO_2 equilibration conditions, the load of sediment in the river system post-river disposal of tailings and empirical relationships associated with tailings weathering.

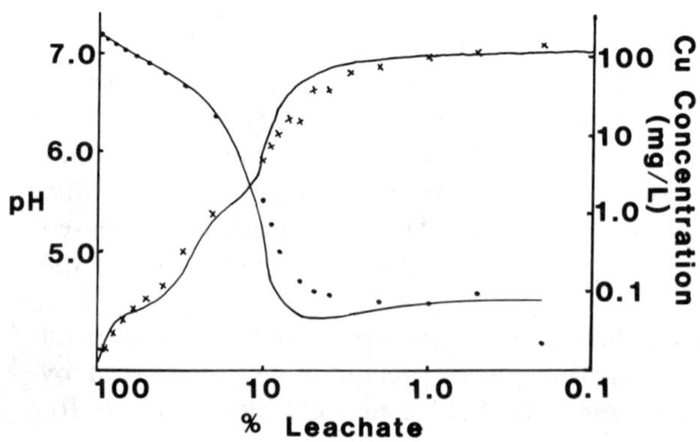

Fig. 6. Observed and predicted relationships between soluble Cu concentration (•), pH (X) and the proportion of leachate in a dump leachate/Jaba River water mix. Experimental data are shown by individual *points* and predicted results by the *solid lines*

Table 3. Comparison of model-predicted results for copper (mg l^{-1}) and pH in the river with field measurements taken during a shutdown of mining and milling operations

Site	River discharge ($m^3 s^{-1}$)	Model predictions		Field measurements		
		Cu	pH	Cu	pH	CO_2^a
Dump leachate	0.65	–	–	70	5.1	–
XS-5	2.95	6.0	6.30	5.1	6.5	-2.37
XS-6	3.12	3.8	6.44	2.8	6.7	-2.52
XS-7	3.62	1.5	6.74	1.8	7.0	-2.66
XS-9	4.09	0.81	6.97	0.88	7.1	-2.76
XS-11	4.12	0.30	7.37	0.11	7.6	-3.17
XS-18	16.92	0.29	7.85	0.04	8.0	-3.20
XS-27	17.12	0.23	8.03	0.03	8.2	-3.38

[a] Log (partial pressure CO_2); value in model set at field measurement; – not determined.

In these cases realistic bounds are set and those that markedly affect model predictions are currently being further investigated. The second form of variation is due to actual variation in the parameters themselves due to climatic, temporal or operational conditions. Inputs affected by such variability include river discharges and the future composition of the waste dumps. A range of values are routinely used in model runs to assess the sensitivity of Cu levels to variations in these parameters.

To date only a limited amount of field data is available against which the model can be validated. These data were collected during normal operating conditions and during a 2-week period when the mine stopped production. When the latter comparison was carried out the surface sorption component of the model was not included but the model predictions for the Kawerong River were in good agreement with field measurements (Table 3). However, the model overestimated the concentrations of soluble copper in the Jaba River where sediment load was much greater and adsorption reactions important.

5.3 Model Uses and Implications

The model has been extensively used as part of the Company's environmental management programme. It has been used to assess the factors which are most important in controlling Cu solubility in the river system and to examine both the effect of possible changes in the current tailings disposal method and the effect of various rehabilitation schemes on river chemistry and thus biology.

Partial pressure of CO_2, the load of suspended sediment, and the sulphide oxidation rate and content of the wastes are important parameters affecting Cu solubility. As in most chemical systems many of the parameters interact in a non-linear fashion and often threshold values exist above which small changes in parameter values markedly affect Cu concentrations. In this regard modelling has greatly enhanced our understanding of the system.

Model predictions of after mine life conditions indicate that Cu concentrations in the Kawerong Valley will decline downstream of the waste dump leachate stream but remain above 1 mg l^{-1}. Upon mixing with the Jaba the Cu concentrations will drop to 0.2–0.3 mg l^{-1} by precipitation reactions alone as long as the oxidation zone of the tailings is reasonably well confined. This can be achieved by ensuring that the sediment in the tailings basins is stabilized so that there is not extensive gully erosion of the deposits. Sorption reactions as a result of erosion of the river bed further decrease soluble Cu levels downstream of the Jaba/Kawerong confluence, the decrease being dependent upon sediment loads in the river.

The predictions also indicate that the river needs to be examined as a system and not as a series of discrete sections. For example, treatment of the waste dump leachate stream has little effect on Cu chemistry if the tailings deposits are not adequately stabilized.

6 Conclusions

Since mine start-up an extensive data collection and a research programme have been carried out on the waste disposal system of Bougainville Copper Limited. The system operates in a high rainfall, high seismicity and mountainous zone. Information is available on the inputs to the disposal area and also on the major chemical processes that affect it. This information has been combined with hydrological and published chemical data to develop first a conceptual and then a numerical model of the chemistry of the terrestrial component of the system. The models have assisted our understanding of the system as a whole, in determining where further investigations are necessary and in the effect of environmental management options on the chemistry of the system.

Model development is an ongoing component of the environmental chemistry programme. The model and our understanding of the important processes need to be updated as the data bases and general appreciation of environmental chemistry increase. Currently, there is an active liason with biologists studying the system to ensure that predictions made about the ecology of the system are as accurate as possible.

Acknowledgements. All chemical analyses during the study programme were carried out by the Analytical Services Department of Bougainville Copper Limited. The assistance and advice received from Mr. J. Van der Linden and Mr. F. Grieshaber in regard to chemical analyses and from Mr. W. Meynink in providing hydrological information and advice about numerical modelling is gratefully acknowledged.

References

Lindsay WL, (1979) Chemical equilibrium in soils. John Wiley & Sons New York

Meynick WJCM (1983) Sediment transport and hydraulic sorting in the Jaba River. DB Simons, Symp Erosion and sedimentation, Colorado State Univ

Heavy Metal Transport in Streams–Field Release Experiments

B.M. CHAPMAN, D.R. JONES, and R.F. JUNG[1]

1 Introduction

1.1 General

The disposal on land of solid wastes from either mining or dredging operations may result in contamination of the surrounding aquatic environment. Percolation of rainwater through dumps can leach out undesirable inorganic solutes (such as heavy metals), harmful organic compounds or generate high levels of particulates which give rise to turbidity. In particular, activity of bacterial species from the genus *Thiobacillus* may result in the oxidation of any pyrite present and lead to the production of sulphuric acid (Lundgren et al. 1972) which not only poses a problem in its own right but can also greatly accelerate the mobilisation of heavy metals. The leachate containing these compounds may drain into nearby watercourses, damaging them ecologically and greatly reducing their value as a resource. Once in the streams the various compounds can be transported (and their undesirable effects persist) for considerable distances downstream.

A thorough analysis of this potential contamination problem involves separate investigations of its two main aspects. Firstly, leaching studies on samples of the solid waste materials should be carried out to determine the nature of the compounds likely to be present in any effluent discharged. Secondly, the transport of the various chemical components of this effluent in the stream should be assessed either directly or indirectly.

In some cases effluents leached from dumps may drain continuously into local streams in an uncontrolled way. In other cases, however, it may be possible to impound the effluent and perhaps also treat it. Even under the latter circumstances accidental or intentional releases to the surroundings can still occur. In the event of such a discharge it is important to know how far any toxic chemical components will be transported before their concentrations are attenuated, by natural purification mechanisms, to below acceptable levels and to know their ultimate fates in the environment. Unfortunately, there are at present insufficient experimental data which would allow a comprehensive understanding of the factors which control the mobility and fate of metals in streams. An a priori quantitative assessment of the likely chemical and biological impact on the aquatic ecosystem of an accidental or planned release is therefore not possible.

---

[1]CSIRO Divison of Fossil Fuels, Hermitage Site, 338 Blaxland Road, Ryde, NSW, 2112, Australia

There are several physical and chemical processes operating in a stream which can give rise, directly or indirectly, to attenuation of pollutants (Chapman 1982, Chapman and Jones 1983). The four main physical processes are advection, dilution, dispersion and sedimentation. Chemical processes include solution reactions, precipitation, coprecipitation and adsorption onto streambed sediments or suspended particulates. Of particular interest are those non-conservative processes which, either alone or together with other processes, can actively remove components from the flowing waters. Such a process is bed adsorption. Similarly, components can be removed from the flowing waters when sedimentation is coupled with precipitation, coprecipitation or adsorption on suspended particulates. It is essential to determine, for any specific industrial site, the relative importance of each of the above processes in the transport of specific leachate components if an accurate determination is to be made of the natural capacity of the local watercourses to purify the effluent prior to discharge from the site.

In order to make a reliable assessment of the likely subsequent behaviour of the metals which might be discharged from solid waste dumps, it is at present necessary to obtain information directly by conducting experiments in which small quantities of these metals are intentionally injected into streams. However, the possibilities for such controlled experimentation have been greatly limited by legal constraints which discourage the intentional release of toxic chemicals into streams. In an attempt to find solutions to this dilemma, a number of different approaches have been followed which avoid such experiments. It is possible, for example, to release into streams relatively innocuous elements such as Sr or K (Zand et al. 1976; Bencala et al. 1983). The likely behaviour of the toxic metal of real interest (e.g. Cd or Cu) is then inferred from the results. Another approach has been to construct artificial flumes with associated ecosystems (Gloyna et al. 1966; Yousef and Gloyna 1977) which are then subjected to the release of heavy metals. Useful results can be obtained, but such equipment is very expensive, limited in size and may not always accurately represent the real system. A third way to obtain data is to study actual pollution events. For example, Neely et al. (1976) studied chloroform contamination of the Mississippi River following the accidental rupturing of a storage tank. Unfortunately, the location and timing of such spills are totally unpredictable and so the changes occurring in the river or stream immediately after the event are usually very difficult to investigate. The detailed study of already chronically polluted streams can also provide considerable information about processes which control the attenuation of pollutants. Investigation of the aquatic chemistry and hydrology of acid mine drainage streams (Boyles et al. 1974; Chapman et al. 1983) provides a good example of this type of approach.

A limited number of studies have been carried out which involve intentional injection of toxic chemicals into freshwater systems. Hall et al. (1980) artificially acidified a stream for a period of 5 months to a pH of about 4. In the stream water they found elevated levels of Al, Ca, Mg and K (and possibly also Mn, Fe and Cd). Hesslein et al. (1980) have added metal radiotracers to a whole lake and monitored the loss from the water column. Airey et al. (1983) released ^{65}Zn and ^{54}Mn into Magela Creek in northern Australia and observed very rapid adsorption onto the sediments and vegetation in the floodplain. More recently Kuwabara et al. (1984)

have injected a mixture of $CuSO_4$ and NaCl into a stream in the form of a square wave pulse of 9-h duration and monitored the response at two stations downstream. Physico-chemical transport modelling indicated that biological and sediment uptake played an important role in attenuating the metal. Kennedy et al. (1984–1985) and Jackman et al. (1984–1985) have studied the transport of and concentration controls on strontium, potassium and lead injected into a small mountain stream in California. It was shown that considerable underflow of water took place through the gravel beds, giving a false impression of groundwater infiltration at some points. Lead was strongly adsorbed by the sediments.

Because the above studies are few in number and relatively restricted in scope there is still a paucity of data and understanding of metal transport in previously uncontaminated streams. In addition, some of the results obtained thus far may be site-specific and hence it may not be possible, a priori, to estimate the likely effects of a metal release into a stream at a different location. In order to fill these substantial gaps in our knowledge, it is necessary to identify the key hydrodynamic and chemical variables controlling metal transport in streams. The fulfillment of this aim requires extensive investigation of the transport of a number of metals under different hydrodynamic and chemical conditions. This chapter presents an overview of such a study which, of necessity, has been much more comprehensive than any which have previously been attempted in a natural stream channel. A much more detailed account of these investigations, including physico-chemical modelling, is to be published in a forthcoming series of papers.

1.2 The Rundle Project

The work to be described in this chapter was carried out as part of a study of the likely effects on the local aquatic environment if an oil shale industry were established near Gladstone in Queensland, Australia (Chapman et al. 1984a). The Rundle deposit has reserves estimated to be greater than 2 billion bbl of fuel oil equivalent and covers a total area of 25 km^2 (Graham 1983). It has been suggested that a final production target of 100–120000 bbl day^{-1} would involved the removal, processing and disposal in surface dumps, of very large quantities of material ($\sim 10^6$ t day^{-1}). Particularly if conditions in the dump favour the establishment of colonies of metal-leaching bacteria, the effluents could contain substantial quantities of acids and heavy metals (as well as other dissolved and particulate organic and inorganic constituents). This might represent a potential source of surface water and groundwater contamination affecting local freshwater streams and ultimately the nearby estuarine and marine environments.

In the first part of this study both batch and column leaching experiments were carried out to determine the identities and possible concentrations of metal contaminants likely to be discharged in the dump effluents (Jones and Chapman 1983; Chapman et al. 1984a). Distilled water was used as an irrigant to simulate rainwater and dilute sulphuric acid solution used to approximate the pH conditions which might occur if colonies of *Thiobacillus* became established. A number of elements were identified as being appropriate for further work in the second part of the study which involved releasing small quantities of dissolved inorganic salts of these elements into a stream on the Rundle site to determine their relative mobilities (Chapman et al. 1984a,b). It is this aspect of the project which will be described in detail below.

1.3 Introduction to Field Release Experiments

The batch and column leaching studies referred to above had identified Cu, Zn, Ni, As and sulphuric acid as being possible contaminants in Rundle dump leachate (Chapman et al. 1984a). The series of experiments to be described here was designed to assess the mobility of simple ionic forms of these elements in a natural stream channel on the Rundle site. In addition to the metal salts and acid, it was necessary to release water tracers to determine the basic hydrodynamic characteristics of the creek. The study provided a convenient opportunity to compare the behaviour of three different water tracers, rhodamine WT, Li and Br.

The metals were injected into the flowing waters as instantaneous pulses to allow dynamic behaviour to be observed. In order to obtain maximum information, densely-spaced sampling in space and time, together with continuous pH monitoring of the stream and detailed chemical analysis of the water samples, was carried out. Stream pH was of special interest because it is recognised as one of the most critical parameters (Leckie et al. 1979; Farrah et al. 1980) controlling adsorption reactions. Information was also sought on the effects of sediment chemistry, and water chemistry, and especially aqueous speciation of chemical entities including solution/particulate partitioning. In addition, it was important to establish whether any ion exchange might be occurring as a result of sorption reactions.

Munduran Creek, the only perennial stream on the Rundle prospect, was not used because its considerable size would have required the release of unacceptably large quantities of metal to achieve measurable levels. Since flow rate was unpredictable and variable in the other streams on the site, it was found to be impractical to conduct these releases under natural flow conditions. Instead, a stream was created by pumping water from Munduran Creek to induce artificial flow in the normally dry bed of one of its ephemeral tributaries. Because this system allowed flow rate to be adjusted over a wide range and to be held constant for long periods, the effect of flow rate as a variable on metal transport could also be studied.

When a pulse of dissolved metal ions travels downstream and is diluted as a result of dispersion, an opportunity is provided to study the effects of concentration which may vary over several orders of magnitude. The various metals were expected to behave differently and it was desired to investigate any competitive effects between dissimilar metals released as a mixture. It was also important to ascertain whether the response of the stream altered as a result of repeated releases over the same reach, since such a change would have implications for the ultimate metal-buffering capacity of the sediments. Finally, it was desired to establish the role of the key hydrodynamic variables of flow rate, hydraulic radius (defined below), velocity and time of travel, in influencing metal attenuation.

2 Experimentation

2.1 Field Site and Stream

Figure 1 shows a schematic plan of the site. A diesel centrifugal pump with a maximum capacity of 750 l min^{-1} delivered water from Munduran Creek through 500 m of 100 mm diameter collapsible plastic tubing to a calibrated V-notch weir box. This box enabled flow to be accurately monitored and controlled. Water from the box overflowed onto an energy dissipater consisting of a sheet of corrugated fibreglass on the creek bed, supporting a 50-mm layer of 17-mm crushed basalt. Instantaneous releases of chemicals were made at R, 21.5 m downstream of V, the position of the V-notch, and monitored at six sampling stations designated A-F further downstream within the 565-m study reach. The watercourse joined Munduran Creek about 1 km below station F. Table 1 lists important creek dimensions including curvilinear distances and heights between the marker points. The change in height with distance downstream indicates that the slope decreases along the study reach. Three flow rates were employed, viz. low, medium and high, with values of 200, 400 and 700 l min^{-1} respectively. Travel times over the study reach varied between 3.5 h for high flow and 11 h for low flow.

There were two main concerns prior to the tests. Firstly, it was feared that flow might be lost through subterranean seepage from the creek bed and banks and accordingly the creek was allowed several days of flow to become "saturated" before the first experiments were commenced. In order to check for this possibility, flow rates at the beginning and end of the study reach were measured by instantaneously releasing dye at station F and shortly thereafter at R. Integration of the areas under the resulting concentration-time curves allowed the two flow rates to be calculated and these were shown to be equal to within experimental error (~5%) indicating insignificant loss of flow along the reach. Secondly, because consecutive

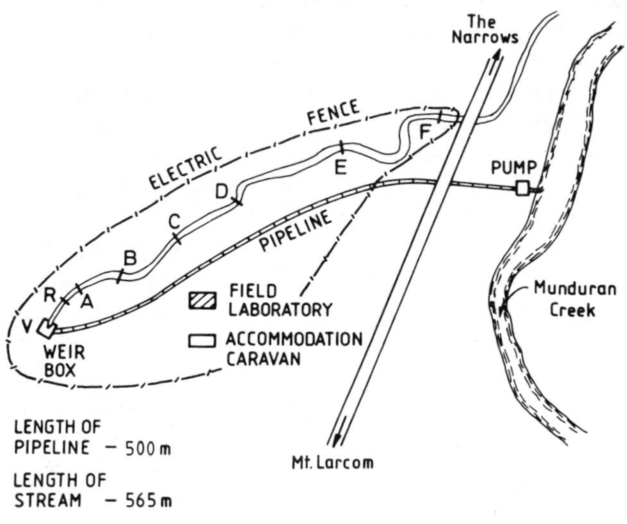

Fig. 1. Schematic plan of Rundle field site

Table 1. Creek length and levels for stations

Station	Cumulative D/S distance (m)	Incremental distance (m)	Cumulative height (m)	Incremental height (m)
V	-21.5		0.38	
		21.5		-0.38
R	0		0	
		35.7		-0.37
A	35.7		-0.37	
		44.9		-0.54
B	80.6		-0.91	
		53.8		-0.38
C	134.4		-1.29	
		125.0		-0.95
D	259.4		-2.24	
		124.9		-1.00
E	384.3		-3.24	
		181.1		-1.05
F	565.4		-4.29	

Note: V is 10.2 m above water level in Munduran Creek.

releases of metals would be made into the same creek, there was a possibility that the response of the creek to later releases might be altered as a result of adsorbed metal remaining from prior releases. Therefore, at least 24-h clean flow was allowed between each release. For the releases involving metals only, all evidence suggested that prior releases had an inconsequential effect on subsequent releases, although it was found that background levels of the elements released in the previous experiment were sometimes slightly elevated for the subsequent release. In contrast, however, it will be seen that during the final experiments when acid was released, significant levels of Cu and Zn were scavenged from the sediment. This Cu, and perhaps also the Zn, were probably incorporated into the sediments during prior releases.

During the series of tests there was no obvious biological activity, such as coatings of filamentous algae developing on sediments, which could have materially affected metal transport. For this reason, and in the absence of any other evidence to the contrary, all results have been interpreted in terms of inorganic physico-chemical processes.

Finally, in order to ascertain whether metal concentrations in the water downstream from the study reach remained below acceptable levels, regular samples were taken in the stream before it joined Munduran Creek. It was found that all of the added metals remained well below the World Health Organisation's recommended limits for drinking water (WHO 1971) throughout the test program.

2.2 The Releases

The aim of this work was to study the relative mobility (attenuation and retardation) of Cu^{2+}, Zn^{2+}, Ni^{2+}, AsO_4^{3-} and H^+ released individually or as mixtures into the creek. However, as a prerequisite to the metal releases, it was essential to measure the hydrodynamic parameters of the creek. This was done by conducting a series of experiments with non-reactive tracers, at the three flow rates, to determine times of travel and dispersion coefficients. From these fundamental parameters, values of velocity, cross-sectional area and hydraulic radius were derived. By comparing the results for the non-reactive tracers with those for reactive metal ions, attenuation and retardation factors for the reactive entities could be calculated. Three comparative tracers where chosen for this experiment: rhodamine, and Li and Br as LiBr. Table 2 briefly summarises the experiments carried out. Table 3 shows a typical analysis for Munduran Creek water, the composition of which changed only slightly over the duration of the experimental series.

The metals Cu, Zn and Ni were released as their sulphate salts, As in the form of Na_2HAsO_4 and H^+ as sulphuric acid. All chemicals were injected as effectively instantaneous pulses. The absolute quantities of each chemical released were proportional to the nominal flow rate. Cu, Zn, Ni and As were injected as individual components at the nominal rate of 10^{-3} mol $(1\ min^{-1})^{-1}$ flow rate, i.e. 10^{-3} mol min 1^{-1}, Li and Br at 2.5×10^{-3} mol min 1^{-1} and rhodamine at 0.05 ml stock solution per 1 min^{-1}. Acid experiments at high levels and low levels required 60×10^{-3} and 16×10^{-3} mol min 1^{-1} respectively. However, when Cu, Zn and Ni were injected

Table 2. Release experiments

Release	Substance	Amount released for each 1 min^{-1} of flow rate	Flow conditions (1 min^{-1})
TRACERS			
Rhodamine WT	0.84 M Rhodamine solution	0.05 ml	200, 400, 700
Li, Br	LiBr	2.5 mmol	
METALS			
Cu	$CuSO_4$	1 mmol	200, 400, 700
Cu + low acid	$CuSO_4$	1 mmol	400
	H_2SO_4	16 mmol	
Cu + high acid	$CuSO_4$	1 mmol	400
	H_2SO_4	60 mmol	
Zn	$ZnSO_4$	1 mmol	200, 400, 700
Ni	$NiSO_4$	1 mmol	400
As	Na_2HAsO_4	1 mmol	400
Brew Cu-Ni-Zn	Cu-, Zn-, $NiSO_4$	0.33 mmol each	400
Brew + low acid	Cu-, Zn-, $NiSO_4$	0.33 mmol each	400
	H_2SO_4	16 mmol	
NO METALS			
High Acid	H_2SO_4	60 mmol	400

Table 3. Analysis of Munduran Creek water

Ion	Concentration
Ca^{2+}	2.8 mM
Mg^{2+}	6
Na^{+}	20
K^{+}	0.06
SO_4^{2-}	1.5
Cl^{-}	29
HCO_3^{-}	5.5
Br^{-}	0.025
pH	8.1

together as a mixture, only one-third as much of each metal was used as when they were released individually.

At each station downstream, water samples were collected at regular preset time intervals and, where appropriate, temperature and pH were measured in situ as functions of time. Selected water samples were also filtered through 0.1-μm membranes, within 20 min after collection, to allow differentiation between particulate and soluble metal. The water samples from metal and acid releases were subjected to extensive chemical analysis.

3 Results and Discussion

3.1 Tracers

Similar results were obtained for each of the three tracers but for the purposes of this discussion attention will be confined to rhodamine since its measurement has the lowest associated relative error. Figure 2 shows the results of rhodamine at medium flow rate. The concentration of rhodamine (expressed as a dilution ratio

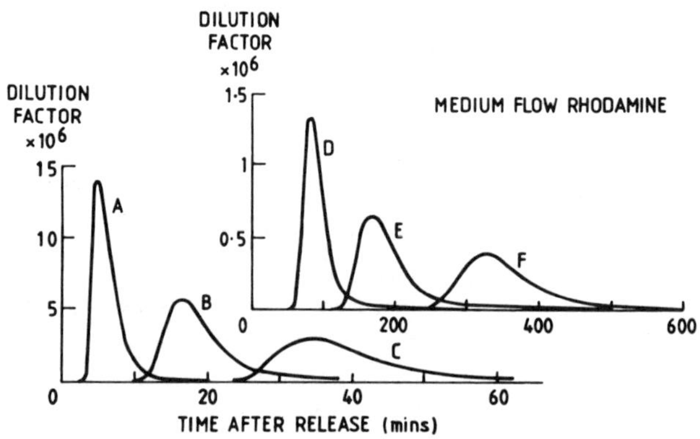

Fig. 2. Data from sampling stations *A* to *F* illustrating the downstream effects of advection and dispersion on rhodamine WT dye released as an effectively instantaneous pulse at R. Flow rate, 400 l min^{-1}

of the stock) is plotted as a function of time after release for the six sampling stations. These results are conveniently arranged as a composite on two sets of axes with different scales. The values of t_{max}, the time at the concentration maximum, can be measured for each of the stations. These values are close to but not equivalent to the times of travel for the stations (Chapman 1979). This arises because the peak of the pulse, as determined by observing concentration changes as a function of time at a fixed point downstream, will appear to arrive slightly earlier than if its position were determined by measuring the concentration profile, as a function of distance, at a number of fixed times. These t_{max} values are plotted against distance downstream in Fig. 3 for each of the rhodamine experiments. With the approximation that t_{max} is equal to the time of travel, the inverse of the slope at any point on the curve for each flow rate is equal to the the velocity at that point. At medium flow the velocities so calculated show about a five-fold drop over the study reach, indicating marked non-uniformity in stream characteristics.

Any loss in rhodamine, for example by adsorption onto bed sediments, will appear as a progressive decrease in the integrated areas under the concentration-time curves for each consecutive station downstream. The integrated area for each station, normalised for total rhodamine added, was plotted against distance for the three flow rates in Fig. 4. Curves for all three flow rates show a small but measurable loss in rhodamine over the study reach. The loss is consistent with the observation that rhodamine can be adsorbed from solution when shaken with samples of the sediments. At low flow there is a large loss in integrated area apparent between stations E and F. Examination of the stream between these two stations revealed a gravel bank through which, at low flow rates, the major portion of the water percolated. Consequently, the more intimate contact between water and sediment would allow greater loss of rhodamine to occur in this region.

The rhodamine pulses in Fig. 2 illustrate the effects of dispersion whereby the pulse simultaneously broadens in width and decreases in height as it travels downstream from station A to station F. However, analysis of the curves shows that the variance of the pulse does not increase linearly with time as with a uniform stream (Fischer 1967), but at an increasingly rapid rate. This is another example of the hydrodynamic non-uniformity noted previously. Moreover, the pulses are not Gaussian as expected but are assymetric. They can be fitted by a skewed Gaussian function to be described below in Section 3.5.

At low flow rates significant discrepancies were observed between the results for a series of trial experiments and those for the subsequent series reported here. Accordingly, the low flow experiment was performed twice, by releasing rhodamine at 11 p.m. as well as at 10 a.m. For the night release, lower dispersion and a time of travel 30 min greater than for the morning release were obtained. The occurrence of variable diurnal stratification was suspected and two further releases commencing at 10 a.m. and 11 p.m. respectively, were carried out specifically to allow any stratification effects to be observed. Detailed investigation of some of the larger pools at the lower end of the study reach revealed a marked thermocline, in excess of 3 °C from bottom to top at certain times of the day. Depth sampling also showed that there were significant gradients in the concentration of rhodamine in these bodies of water. It was concluded that under certain weather conditions

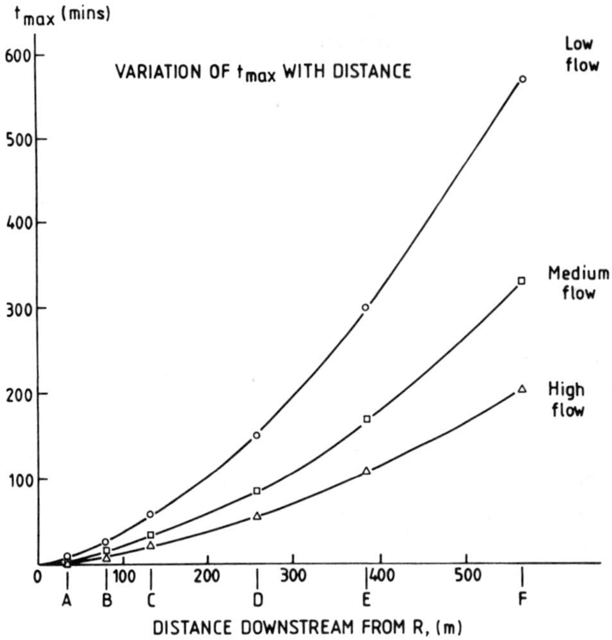

Fig. 3. Dependence of the time of travel (t_{max}) of the concentration maximum for rhodamine as a function of distance downstream for the three flow rates employed (viz. 200, 400 and 700 $l\ min^{-1}$)

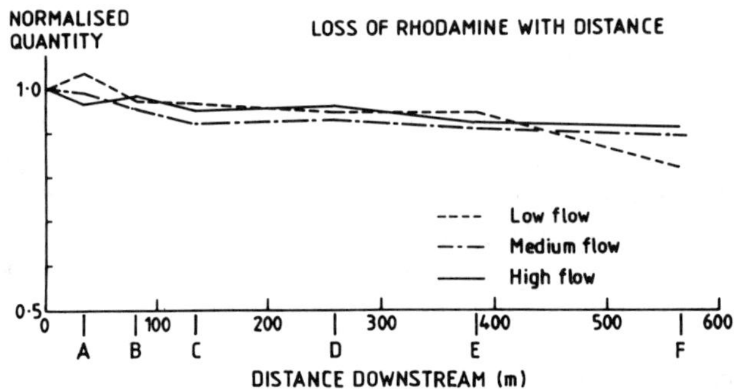

Fig. 4. The normalised integrated areas under the rhodamine pulses are plotted against downstream distance for each of the standard flow rates

stratification occurred in larger pools from about 2 h after dawn to 2 h after sunset. This allowed warm inflowing waters to skim across the colder, lower layers. The resulting short-circuiting of the volume of these pools gave rise to decreased times of travel and increased dispersion. It was also found that a rhodamine pulse travelling through an unstratified pool during the night will become partially trapped as stratification develops after dawn. These effects were observed only at the lowest flow rate since at higher flow rates there was sufficient mixing energy available to prevent the formation of a temperature gradient.

3.2 Metal Releases

When Cu was released into the creek, a fine, stable, pale-blue colloidal precipitate appeared. Presumably, malachite, $Cu_2(OH)_2CO_3$, is formed by the reaction with the high levels of bicarbonate (~5 mM) present in the water (see Table 3). Figure 5 shows the result of a medium flow Cu release, where concentrations of total Cu (solid curve) and soluble Cu (broken curve) are plotted against time after release for the six stations. The concentration of metal in particulate form is equal to the difference between the two curves. In addition to the obvious dispersion effects as noted previously for rhodamine, it is seen that over half the metal at the peak in station A is in particulate form compared with less than 10% at station F. Samples of this colloidal suspension, withdrawn from the creek and left standing in a beaker for long periods, showed the rate of particle sedimentation to be very low. Therefore, the decrease in the proportion of particulate metal is probably due to the particles redissolving as the pulse is diluted by dispersion as it travels downstream. However, it is also possible that particles are removed from the flowing waters by adhering to, or being filtered out by, surfaces projecting from the stream bed or banks. It should also be noted that there is a much greater degree of "tailing" of the pulses than is observed for rhodamine.

Although not shown, the results for Cu releases at low and high flow rates, after allowing for differences in times of travel, are qualitatively similar to those for medium flow. Results for Zn at medium flow appear in Fig. 6. In contrast to Cu, insignificant quantities of this element exist in particulate form although the curves are again qualitatively the same as observed for total Cu. However, a comparison

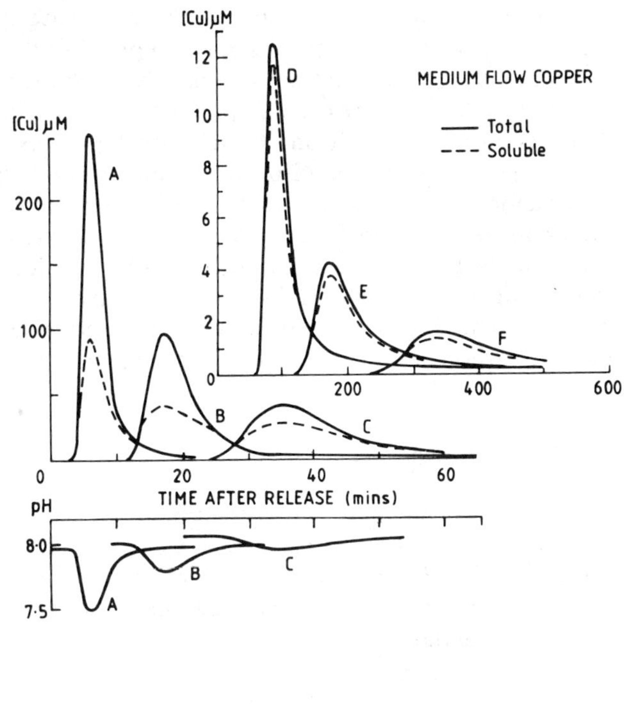

Fig. 5. Concentration of copper and values of pH as a function of time for a release of copper sulphate into the stream flowing at 400 l min^{-1}

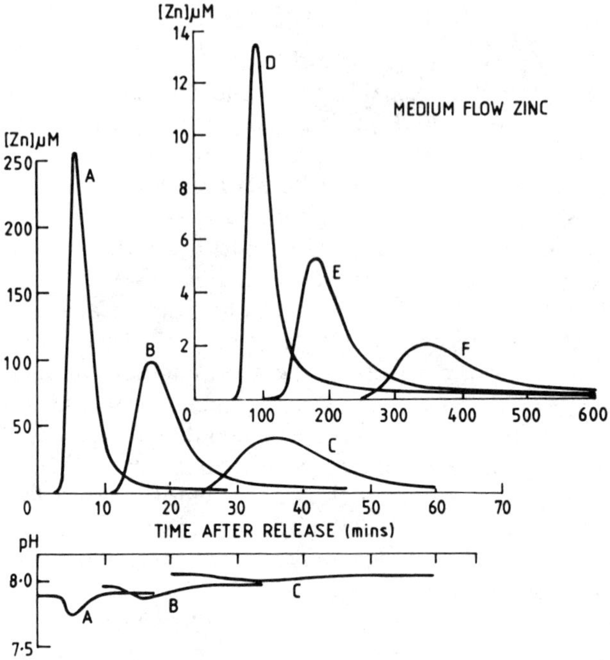

Fig. 6. Plots of zinc concentration as a function of time at each of the six downstream sampling stations. Flow rate, 400 l min^{-1}

of the heights of the curves for the later stations reveal that they decrease somewhat more slowly than for Cu. Zn exhibited a similar dependence on flow rate to Cu under both high and low flow-rate conditions.

Corresponding results for the combined release of Cu, Zn and Ni (hereinafter referred to as a "brew") are again not shown but the individual metals behaved much as they did when released alone, although the percentage of particulate Cu was initially lower in the brew experiment. This is to be expected from mass action considerations since only one-third as much Cu was released.

As a consequence of adsorption of metals on bed sediments it was expected that major cations would have been exchanged into solution and that pulses of these cations, corresponding in size to the loss of metal, would have been observed. However, because the background levels of the major cations were so much higher than any expected perturbation, no such pulses were detectable. It is also worth noting that no significant changes were observed in the concentrations of dissolved organic carbon as a function of time during any of the releases.

3.3 Acid Releases

Following the metal releases described above several experiments involving injection of various quantities of acid were performed (see Table 2). A number of releases of high levels of acid were carried out before those at lower levels. Some of these experiments also involved the simultaneous discharge of metals or mixtures of metals.

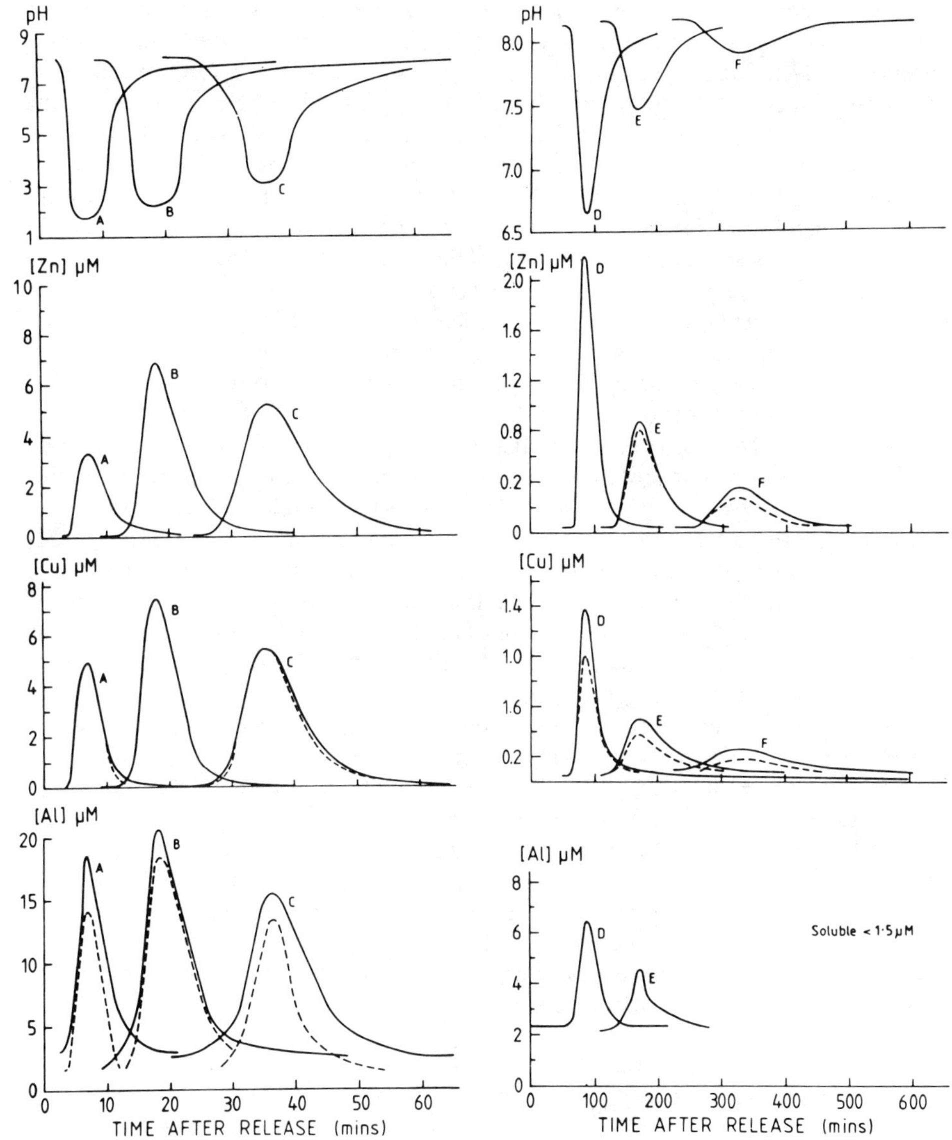

Fig. 7. Concentrations (total) of selected elements scavenged from the sediments together with the corresponding pH profiles following the "high acid" release. Total metal (–); soluble metal (---)

Only selected results for the high acid release (Fig. 7) are shown since the effects for the low acid releases were similar but less marked. Very deep troughs in pH were recorded at the upstream stations but these became shallower as the pulse progressed downstream. Simultaneously, Cu, Zn, Mn, Fe, Al and SiO_2 were scavenged from the sediment. The peak heights of the pulses of each of these elements increase initially as they travel downstream. This occurs even against the effects of dispersion, which on its own, would cause a rapid decrease in peak height. Eventually, however, as the pH pulse becomes shallower and its scavenging ability is progressively reduced, peak height approaches a maximum and then begins to decrease. Since total metal mobilised is proportional to the area under the respective curves it should be realised that in some cases, even when peak height is decreasing, scavenging can still be occurring, if the rate of decrease in peak height is slower than the rate of pulse broadening. It is suspected that the Cu, and perhaps also the Zn, dissolved from the bed sediments in this experiment originated from the previous releases of these metals. All of the Cu is in soluble form at the upstream stations but as the pulse moves downstream and pH increases, it becomes increasingly converted into particulate species, probably by precipitation as malachite and perhaps also by coprecipitation with precipitates of the other scavenged metals. A decrease similar to that observed for Cu, in the proportion of soluble forms of these other metals is evident.

A significant fraction of the zinc also exists in particulate form at stations E and F. This result is in marked contrast to the findings from the medium flow zinc release (Fig. 6) where all zinc remained soluble even though the concentrations at the downstream stations were five times greater, and the values of pH considerably higher. The most probable explanation for the lower solubility observed at E and F in the acid injection experiment is that the increase in the pH minimum from 6.7 at D to 7.7 at F leads to the adsorption of a significant proportion of the initially soluble zinc onto the amorphous iron and aluminium hydroxide precipitates which are also being carried downstream by the flowing water. This hypothesis is supported by detailed studies of zinc uptake onto synthetic precipitates of Fe and Al hydroxides which have shown that the adsorption edge for this metal lies between the pH values of 6 and 8 (Leckie et al. 1979; Millward and Moore 1982).

As expected, when Cu was released together with the above quantity of acid the major difference was the higher levels of Cu measured, which was essentially all in soluble form for the first three stations. Although the pH curves for corresponding stations in the two experiments (i.e. with and without Cu) were virtually identical, the quantities of other elements scavenged were slightly less than for acid alone probably because the more easily accessible forms had been removed in the acid-only release which was performed first.

3.4 Peak Concentration vs t_{max}

3.4.1 General

In order to facilitate comparison between the set of six curves of concentration vs time for each chemical entity in each release, it was desirable to reduce key data in all these sets of curves to a single curve. Accordingly, the point of maximum concentration for each station was selected and the concentration C_{max} at this point was plotted on a log-log scale against the corresponding time t_{max} at which it was observed to occur.

For a conservative tracer released into a hydrodynamically uniform stream of constant flow rate, there is a simple but close approximate relationship which exists between C_{max} at any station and its corresponding t_{max} (Fischer 1967). If the logarithm of the peak concentration is plotted against the logarithm of t_{max}, a straight line of slope −0.5 is obtained under these conditions, but a different relationship would be observed for a non-uniform stream. For metals whose concentrations decrease more rapidly than those of a conservative tracer as a result of attenuation by physico-chemical removal mechanisms from the flowing water (e.g. by adsorption onto bed sediments, sedimentation of precipitates, etc.) plots of steeper negative slope than for the tracer would be obtained.

Accordingly, for each element in each release, C_{max} was plotted against t_{max} on a log-log chart.

3.4.2 Effect of Flow Rate

The data for the rhodamine plots shown in Fig. 8 have been obtained from Fig. 2 and from the results for the releases carried out at high and low flow and show the effect of flow rate on rhodamine transport. Except for their initial portions the curves for the three flow rates are seen to be nearly straight lines which are remarkably close to one another. Corresponding curves for Zn showing similar effects of flow rate are also displayed in Fig. 8.

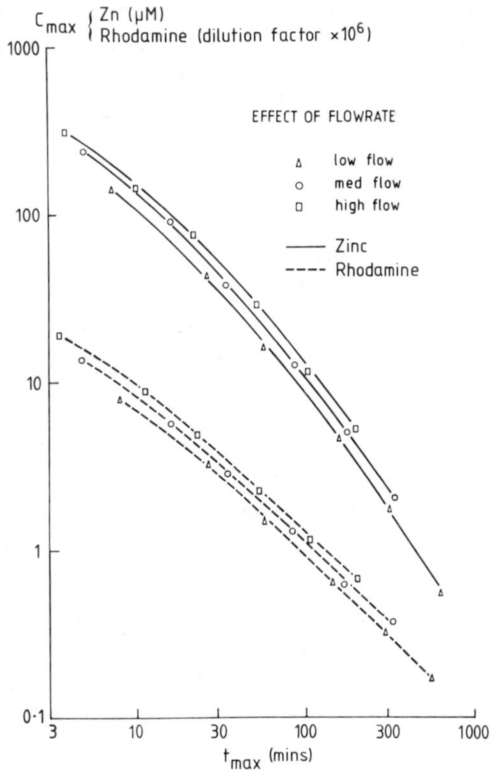

Fig. 8. C_{max} versus t_{max} curves for the rhodamine and zinc releases plotted on a log-log scale

3.4.3 Effect of Chemical Type

The curves for the different chemical entities have different slopes and a comparison can be made by plotting, together in Fig. 9, medium flow-rate curves for the water tracer rhodamine WT and the metals Cu, Zn, Ni and As. Also shown for comparison is the relationship expected for a conservative tracer in a uniform stream. Although it was mentioned previously that a small proportion of rhodamine was lost by adsorption on the stream bed, for the purposes of the present analysis, this tracer can be considered truly conservative. Therefore, the large deviation between the water-tracer plot and that expected for an ideal uniform stream is the result of the hydrodynamic non-uniformities. However, the deviations occurring between the curve for the water tracer and those for the metals are the result of chemical attenuation mechanisms removing metal from the flowing waters. A comparison of the slopes of these lines shows that Cu is the most highly attenuated (i.e. the least mobile) of all the chemical components released, followed by Zn, N and As.

3.4.4 Soluble and Particulate Components

It is of particular interest to note that even in the case of Cu, which was the most highly attenuated of all the metals, no chromatographic retardation effects on the peak of the pulse were apparent (although marked tailing was noticed). This is in contrast to what is usually observed in groundwater studies (Vachand et al. 1976; Valocci et al. 1981) where the velocity of a pulse of reacting solute can be much less than that of a non-reactive tracer and reflects the much less intimate contact

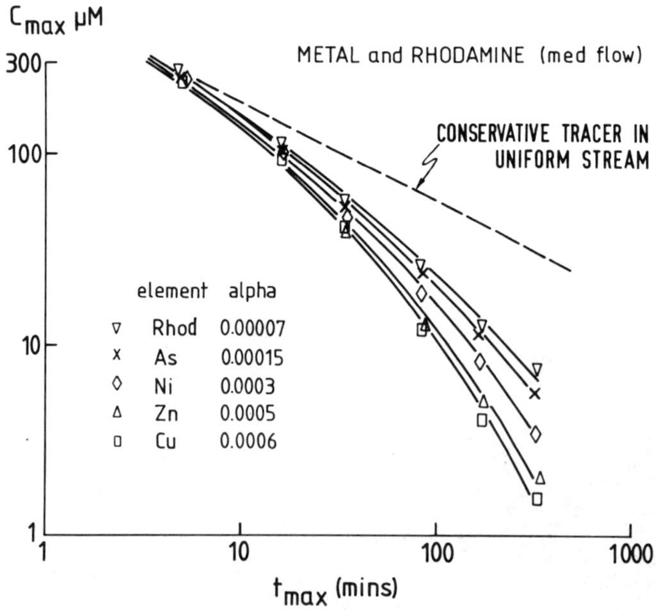

Fig. 9 Experimental points and theoretical curves (calculated by using the equations derived in Sect. 3.5.2) for the tracer and metal release experiments carried out at a flow rate of 400 l min^{-1}. The values of rhodamine shown have been obtained by multiplying dilution factors by 20×10^6 to allow visual comparison with metal concentrations. The line for a conservative tracer in a uniform stream ($C_{max} \propto t^{max-1/2}$) is included for reference

between solid and solution in surface waters. Since insignificant proportions of the metals, apart from Cu, were observed to be in particulate form, it seems probable that the metals are attenuated by adsorption on the bed sediment. The particulate Cu showed very little tendency to sediment (Chapman et al. 1984a) and therefore a major contribution to attenuation of total metal by this mechanism is unlikely. This hypothesis is supported by Fig. 10 which shows the curve for medium flow-rate Cu partitioned into its soluble and particulate components. The particulates are seen to decrease very rapidly in the upstream regions, while the soluble component decreases even more slowly than rhodamine (and hence more slowly than would be required by dispersion alone). It seems certain therefore that particulate Cu is dissolving to replenish the levels of soluble Cu reduced by dispersion and also presumably depleted by adsorption.

3.4.5 Effect of Competition

The "brew" releases were designed to allow the effects of competition to be studied. Although not shown here, C_{max} -t_{max} curves for all three metals in the release were compared to those obtained when the metals were released individually in separate experiments. When the curves for the individual releases were suitably scaled to adjust for the fact that only one-third as much metal of each type was released in the brew, the results for Zn and Ni were found to be identical to those in the brew releases. There was therefore no indication of any competitive effect between these metals. Furthermore, this observation suggests that the attenuation mechanisms are behaving linearly with respect to concentration, over the three-fold concentration range covered by the brew and individual release experiments. These

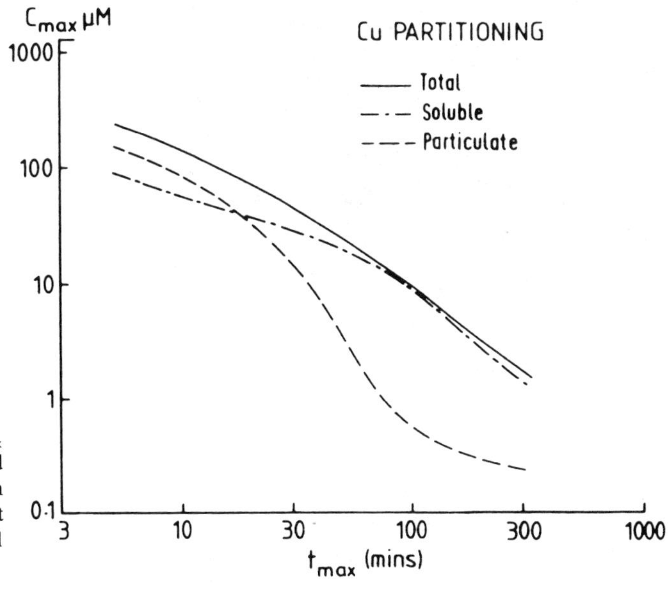

Fig. 10. C_{max} versus t_{max} curves for total, dissolved and particulate forms of copper in the experiment carried out at the medium flow rate 400 l min^{-1}

results are similar to those obtained for adsorption of metals onto model substrates (Benjamin and Leckie 1981a,b) where the fraction of total available surface sites occupied is very low and where each metal type may be binding to its own preferred high affinity sites. Either of these aspects in its own right would prevent competition effects from being observed. Circumstantial evidence for low surface coverage in the present system is provided by the observation that previous releases of a given metal had an insignificant effect on subsequent releases of either the same or another metal.

In contrast to the behaviour for Zn and Ni, Cu was attenuated to a greater degree in the brew release which is in a direction opposite to what would be expected if competition effects were operative. This suggests a slight non-linearity with respect to concentration in the behaviour for Cu and most likely results from the differing proportions of soluble and particulate metal in the two cases.

3.4.6 Effect of Acid

The effects of various levels of acid on the Cu release at medium flow rate are shown in Fig. 11 together with the corresponding plots for pH minima. It is surprising that the curves for the release of Cu without acid and with low acid are virtually identical

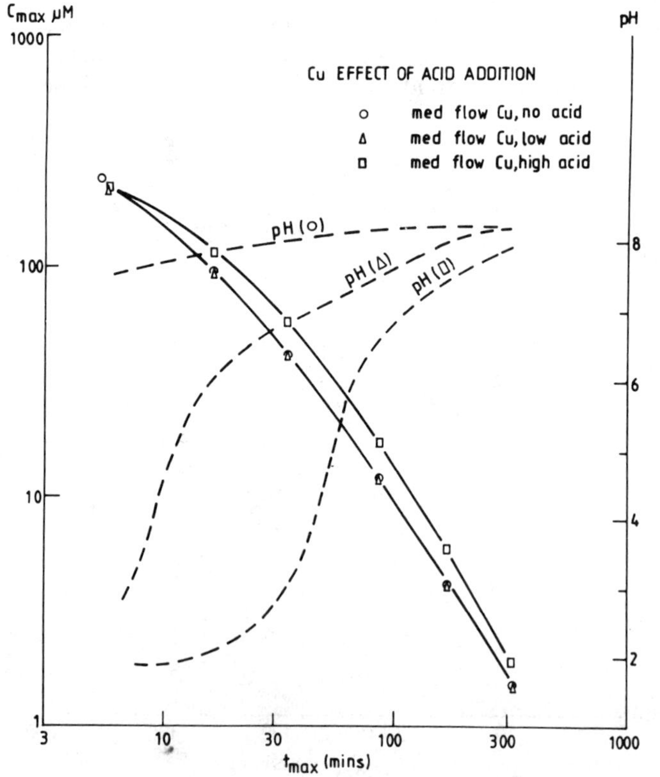

Fig. 11. Plots of pH and C_{max} versus t_{max} illustrate the effects of acid addition on pH and metal attenuation in the stream at medium flow rate

in spite of considerable differences both in pH and in the relative quantities of soluble and particulate metal forms. These results can be explained by the operation of two compensating mechanisms. The addition of acid has the effect, not only of keeping the metal in soluble form and of reducing its complexation with HCO_3^-, but also at the same time of competing with it for available surface sites. These two opposing effects have apparently balanced out so that no net difference in attenuation was observed. Higher levels of acid do appear to reduce Cu attenuation at the upstream stations, as the curves are seen to diverge, but by station F the rate of attenuation appears to have perhaps even exceeded that observed in the absence of acid since the curves are starting to converge.

In Fig. 12 the results for the low acid brew release are compared with those for the brew release with no acid and show that the acid reduces attenuation for all three metals. By station F the peak heights of the pulses are about 20% greater in the presence of acid. Apparently the sulphuric acid, which is 16 times as concentrated as total added metal on a molar basis, reduces the rate of attenuation of metals by competition for the available surface sites. It is perhaps surprising that the effects of such relatively large quantities of acid are not more marked. The results obtained here for Cu are apparently in conflict with those for the releases discussed above where low levels of acid had virtually no effect on Cu attenuation.

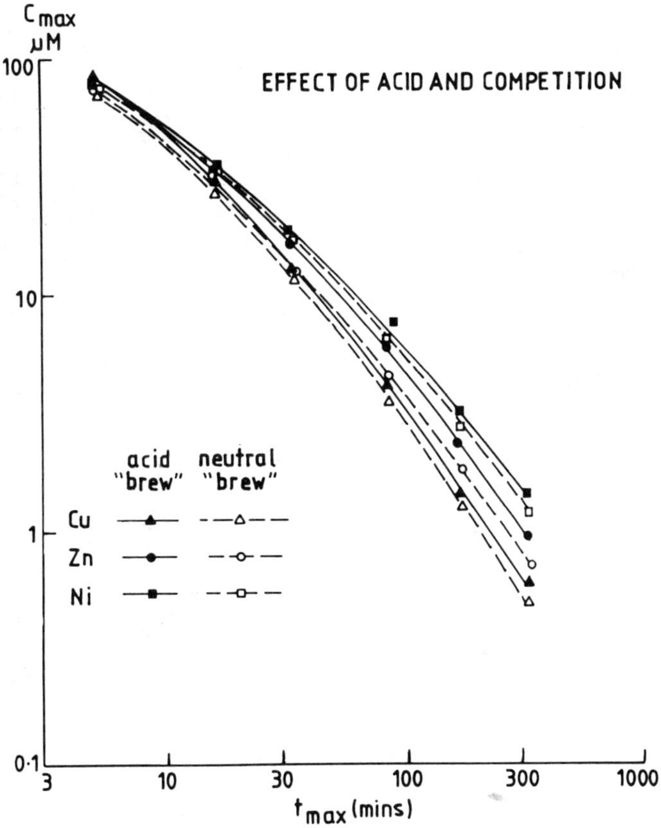

Fig. 12. The influence of low levels of acid on the attenuation of a mixture ("brew") of copper, nickel and zinc at medium flow rate

However, in these former experiments three times as much Cu was released as in the brews. This higher concentration of total Cu gave rise to a much greater percentage of the metal in particulate form (when acid was absent) and therefore a much lower percentage of soluble Cu available for adsorption. However, in the presence of acid the concentration of soluble metal available for adsorption was greatly increased which offset the competition for sites from the acid. The corresponding percentage increase in soluble Cu for the brew in the presence of acid was less and consequently the competition for adsorption sites from the acid was not completely compensated for.

3.4.7 Effect of Acid Scavenging

The growth in the peaks for metals scavenged in the acid release which was described in Section 3.3 are very clearly illustrated in the log-log plot of Fig. 13. The results for Cu show how the pulse height builds up, reaches a maximum and then falls away, eventually giving rise to a slope after station C close to those observed when Cu was released alone, as a brew, and with acid. This behaviour was interpreted above as an initial scavenging at low pH, of previously absorbed Cu,

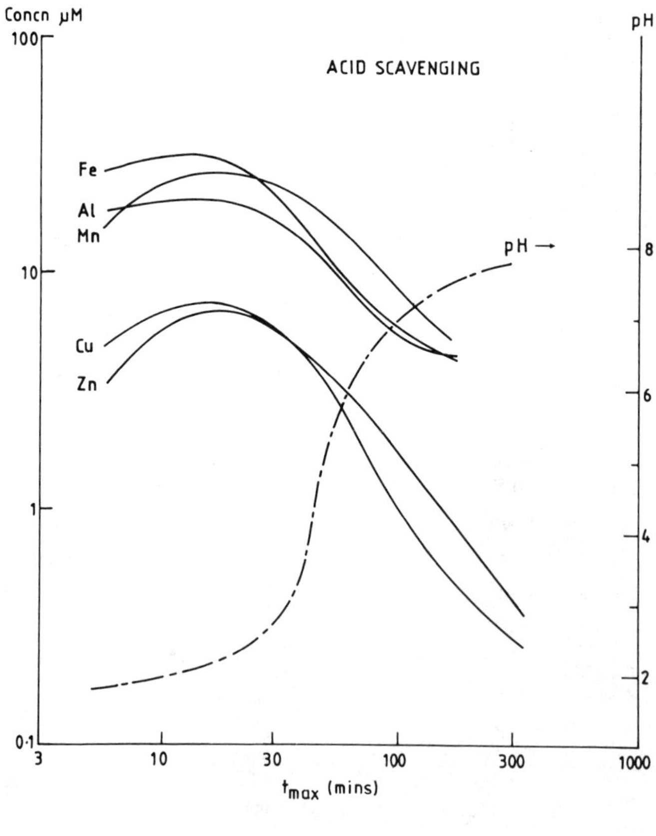

Fig. 13. Plots of pH and C_{max} (total concentrations) versus t_{max} for selected elements scavenged from the sediments following the "high acid" release

followed by re-adsorption onto sediments further downstream as the severity of the pH perturbation is reduced by dispersion and neutralisation.

On the other hand, the metals Fe and Al and perhaps Mn would be scavenged at least partly by acid dissolution of the ubiquitous metal oxides and fine clay particles. The presence of an accompanying pulse of SiO_2 of about twice the height of that for Al supports the suggestion of clay dissolution. Although the concentration data for Fe and Al are relatively imprecise because the detection limits of analysis are being approached, at stations D and E the curves for these metals, which are essentially all in particulate form, appear to be decreasing at a rate greater than can be accounted for by dispersion alone. This probably means that some loss of these metals is occurring by sedimentation.

3.5 Modelling of the Results

In the analysis to be outlined below we have attempted to determine how far the salient features of the results can be explained in terms of very simple concepts and approximate analytical expressions. This approach is particularly useful as it can quickly identify the broad dependencies of the results on the important variables of interest.

3.5.1 Behaviour of a Conservative Tracer

As was mentioned in Section 3.1, the rhodamine pulses shown in Fig. 2 were not Gaussian as expected for a uniform stream. However, they can be fitted by the following skewed Gaussian function:

$$g(\xi) = 1/[\sqrt{\pi}(\sqrt{2\sigma\text{-}k\xi})]\exp\{-(1/k\ \ln[1\text{-}k\xi/(\sqrt{2\sigma})])^2\}, \tag{1}$$

where σ is the standard deviation and k the skewness constant. The source of this skewness is undoubtedly the non-uniformity in the dispersion coefficient which causes the pulses to tail. (The dispersion coefficient is a quantity, analogous to a diffusion coefficient, which is a measure of the spreading and mixing of a pulse of solute caused by fluid shear and eddy currents in the water column.)

It was found after careful analysis employing the expression above (and neglecting the small observed attenuation in the amount of dye) that the response of the system exhibited in Fig. 2 and those for the other flow rates could be described by the following analytical equation relating concentration C_c to time t, downstream distance x, flow rate r and volume of rhodamine added V.

$$C_c = V/\{r\sqrt{\pi}[\sqrt{(\bar{t}^*\ t)}\text{-}k(y\text{-}t)]\}\exp\{-(1/k\ \log[l\text{-}k(y\text{-}t)/\sqrt{(\bar{t}^* t)}])^2\}, \tag{2}$$

where:

C_c is the concentration of rhodamine if it were conservative:

$y = ar^N x^n$,

$\bar{t}^* = (p + q\ x^m)r^M$ and

k, a, N, n, p, q, m and M are constants.

The skewness constant, k, for each pulse varied only slightly with flow rate and distance downstream, and had an average value of 0.325.

3.5.2 Behaviour of a Non-Conservative Component

To obtain an approximate analytical expression to describe the behaviour of a non-conservative reacting solute the following assumptions were made:

1. The hydrodynamics of the stream can be represented by Eq (2).
2. The reacting solute is attenuated by adsorption onto the stream sediments with a mass transfer coefficient β across the sediment-water interface which depends on the average velocity, v, according to the following relationship (Shull and Gloyna 1968):

$$\beta = \alpha v^{Q},$$

where α and Q are constants.

3. The properties of the sediments, including roughness, are constant along the stream.
4. In calculating the peak height of the pulse the effect of desorption can be neglected. (Some justification for this is provided by the observation that chromatographic retardation effects were not observed for the peaks of the pulses of released metals.)
5. The hydraulic radius R, which is defined as the cross-sectional area of the stream divided by its wetted perimeter, was approximated by the relationship (Chapman, unpublished results):

$$R \doteqdot 1/3 \sqrt{(r/v)}.$$

On the basis of the above assumptions, an equation was derived for the concentration C of non-conservative solutes as a function of time and distance downstream:

$$C = C_c \exp(-\bar{k}t), \tag{3}$$

where:

C_c is given by Eq. (2) above and

$$\bar{k} = \frac{3\alpha x^{(1-n)(1/2+Q)}}{n^{(Q-1/2)}\, a^{(Q+1/2)}\, r^{N(Q+1/2)+1/2}\,(1+(n-1)(1/2-Q))}.$$

From an analysis of the data a value for Q [see point (2) above] of 0.6 was found to fit the results for all components released. This compares with a value of 0.8 used by Shull and Gloyna (1968) in flume studies. The relationship above was used to obtain the theoretical C_{max} -t_{max} curves for each of the components released which are shown inFig. 9 together with their corresponding values of α.

4 Conclusions

The results obtained in this work represent, chemically and hydrodynamically, the most complete and comprehensive data set available on the transport of metals in streams. The behaviours of four elements, as well as acid, have been extensively investigated in a natural watercourse which was artificially pumped at three different flow rates using water from a nearby stream. A number of important conclusions can be drawn from the study, many of which will be of general applicability to streams elsewhere.

The stream chosen for this work was hydrodynamically very non-uniform. For an instantaneous release of solute, this gave rise to concentration pulses downstream which exhibited a skewed Gaussian form. However, the skewness was found to be independent of flow rate and distance downstream. Logarithmic plots of C_{max} versus t_{max} for conservative tracers were almost straight lines but had much greater negative slopes than the value of -0.5 expected for a uniform stream. In practice this means that in such a non-uniform stream, pulse concentrations will be reduced as result of dispersion much more rapidly than in a uniform stream. Also, at any given distance downstream, lower flow rates were comparatively more effective in dispersing the pulse than higher flow rates. Furthermore, this trend was also found to be true for the chemical process of attenuation by adsorption.

The percentage losses due to adsorption of components over the study reach (based on peak height) increase in the order rhodamine(10%)<As(35%) <Ni(58%)<Zn(74%)<Cu(79%). When metals were released together as a mixture there was no evidence of competition for available surface sites and moreover Zn and Ni attenuated linearly with respect to concentration under the conditions investigated. However, some of the Cu formed a colloidal suspension which did not sediment and inhibited adsorption by reducing the concentration of Cu in solution in the flowing water. Partial redissolution of this colloid at stations further downstream tended to buffer the soluble Cu concentrations which would otherwise have decreased rapidly due to dispersion and adsorption.

During the first high acid release Cu, Mn, Fe, Al and Si were scavenged from the sediments near the upper stations, but they subsequently proceeded to attenuate (for Cu, at the same rate as seen before) by sediment adsorption (and possibly by sedimentation of Fe and Al hydroxide precipitates) in the lower reaches. In the high acid release which followed, lesser quantities of these elements were scavenged. For the brew releases, additions of low levels of acid moderately reduced the degree of attenuation of all three metals, presumably by competition with them.

By employing some reasonable assumptions and approximations, a simple and useful semi-empirical equation has been derived to describe the effects of the processes of advection, dispersion and adsorption in this system. This equation has been used successfully to fit the changes observed in peak height with time and distance downstream for the five components released and will form the basis for a more complex simulation study of this data set using a combined transport model such as RIVEQL(Chapman 1982), which couples detailed aquatic chemistry and hydrodynamics. Although the work described here has been carried out in response to a specific problem, the methodology is applicable to the study of contaminant transport in streams generally.

The release of small quantities of metals into streams, to enable the response of the aquatic environment to be determined, remains an essential part of the process of assessing the impact of proposed mining and other industrial activities. When experiments of this nature are undertaken it is most important to ensure that maximum information is obtained. Sampling should be carried out sufficiently frequently, in both space and time, both "total" and filtered samples should be taken and key variables such as pH continuously monitored. Exhaustive chemical analysis should be carried out on all components likely to be of interest, including major cations and anions.

Simultaneously with this work, numerical simulation models are being developed with the ultimate aim of being able to predict the consequences of waste discharge to a watercourse using input data obtained only from the release of innocuous tracers and from laboratory studies on the kinetics of uptake of metals by samples of the sediments from the stream in question. It is intended that, eventually, this approach will greatly reduce the necessity for experimental discharges of metals to the environment. In the meantime, however, such experiments will continue to be required to directly obtain information on metal transport and also to calibrate and validate the models under development. These models will allow a quantitative assessment to be made of the likely adverse consequences of a release to the environment and may suggest possible remedial procedures in the case of such a release. As a result of this work it should be possible to achieve a more systematic planning of the location and construction of industrial facilities such as waste dumps and the type and extent of treatment which may be necessary for contaminated water together with the size of any natural buffer zone required.

Acknowledgements. This project was partially funded by the National Energy Research Development and Demonstration Program (NERDDP) of Australia and by Esso Australia Ltd.

References

Airey PL, Calf GE, Davison A, and Easey JF (1983) The transport of zinc and manganese tracers across the Magela Creek system Northern Territory. In: Proc Int Spec Conf Water regime in relation to milling, mining and waste treatment including rehabilitation with emphasis on uranium mining. Darwin, 1983, pp 811–818

Bencala KE, Jackman AP, Kennedy VC, Avanzino RJ, Zellweger GW (1983) Kinetic analysis of strontium and potassium sorption onto sands and gravels in a natural channel. Water Resour Res 19:725–731

Benjamin MM, Leckie JO (1981a) Multiple-site adsorption of Cd, Cu, Zn and Pb on amorphous iron oxyhydroxide. J Colloid Interface Sci 79:209–221

Benjamin MM, Leckie JO (1981b) Competitive adsorption of Cd, Cu, Zn and Pb on amorphous iron oxyhydroxide. J Colloid Interface Sci 83:410–419

Boyles JM, Cain D, Alley W, Klusman R (1974) Impact of Argo Tunnel acid mine drainage Clear Creek County Colorado. Proc Am Water Resour Assoc 18:41–53

Chapman BM (1979) Dispersion of soluble pollutants in non-uniform rivers, Part I: Theory. J Hydrol 40:139–152

Chapman BM (1982) Numerical simulation of the transport and speciation of non-conservative chemical reactants in rivers. Water Resour Res 18:155–167

Chapman BM, Jones DR (1983) Simulation of metal transport in streams. In: Proc ASCE Conf Environmental Engineering, Boulder, Col, pp 635–642

Chapman BM, Jones DR, Jung RF (1983) Processes controlling metal ion attenuation in acid mine drainage streams. Geochim Cosmochim Acta 47:1957–1973

Chapman BM, Jones DR, Jung RF (1984a) Environmental aspects of oil shale mining at Rundle. Final project report to the Dep Resourc Energ, Australia. Rep NERDDP/EG/85/383

Chapman BM, Jones DR, Jung RF (1984b) Transport of metals of environmental interest released into a creek on the Rundle prospect. In: Proc 2nd Australian Worksh Oil shale, Brisbane, 1984, pp 270–275

Farrah H, Hatton D, Pickering WF (1980) The affinity of metal ions for clay surfaces. Chem Geol 28:55–68

Fischer HB (1967) The mechanisms of dispersion in natural streams. Hydrol Div Proc ASCE 93(HY6):187–216

Gloyna EF, Bhagot SK, Yousef Y, Shih C (1966) Transport of radionuclides in a model river. In: IAEA Symp Disposal of radioactive waters into seas, ocean and surface waters. Int Atom Energ Ag, Vienna, pp 11–32

Graham BK (1983) The environmental implications of oil shale development and their relevance to energy research and development programs. In: The environmental implications of three synthetic liquid fuel technologies. Aust Gov Publ Serv, Canberra, pp 64–127

Hall RJ, Likens GE, Fiance SB, Hendry GR (1980) Experimental acidification of a stream in the Hubbard Brook experimental forest, New Hampshire. Ecology 61:977–989

Hesslein RH, Brocker WS, Schindler DW (1980) Fates of metal radiotracers added to a whole lake: sediment-water interactions. Can J Fish Aquat Sci 37:373–377

Jackman AP, Walters RA, Kennedy VC (1984–1985) Transport and concentration controls for chloride, strontium, potassium and lead in Uvas Creek, a small cobble-bed stream in Santa Clara County, California, USA. 2. Mathematical modelling. J Hydrol 75:111–141

Jones DR, Chapman BM (1983) An investigation of the leaching behaviour of solid wastes from Rundle oil shale. In: Proc 1st Aust Workshop Oil shale, Lucas Heights, pp 183–186

Kennedy VC, Jackman SM, Zand SM, Zellweger GW, Avanzino RJ (1984–1985) Transport and concentration controls for chloride, strontium, potassium and lead in Uvas Creek, a small cobble-bed stream in Santa Clara County, California, USA, 1. Conceptual model. J Hydrol 75:67–110

Kuwabara JS, Leyland HV, Bencala KE (1984) Copper transport along a Sierra Nevada stream. J Environ Eng 110:646–655

Leckie JO, Benjamin MM, Hayes K, Kaufman G, Altman S (1979) Adsorption coprecipitation of trace elements from water with iron oxyhydroxide, EPRI-RP-910, Electric Power Res Inst, Palo Alto, Cal

Lundgren DG, Vestal JR, Tobita FR (1972) The microbiology of mine drainage pollution. In: Mitchell R (ed) Water pollution microbiology. John Wiley, New York, pp 69–88

Millward GE, Moore RM (1982) The adsorption of Cu, Mn and Zn by iron oxyhydroxide in model estuarine solutions. Water Res 16:981–985

Neely WB, Blau GE, Alfrey T (1976) Mathematical models predict concentration-time profiles resulting from chemical spill in a river. Environ Sci Technol 10:72–76

Shull RD, Gloyna EF (1968) Radioactivity transport in water – Simulation of sustained releases to selected river environments. Tech Rep US Atom Energ Comm, Contract At-(11-1)-490. Centre for research in water resources, environmental health engineering research laboratory, civil engineering department. Univ Texas, Austin

Vachaud G, Wierenga PJ, Gaudet JP, Jegat H (1976) Simulation of miscible displacement in unsaturated porous media. In: Vansteenkiste GC (ed) System simulation in water resources. Elsevier/North Holland Biomedical Press, Amsterdam, pp 129–140

Valocci AJ, Street RL, Roberts PV (1981) Transport of ion-exchanging solutes in groundwater: chromatographic theory and field simulation. Water Resour Res 17:1517–1527

WHO (1971) International standards for drinking water, 3rd edn. WHO, Geneva

Yousef YA, Gloyna EF (1977) A transport model for long term release of low level radionuclide solutions into a stream ecosystem. In: Suffet I H (ed) Fate of pollutants in the air and water environments, pt 2: Chemical and biological fate of pollutants in the environment. John Wiley, New York, pp 239–259

Zand SM, Kennedy VC, Zellweger GW, Avanzino RJ (1976) Solute transport and modelling of water quality in a small stream. Res US Geol Surv 4:233–240

Subject Index